Advances in Electrochemical Science and Engineering

Volume 1

Advances in
Electrochemical Science
and Engineering

Volume 1

Advances in Electrochemical Science and Engineering

Volume 1

Edited by Heinz Gerischer
and Charles W. Tobias

Contributions from

V. Brusic, J. Horkans, D. J. Barclay,
Yorktown Heights, Winchester
Der-Táu Chin, Potsdam
G. P. Evans, Cambridge
T. Iwasita-Vielstich, Bonn
R. Kötz, Baden
J. Winnick, Atlanta

VCH

Editors:
Prof. Dr. Heinz Gerischer
Fritz-Haber-Institut der MPG
Faradayweg 4 – 6
D-1000 Berlin 33

Prof. Charles W. Tobias
Dept. of Chemical Engineering
University of California
Berkeley, California 94720, USA

Published jointly by
VCH Verlagsgesellschaft mbH, Weinheim (Federal Republic of Germany)
VCH Publishers Inc., New York, NY (USA)

Editorial Director: Dr. Michael G. Weller
Production Manager: Claudia Grössl

British Library Cataloguing-in-Publication Data:
Advances in electrochemical science and engineering. Vol. 1
1. Electrochemistry
I. Gerischer, Heinz. II. Tobias, Charles W.
541.37
ISBN 3-527-27884-2

Deutsche Bibliothek Cataloguing-in-Publication Data:
Advances in electrochemical science and engineering
ed. by Heinz Gerischer and Charles W. Tobias
Weinheim; New York, NY: VCH
NE: Gerischer, Heinz [Hrsg.]
Vol. 1. Contributions from V. Brusic ... – 1990
ISBN 3-527-27884-2 (Weinheim)
ISBN 0-89573-892-9 (New York)
NE: Brusic, Vlasta [Mitverf.]

Composition: Macmillan India Ltd., Bangalore 560 025, India
Printing: Zechnersche Buchdruckerei GmbH, D-6720 Speyer
Bookbinding: Großbuchbinderei Josef Spinner, D-7583 Ottersweier
Printed in the Federal Republic of Germany

Distribution:
VCH Verlagsgesellschaft, P.O. Box 10 11 61, D-6940 Weinheim (Federal Republic of Germany)
Switzerland: VCH Verlags-AG, P.O. Box, CH-4020 Basel (Switzerland)
United Kingdom and Ireland: VCH Publishers (UK) Ltd., 8 Wellington Court, Wellington Street,
        Cambridge CB1 1HZ (Great Britain)
USA and Canada: VCH Publishers, Suite 909, 220 East 23rd Street, New York, NY 10010-4606 (USA)

ISBN 3-527-27884-2 (VCH, Weinheim)
ISBN 0-89573-892-9 (VCH, New York)

Introduction

The first volume of this series continues, with a somewhat modified title, the previous series founded in 1961 by Paul Delahay and Charles W. Tobias under the name "Advances in Electrochemistry and Electrochemical Engineering". Thirteen volumes of this series have appeared in irregular sequence. The aim was "to publish authoritative reviews in the area of electrochemical phenomena and to bridge the gap between electrochemistry as part of physical chemistry and electrochemical engineering" as stated in the first issue of the previous series by the editors. After the resignation of Paul Delahay in 1976, Heinz Gerischer took over his responsibilities as editor.

Our request, in 1987, for transfer of the series to another publishing house was graciously granted by Wiley-Interscience. We have adopted a new title: *Advances in Electrochemical Science and Engineering,* and the new publisher, VCH Verlagsgesellschaft, is committed to provide type-set volumes at regular intervals.

The scope of this series will be expanded to reflect the extension of electrochemical science and its applications in new areas. We are committed to continue to seek highly competent authors, and to maintain the high standards set in the past for this series. We sincerely hope that the forthcoming volumes of the Advances will be received as favorably as were those during the past decades.

Heinz Gerischer
Charles W. Tobias

Preface

This volume combines chapters oriented towards new materials with chapters on experimental progress in the study of electrochemical processes. G. P. Evans reviews the electrochemical properties of conducting polymers, materials which are most interesting from a theoretical point of view and promise to open up new fields of application. His approach gives a survey of the main classes of such polymers, describing their synthesis, structure, electronic and electrochemical properties and, briefly, their use as electrodes.

R. Kötz reviews the application of the most powerful surface physics technique, photoelectron spectroscopy, for the elucidation of the composition of electrodes. He exemplifies the potential of this technique for materials which play a key role in electrochemical oxidation processes or are used in some other electrochemical process.

T. Iwasita-Vielstich shows how modern spectroscopic techniques enable us to analyze the mechanism of catalyzed multi-step electrode reactions of organic molecules by detecting intermediates. This demonstrates the current general trend in electrochemical research involving the development of techniques that provide information on the atomic or molecular scale.

With its axisymmetric transport and current distribution, the rotating hemispherical electrode complements the rotating disk as a tool for studying electrode processes. Der-Tau Chin provides a valuable overview and summary of the fundamental theory and applications of this interesting device.

A timely chapter on the theory and applications of electrochemical gas separation processes is presented by Jack Winnick. These alternatives for the removal of dilute components from gas streams in pure form are characterized by high selectivity, simplicity, and favorable economics.

V. Brusic, J. Horkans, and D. J. Barklay offer an authoritative review of the electrochemical aspects of fabrication and of evalution of the stability of thin film storage media. Recent developments have led to improved thermodynamic stability, thereby rendering thin film disks suitable for high density storage applications.

Contents

List of Contributors

Donald J. Barclay
IBM Laboratories, Ltd.
Hursley House, Hursley Park
Winchester, Hampshire S021 2 JN
United Kingdom

Vlasta Brusic
IBM, Thomas Watson Research Center
P. O. Box 218
Yorktown Heights, New York 10598
USA

Der-Tau Chin
Department of Chemical Engineering
Clarkson University
Potsdam, New York 13676
USA

Gary P. Evans
Cambridge Life Sciences
Science Park, Milton Road
Cambridge CB4 4GN
United Kingdom

Jean Horkans
IBM, Thomas J. Watson Research Center
P. O. Box 218
Yorktown Heights, New York 10598
USA

Teresa Iwasita-Vielstich
Institut für Physikalische Chemie
der Universität
D-5300 Bonn 1
Federal Republic of Germany

R. Kötz
Paul Scherrer Institut
Sektion Elektrochemie
CH-5232 Villigen PSI
Switzerland

Jack Winnick
Georgia Institute of Technology
School of Chemical Engineering
Atlanta, Georgia 30332
USA

The Electrochemistry of Conducting Polymers

Gary P. Evans

Cambridge Life Sciences c/o Walker Laboratories, Cambridgeshire Business Park, Ely, Cambridgeshire, CB7 4DT, United Kingdom

Contents

1   Introduction

During the previous ten years few other areas in polymer research can have generated as much interest among such a wide variety of disciplines as that of the so-called 'conducting polymers'. These materials, whilst being organic polymers, have the unusual property of possessing high electrical conductivity, and can exhibit a range of properties from semiconducting to near-metallic behaviour. In view of this, these materials offer the electrochemist a whole new range of potential electrode materials, and a significant body of research has been built up in recent years covering many aspects of their behaviour in electrochemical systems.

Although the advent of large-scale interest in conducting polymers is a relatively recent occurrence, (notwithstanding the existence of well-known examples of materials which might be considered electrically-conductive polymers, such as graphite), the most widely studied material, polyacetylene, was first synthesised as long ago as 1958, and indeed prior to this many other materials were known which now fall into the category of conducting polymers, such as pyrrole and aniline blacks which were known at the beginning of this century. The current interest in such materials began in the 1970s, however, when it was found that the electrical properties of polyacetylene, a semiconductor when pristine, could be radically altered by treatment with oxidising agents such as iodine. It is the intention with this review to cover the main areas of interest concerning the use of these materials in electrochemistry, and also to provide a basic theoretical background in which such work can be incorporated.

The term 'Synthetic Metals' is now generally employed when discussing materials with high electrical conductivity which do not consist primarily of metallic elements, and this includes a range of materials from charge-transfer salts such as TTF-TCNQ, to so-called molecular wire type materials containing stacks of metal pthalocyanine molecules. The sheer volume of published literature makes it beyond the scope of a single review to consider all the materials which fall under this heading, and the scope of this review will be restricted to those materials which contain an organic polymeric backbone, the conductivity of which can be altered by electrochemical methods, and their application and study, in or using electrochemistry. This can be divided into three broad areas; the use of electrochemistry for the synthesis of conducting polymers, the electrochemistry the materials themselves (i.e., doping reactions, intercalation etc.) and the electrochemistry occurring at electrodes made from or coated with these materials (e.g., the electrochemistry of couples such as ferrocene/ferrocinium at polymer electrodes). Discussion of the structure and morphology of the materials is also included as this has a significant effect on both the electrical and electrochemical properties. The polymer poly(sulfur nitride) (polythiazyl) is not included as the conductivity of this material has a completely different basis to that of the other conducting polymers, involving a genuinely metal-like conduction band (for an account of the electrochemistry of this material see [1]). In-depth discussion of the applications of conducting polymers is also left for subsequent contributions.

2 Theoretical Aspects of Charge Conduction

2.1 The Concept of Doping as Applied to Conjugated Polymers

Doping is the term applied to the process of changing the state of oxidation or reduction of conjugated polymers with a concomitant change in the electronic properties of the material (e.g., to increase their conductivity), and arises from the initial interest in these materials from semiconductor physicists. Although the term 'doping' as applied to these processes is strictly correct, in that small quantities of the dopant give rise to disproportionately large changes in the properties of the doped material, its use implies a similarity between the doping of conjugated polymers and the doping of semiconductor materials such as silicon, which can be misleading. In the latter case, the dopant species occupies positions within the lattice of the host material, resulting in the presence of either electron-rich or electron-deficient sites, with no charge transfer occurring between the two species. The effect of the dopant boron on silicon for example, arises from the fact that Si has four valency electrons and B has only three, resulting in the creation of electron deficient sites within the lattice i.e., the creation of positive 'holes'. Similarly replacing Si with phosphorus results in electron rich lattice sites due to the five valency electrons of the phosphorus. The doping reaction in conjugated polymers however, is essentially a charge-transfer reaction resulting in the partial oxidation or reduction of the polymer, rather than the creation of holes etc.

The partial oxidation of conjugated polymers is generally referred to as p-doping, again in analogy to other semiconductor materials, but the basic process is the removal of electrons as in any other branch of chemistry, i.e.,

$$(\text{Monomer Unit})_x \rightarrow [(\text{Monomer Unit})^{y+}]_x + xy\,e^- \qquad (2.1)$$

Similarly n-doping is a partial reduction of the polymeric material, i.e.,

$$(\text{Monomer Unit})_x + xy\,e^- \rightarrow [(\text{Monomer Unit})^{y-}]_x \qquad (2.2)$$

These processes can be made to occur in a number of ways, for example with gas phase reagents such as AsF_5 and I_2, solution species such as $FeCl_3$ or using electrochemical oxidation and reduction, but regardless of the method used the basic process is the ·same. If the material is to maintain overall electrical neutrality during and after doping, a counter ion is required, i.e., for p-doped materials,

$$[(\text{Monomer Unit})^{y+}]_x + xy\,A^- \rightarrow [(\text{Monomer Unit})^{y+}\,A_y^-]_x \qquad (2.3)$$

and for n-doped materials,

$$[(\text{Monomer Unit})^{y-}]_x + xy\,M^+ \rightarrow [M_y^+(\text{Monomer Unit})^{y-}]_x \qquad (2.4)$$

It is a convenient aspect of electrochemical doping of the polymer that the electrolyte can provide the counter ion, although a similar result can be achieved *via* chemical doping under certain circumstances, e.g., using a radical ion containing species such as sodium naphthalide, Na^+ npth$^-$, i.e.,

$$(\text{Monomer Unit})_x + xy\,\text{npth}^- \rightarrow [(\text{Monomer Unit})^{y-}]_x + xy\,\text{npth} \qquad (2.5)$$

$$[(\text{Monomer Unit})^{y-}]_x + xy\,Na^+ \rightarrow [Na_y^+(\text{Monomer Unit})^{y-}]_x \qquad (2.6)$$

A similar result can also be achieved by using an alkali metal in an amalgam, where the metal serves both to reduce the polymer and to provide the counter ion, i.e.,

$$(\text{Monomer Unit})_x + xy\,Na \rightarrow [(\text{Monomer Unit})^{y-}]_x + xy\,Na^+ \qquad (2.7)$$

followed by,

$$[(\text{Monomer Unit})^{y-}]_x + xy\,Na^+ \rightarrow [Na_y^+(\text{Monomer Unit})^{y-}]_x \qquad (2.8)$$

although it is debatable whether these processes occur as discrete steps rather than by a single concerted process.

Although such chemical doping methods have often been used, electrochemical doping is emerging as the preferred technique in many applications as it provides a potentially highly controllable and reproducible method for investigation of the doping process, in which the transfer of charge can be accurately monitored and regulated, giving a degree of control which is beyond the scope of gas or solution phase chemical doping.

2.2   The Nature of Charge Carriers in Conjugated Polymers

Both theoretical and experimental evidence suggest that the precise nature of the charge carriers in conjugated polymer systems varies from material to material, and it is still a subject of debate in many cases. A discussion of the various theoretical models for the electronic structure of conjugated polymers is given below, using polyacetylene and poly(paraphenylene) as examples. More detailed information on these materials and the applicability of these theoretical models is given in subsequent sections.

The material which has received most attention to date is polyacetylene (Section 4.2), as it presents the most simple of these polymers from the point of view of its structure. Two of the three p-orbitals of the carbon atoms in polyacetylene are in the form of sp^2 hybrid orbitals, two of which give rise to the sigma bond backbone of the polymer (an essentially 1-D lattice), with the third forming a bond with the s-orbital of the hydrogen atom. The third p-orbital, the p_z forms an extended π-system along the carbon chain, which would in principle produce a metallic material with a half-filled conduction band, in which all carbon-carbon bonds were identical. In fact a lower energy structure is achieved by the presence of a band gap at the Fermi level (the formation of π and π^*, bonding and antibonding orbitals), resulting in a lowering of the energy of the occupied levels, and also resulting in a periodic bond alternation along the carbon chain [2]. That is to say that the spacing of the carbon atoms in the polymer backbone is altered to give a system of alternating long and short bonds, and although this approximates to a sequence of alternating double and single bonds, the p_z electrons are not completely localised.

A number of structures are possible for polyacetylene with this periodic bond alternation, two trans-forms, and two cis-forms as shown in Fig. 1. It should be noted that the two trans-forms are essentially equivalent, and are both thermodynamically stable, giving trans-polyacetylene a degenerate ground state. In contrast, of the two possible cis-forms, the trans-cis structure is of higher energy than the cis-trans, with the result that only the latter is thermodynamically stable, and cis-polyacetylene has a non-degenerate ground state. The degeneracy of the ground state of the trans-material gives rise to the possibility of structural defects (kinks) in chains, where there is a change in the sense of the bond alternation (Fig. 1e). It is clear from a consideration of the bonding of the carbon atoms on the chain that where the defect occurs, a single unpaired electron exists (although the overall charge remains zero) with the result that a new state (energy level) is created at mid-gap (i.e., the unpaired electron resides in a non-bonding orbital) (Fig. 2). This neutral defect state known as a *soliton* is singly occupied and therefore has spin 1/2, and has been calculated to be delocalised over about fifteen carbon atoms [3]. The soliton energy level can accommodate either 0, 1 or 2 electrons and thus the soliton may also be positively or negatively charged, giving the unusual property of seperating spin and charge (i.e., neutral solitons have spin but no charge, whereas charged solitons have no spin) (Fig. 2). Theoretical calculations also show that the formation of charged solitons on doping gives rise to a lower energy configuration for the polymer than the formation of electron-hole pairs [4–7],

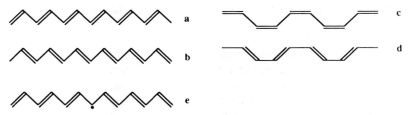

Fig. 1. Possible structures for polyacetylene chains showing the two degenerate trans-structures (a) and (b), and the two non-degenerate cis-structures, (c) cis-transoid and (d) trans-cisoid and (e), a soliton defect at a phase boundary between the two degenerate trans-phases of polyacetylene, where the bond alternation has been reversed.

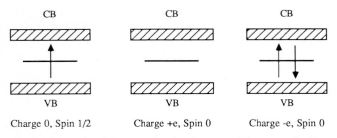

Charge 0, Spin 1/2   Charge +e, Spin 0   Charge -e, Spin 0

Fig. 2. Energy level diagram for the three possible types of soliton.

and when charge is added to the polymer chain by doping it will be located in the mid-gap states, as these provide the highest occupied molecular orbital (HOMO) for charge removal and the lowest unoccupied molecular orbital (LUMO) for charge injection. There is translational symmetry in the system (the kink can occur anywhere along the chain), thus the soliton should be mobile along the chain, giving rise in principle to the inherent conductivity of the material. It has been shown that two neutral soliton defects would be expected to recombine leaving no deformation, although single soliton defects can arise on chains with imperfections [3, 8, 9]. In fact there is evidence to suggest that trans-polyacetylene as prepared has a higher incidence of neutral solitons than would be expected from theoretical considerations [10], which are likely to have been trapped by crosslinking (which prevents their recombination).

As stated above two neutral solitons on the same chain will tend to recombine, but two charged solitons will repel each other and lead to two isolated charged defects [3]. A neutral soliton and a charged one can however achieve a minimum energy configuration by pairing [3, 11] (when they occur on the same chain) to give a *polaron* [3, 12] (which is in effect a radical cation), and this polaron defect gives rise to two states in the band gap (a bonding and an antibonding orbital), symmetrically placed about the mid-gap energy. If the number of charges on the chain is increased (i.e., if more electrons are located in the polaron energy levels), a stage is reached where the polaron states will begin to interact, and if a sufficiently high level is achieved, they would be expected to recombine to form two charged solitons which would then separate [3].

Trans-polyacetylene is unique in the degeneracy of its ground-state, and all other conjugated polymers, and indeed cis-polyacetylene, possess non-degenerate ground states, and this affects the nature of the charges which they can support. In such polymers, where two regions separated by a topological defect are not degenerate, the formation of single solitons, whether as the result of doping or from inherent defects, is energetically unfavourable [3], and the energetically preferred configurations involve paired sites [11]. This can be illustrated with reference to poly(paraphenylene), which can be drawn with either a benzenoid (Fig. 3a) or quinoid (Fig. 3b) structure, the latter of which is the higher energy configuration. If a pair of radicals exist on the chain and they are both neutral they will recombine to eliminate the structural defect [3], but if one is charged, a polaron is formed which in this case is delocalised over about five

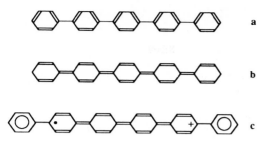

Fig. 3. Structural diagrams for polyparaphenylene showing the non-degenerate benzenoid (a) and quinoid (b) configurations and (c) a polaron defect.

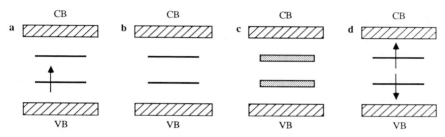

Fig. 4. Energy level diagrams showing possible electronic configurations for positively-charged polaron (a) and bipolaron (b) defects and (c) a schematic 'bipolaron band' model. The negatively-charged polaron would carry three electrons and the bipolaron four. Also shown is the neutral polaron-exciton (d) which would decay to restore the chain structure.

rings (Figs. 3c and 4a). In contrast to what is predicted for polyacetylene, when both defects are charged, rather than seperating along the chain they are predicted to pair up to form a *bipolaron* [3, 13, 14] (Fig, 4b), a doubly-charged defect which extends over a similar number of rings to the polaron and again has no spin. At high doping levels bipolarons may also combine to form 'bipolaron bands' within the gap [3, 13–16]. A more detailed discussion of these topics is given in [13].

These charge carriers will give rise to conductivity in the materials in which they occur and in subsequent sections experimental evidence is given supporting their existence, but it should be borne in mind that any polymer system consists primarily of chains of finite length which will generally be randomly oriented and are unlikely to be either regular or fully conjugated along their whole length, with a high likelihood of large numbers of structural imperfections such as sp^3-bonded carbons interrupting the conjugation. Thus any consideration of conductivity or of the electrical properties of these materials must include charge transfer between chains (and between different conjugated segments on the same chain) and, in doped conducting polymers, the pinning effect of the relatively immobile counterions on the charged defects. In addition to these effects which may act at a molecular level, the actual values measured are also likely to be dominated by grain boundaries and other morphological effects [17–21].

2.3   Experimental Methods and their Relation to Theoretical Models

Electrochemistry has proved an invaluable tool for the investigation of the properties of conducting polymers, not only *via* the use of standard techniques such as cyclic voltammetry (CV), and constant current charge/discharge, but through its suitability to be coupled with other experimental methods in order to achieve a more precise control over the polymer system (with the degree of doping controllable and calculable directly *via* the charge passed). From the earlier theoretical discussion it is apparent that many of the proposed charge carriers or defects in conducting polymers can contain unpaired spins, and would therefore be expected to be ESR active, and a number of workers have successfully combined *in situ* electrochemical doping with ESR measurements. The nature of the majority of conducting polymers is also such that the electronic transitions between the valence and conduction bands (the π and π^* orbitals), and to and from states within the gap (e.g., Figs. 2 and 4), have energies which fall in the range 1–4 eV (1240–310 nm) and thus these transitions are accessible using UV, visible and IR spectroscopy. The use of optically transparent electrodes (OTEs) allows these spectra to be obtained from samples which are incorporated in electrochemical cells, allowing the charge state of the material (i.e., the doping level) to be electrochemically controlled. OTEs are easily obtained by treating glass to give it a conducting coating, either by annealing to incorporate indium-tin oxide (ITO) or by evaporation of a thin layer of a metal such as platinum, and the polymer can then be cast or synthesised directly onto the OTE. ITO slides are sufficiently transmitting to allow measurements to be made into the UV, with a cutoff at around 4.0 eV.

In addition to these experimental techniques which employ electrochemistry largely as a means of controlling the state of the system, new methods have also been developed which are particularly suited to the study of conducting polymers, which are less widely known. Of these techniques electrochemical voltage spectroscopy (EVS) [22–24], and the associated technique of gravimetric electrochemical voltage spectroscopy (GEVS) [25] have proved especially useful for the investigation of the

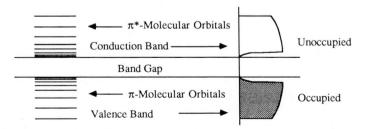

Fig. 5. A schematic band structure and molecular orbital diagram for a conjugated polymer containing no mid-gap states.

electrochemistry of these materials. EVS was first developed by Thompson [22] in order to accurately measure the charge/potental relationships of electrochemical systems under conditions of quasi-equilibrium. The system is initially allowed to reach equilibrium and the equilibrium potential V, is measured. Under computer control the potential is then stepped to a new value $V + dV$ and the subsequent flow of current is simultaneously monitored and integrated to give the charge passed, whilst the system re-equilibrates at the new potential. A suitable threshold level must be selected for the current, below which it is considered that the system has reached its new equilibrium, and once the current has fallen below this level, the potential is stepped again to $V + 2dV$ and the process repeated. By stepping the potential in this way to some maximum value and then reversing the process to return to the equilibrium value the whole redox cycle can be investigated, and a plot of V vs. Q can be obtained by integration of dQ/dV (as measured in the experiment). Thus the overall process can be likened to cyclic voltammetry performed at an infinitely slow sweep rate [24].

The oxidation or reduction of the polymer is more clearly understood with reference to the energy level diagram for the material (Fig. 5). Between the top of the valence band (the highest occupied π bonding orbital) and the bottom of the conduction band (the lowest unoccupied π^* antibonding orbital), the density of states (i.e., the number of molecular orbitals) is extremely small. That is to say that the number of available energy levels is effectively near zero, with the result that these orbitals can be rapidly filled or emptied even if only a relatively small amount of charge is transferred. The measured open-circuit potential is determined by the energy of the highest occupied orbital, so in this region where the polymer is close to being neutral, small amounts of charge-transfer will markedly affect the potential and its precise measured value will depend on the history of the given sample.

The small density of states in the band gap means that the removal of electrons from the polymer during oxidation can therefore effectively only occur in significant numbers when the potential of the electrode has been made sufficiently positive (relative to the counter or reference electrode) for the electrons in the valence band to be extracted into the substrate material and external circuit (i.e., when it becomes energetically more favourable for the electron to reside in the electrode rather than the polymer). Similarly, addition of electrons to the polymer during reduction can only occur when the energy of the electrons in the substrate is raised (by taking the potential sufficiently negative), for it to become energetically more favourable for them to reside in the levels of the conduction band of the polymer. This is directly analogous to the oxidation or reduction of a species in solution at a metal electrode, and in a similar way, the potentials at which these redox processes occur define the energies of the HOMO (in this case, the top of the valence band), and the LUMO (the bottom of the conduction band). Measurement of the potentials at which these two processes begin to occur is therefore a measurement of the bandgap of the material (the separation between the π and π^* orbitals) and the use of EVS allows this to be accurately determined, although there may be practical restrictions to the applicability of this approach in that suitable solvent systems must be found for both oxidation and reduction of the polymer.

3 Synthesis

The synthesis of conducting polymers can be divided into two broad areas, these being electrochemical and chemical (i.e., non-electrochemical). Whilst the latter may be considered to be outside the scope of this review, it is worth noting that many materials which are now routinely synthesised electrochemically were originally produced *via* non-electrochemical routes, and that whilst some may be synthesised by a variety of methods many, most notably polyacetylene, are still only accessible *via* chemical synthesis. In view of this it is useful to have an appreciation of the synthesis of these materials *via* routes which do not involve electrochemistry.

3.1 Chemical Synthesis

A large number of conducting polymers can be synthesised *via* the use of catalysts [28–30], but generally little control can be exercised over the morphology of the product and purification of the material obtained can be problematical. In recent years however, a number of alternative synthetic routes have been devised which involve soluble precursor polymers which can be more easily purified and cast onto substrates, with subsequent conversion (usually by heating) to the desired product, and

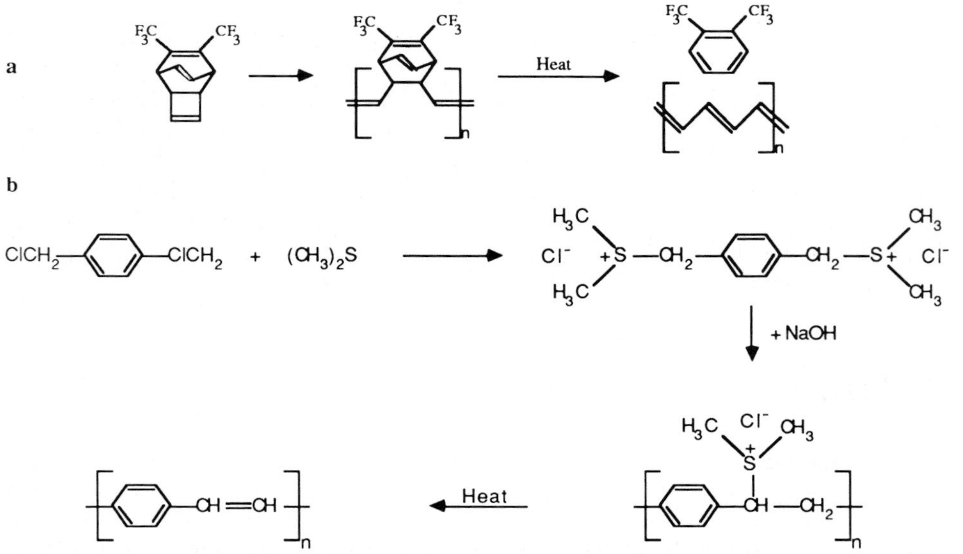

Fig. 6. Synthetic routes to conjugated polymers *via* precursor polymers for (a) polyacetylene, and (b) arylene vinylenes.

this approach has been successfully applied to the synthesis of polyacetylene [31, 32] and the arylene vinylene polymers [33] (Fig. 6).

A non-electrochemical technique which has been employed to alter the physical characteristics of a number of polymers is that of stress orientation [26, 27], in which the material is stressed whilst being converted to the desired form. This has the effect of aligning the polymer chains and increasing the degree of order in the material, and is obviously most applicable to materials which can be produced *via* a precursor polymer. With Durham polyacetylene (Section 4.2.1) increases in length in excess of a factor of twenty have been achieved, with concomitant increases in order, as shown by X-ray diffraction and by measurements of the anisotropy of the electrical conductivity perpendicular and parallel to the stretch direction.

3.2   Electrochemical Synthesis

Many conducting polymers can in principle be electrochemically synthesised, the main requirements being that the monomer (a) has an oxidation potential which is accessible *via* a suitable solvent system, (b) will produce a radical cation which reacts more quickly with other monomers to form the polymer than it will with other nucleophiles in the electrolyte solution [34], and (c) produces a polymer with a lower oxidation potential than that of the monomer (if the material is to be produced in a state which has a higher conductivity). The specific details of electrochemical synthesis are discussed in subsequent sections in relation to specific materials, but in general the method is the same for all polymers – the monomer is simply dissolved in a suitable solvent and a simple cell is used to anodically polymerise the materials onto one of the electrodes (Pt metal, ITO glass etc.) under either potentiostatic or galvanostatic conditions. Electrochemical synthesis also has the advantage of producing the material *in situ* on an electrode already located in a cell, on which further experiments may be performed without necessity for any intermediate treatments such as heating. The fact that a large number of materials can be produced in this way, and the method is so simple to perform, has tended to focus attention onto these materials, with others that require chemical synthesis being less widely studied (with the obvious exception of polyacetylene).

4   Materials

4.1   General

A summary of materials which have been electrochemically synthesised is given in Table 1, and in subsequent sections specific materials are discussed separately in relation to the published literature, but it should be noted that this separation is

Table 1. Electrochemical synthesis of conducting polymers.

Monomer (concn.)	Electrode/solvent[a]	Electrolyte (concn.)	E_{pol}(V)[b]	Counter ion[c]	σ(S cm^{-1})[d]	Ref.
anthracene	ITO/odcb	$CuCl_2$ (0.1 M) + Bu_4NClO_4 (0.1 M)	+ 20 vs. Ni	ClO_4^-	10^{-3}	38
Polyazulene						
azulene (0.01 M)	acn-aq/Ar	Bu_4NClO_4 (0.1 M)	+ 0.90 cv	ClO_4^-	10^{-2}–10^{-1}	39
azulene (0.05 M)	bzn	$LiClO_4$ (1 M)	+ 4.50 vs. Ni	ClO_4^-	1	40
azulene (1–10 mM)	–	Et_4NBF_4 (0.1 M)	+ 0.91 cv	–	–	41
4,6,8-trimethyl-(1–10 mM)	acn/Ar	Et_4NBF_4 (0.1 M)	+ 0.90 cv	–	–	41
Polyfuran						
furan (0.01 M)	acn-aq (0.01 M)/Ar	Bu_4NBF_4 (0.1 M)	+ 1.85 cv	BF_4^-	10–80	39
furan (0.2–0.3 M)	bzn (anhyd.)	$AgClO_4$, $LiClO_4$, Bu_4NBr, $NaAsF_6$, $NaPF_6$, $LiBF_4$ (0.1–0.2 M)	+ 1.80 to + 2.50	ClO_4^-	10^{-4}	42
Polyindole						
indole (0.01 M)	acn-aq (0.01 M)/Ar	Bu_4NClO_4 (0.1 M)	+ 0.9 cv	ClO_4^-	5.0×10^{-3}	39
Polyisothianapthene (PITN)						
isothianapthene (0.23 M)	acn, N_2atm	Bu_4NPF_6, LiBr, Bu_4NBr, Ph_4AsCl (0.3 M)	–	–	–	43
isothianapthene	acn (anhyd.)	Ph_4PCl	–	Cl^-	10–50	44
Polynapthalene						
napthalene	ITO/nbz	$CuCl_2$ (0.1 M) + $LiAsF_6$ (0.1 M)	+ 20 vs. Ni	AsF_6^- neutral	10^{-3} 10^{-9}	45 45

Poly(para-phenylene)

	Electrode/Solvent	Electrolyte		Counter ion	Conductivity	Ref
benzene (1.2 M)				AsF_6^-	100	46
benzene (5.4% v/v)	ITO/nbz $MeNO_2/AlCl_3$ (4g)	$CuCl_2$ (0.1 M) + $LiAsF_6$ (0.1 M) H_2O, Et_3N, Bu_4NCl, Bu_3N, Bu_4NClO_4	$2\,mA\,cm^{-2}$ $12\,mA\,cm^{-2}$	$AlCl_3^-$ (Et_3N^+) $AlCl_3^-$ (Bu_4NCl)	1.1×10^{-5} 2.4×10^{-4}	47 47 47
benzene (5.4% v/v)	$nbz/AlCl_3$ (4g)	Bu_4NCl	$4\,mA\,cm^{-2}$	$AlCl_3^-$	5.9×10^{-4}	48
benzene	Pt/HF (93%)/7°C	—	$0.5-10\,mA\,cm^{-2}$	AsF_6^- (chem.) neutral	10^{-2} 10^{-10}	48 48
benzene (0.02 M)	$SO_2/-40\,°C$	Bu_4NPF_6	$+2.3$ vs. AgCl	neutral	—	49
dihalobenzene	Hg/Ar atm	$LiClO_4$, $NiCl_2$-P-Ph_2-$(CH_2)_2$-P-Ph_2	-2.6 vs. $AgClO_4$	neutral	10^{-18}	50

Polypyridazine

	Electrode/Solvent	Electrolyte		Counter ion	Conductivity	Ref
pyridazine	acn, bzn, pc, nbz	$LiClO_4$, $LiBF_4$, $LiAsF_6$ (0.2 M) Bu_4NClO_4	$+4.0$ to $+8.0$	ClO_4^-	10	51

Polyselenophene

	Electrode/Solvent	Electrolyte		Counter ion	Conductivity	Ref
selenophene (0.1–1 M)	acn, bzn, pc, nbz	$LiClO_4$, $LiBF_4$, $LiAsF_6$, $LiPF_6$, Bu_4NBF_4, Bu_4NClO_4, $AgClO_4$, (0.1–1 M)	$+3.0$ to $+10.0$ vs. Ni	ClO_4^- neutral	$10^{-4}-10^{-3}$ 10^{-10}	52 52
selenophene (0.2 M)	ITO/bzn/Ar	$AgClO_4$ (0.3 M)	$+20$ vs. Ni	ClO_4^- neutral	10^{-4} 10^{-9}	53
3-methyl-(0.1–1 M)	acn	$LiClO_4$ (0.1 M or saturated)	1.48 cv	ClO_4^-	—	54
3-methoxy-(0.1–1 M)	acn	$LiClO_4$ (0.1 M or saturated)	1.20 cv	ClO_4^-	—	54
diselenophene	acn	Et_4NBF_4 (0.5 M)	0.88 cv	BF_4^-	10^{-5} (f)	55

[a] Electrode/Solvent:
acn – acetonitrile, aq – water, bzn – benzonitrile, mc – methylene chloride, nbz – nitrobenzene, nmp – N-methyl-2-pyrrollidone, odcb – o-dichlorobenzene, pc – propylene carbonate, thf – tetrahydrofuran.

NB. Mixed solvent systems are shown as e.g. acn-aq (0.01 M) where the number in parentheses indicates the concentration of the lesser constituent
ITO – Indium/tin oxide-coated glass, Ar – Solutions purged with argon, Ar atm – Experiment performed under an argon atmosphere, N_2 atm – Experiment performed under a nitrogen atmosphere

[b] All potentials are measured vs. SCE unless otherwise stated:
E_{pol} – Potentials are as quoted in the original Ref, cv – The potential given was obtained from cyclic voltammetry.

NB. Where only a current density was given in the original Ref. this is quoted in place of the polymerisation potential.

[c] Counter ions are as incorporated during the electrochemical polymerisation process or by subsequent electrochemical doping unless suffixed with (chem.) which indicates the use of chemical doping.

[d] Conductivities: f – film, p – pressed pellet.

mainly for clarity, and that there are similarities between many of the materials discussed, such as polypyrrole and its analogues, which have similar band structures [35] and which can in some respects even be treated as substituted polyacetylenes with respect to their conjugated backbone [36, 37].

The stability of the materials with respect to reaction with species such as oxygen can be explained by considering the redox potentials for a range of polymers [56] (Fig. 7). If the potential of the material in a given state (i.e., at a given level of doping) lies more negative than the reduction potential of oxygen (for example), the polymer will be oxidised and the oxygen reduced. Similarly, if the potential of the polymer lies more positive than the oxidation potential of a species with which it is in contact, a reaction will occur and the polymer will be reduced. Polypyrrole is therefore stable to oxygen when in its doped (oxidised) form, but not when in its neutral (reduced) form, whereas poly(N-methyl pyrrole) is less easily oxidised and consequently less sensitive to oxygen [57]. In fact the low oxidation potential of polypyrrole means that a wide variety of oxidising agents can be used to dope the material [58]. On the same basis it would be expected that poly(paraphenylene) will be reduced by water when in its doped state, but not when neutral (undoped) [56]. From Fig. 7 it can also be seen that polyacetylene is stable to oxidation by O_2 when neutral and reduction by water when doped, although this does not exclude any chemical (i.e., non-electrochemical) processes which may occur.

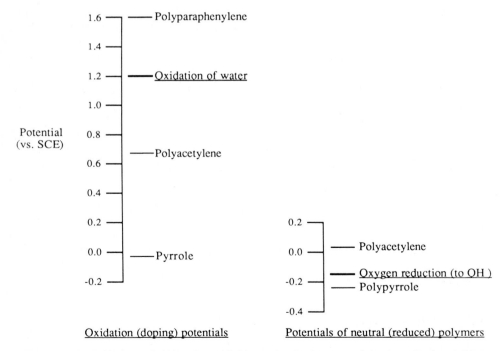

Fig. 7. Schematic diagram showing the oxidation and reduction potentials of conducting polymers relative to oxygen reduction and water oxidation.

The study of the electrochemistry of all conjugated polymers is subject to the same restrictions, imposed by the necessity for a suitable solvent/electrolyte system in which the experiments can be performed. Choice of the solvent is critical if side reactions such as solvent polymerisation [59, 60] are to be avoided, particularly in battery systems where repeated charge and discharge cycles will exacerbate the problem. The two most common solvents used for doping experiments are propylene carbonate for p-doping, and tetrahydrofuran for n-doping, as these have low nucleophilicity and are stable in the range of potentials at which the doping reactions of most polymers occur [61], but it should be noted that the working potential limits for a given system will depend on a number of factors including the electrolyte salt, the current density employed and the nature of the electrode materials used [62, 63]. The study of electropolymerisation reaction is also complicated by the fact that doping of the product will occur simultaneously as soon as polymer is deposited, but it should be noted that all the polymers are produced stoichiometrically (with two electrons per monomer unit) once the charge-transfer due to the doping reaction is accounted for.

4.2 Polyacetylene

Polyacetylene is the one of the most widely studied of the conducting polymers both from the point of view of the physicist and also the chemist, with conductivities as high as those of metals being achieved [64]. Whilst it is an attractive material to study as the structurally most simple example, it is not the most easy material to work with, being extremely air sensitive and only accessible *via* chemical routes. The reaction with oxygen is essentially irreversible and occurs rapidly at room temperature (and even faster in the presence of water), but if the exposure is very brief the oxygen appears to act solely as a dopant and the reaction is reversible [65, 66]. The degradation is inhibited at lower temperature [67] and by the presence of dopants, notably BF_4^- [66, 68], and it has been suggested that this latter effect is due to reaction between the BF_4^- and the oxygen [66] i.e.,

$$(CH)_x + 1/2\ xy\ O_2 \rightarrow [(CH)_x\ (O_2)_{y/2}^{2-}]_x \qquad (4.1)$$

$$4(CH)_x + xy\ O_2 + 4xy\ HBF_4 \rightarrow 4[CH_y^+(BF_y^-)_y]_x + 2xy\ H_2O \qquad (4.2)$$

Irreversible reaction can occur with other dopants such as iodine (at elevated temperatures) [69] and $FeCl_3$ [70], and a loss of conductivity is seen over a period of months for p-doped material even at room temperature, although n-doped material appears to have a higher degree of thermal stability [71].

4.2.1 Polymerisation

The most commonly used form of polyacetylene is produced by the Shirakawa method, which involves the direct polymerisation of acetylene gas onto a substrate at low temperature ($-78\,°C$) using a Ziegler–Natta catalyst [28, 29] (many substituted polyacetylenes have also been prepared [72–74]), although the direct electrochemical polymerisation of acetylene to give a powdery product similar to the Shirakawa material has been reported [75]. An alternative route has been developed by Feast et al., generally known as the 'Durham route', which involves the use of a precursor polymer, which can be purified prior to conversion to the final product – Durham polyacetylene [31, 32] (Section 3, Fig. 6). This method has several advantages over the Shirakawa route, in that the leaving group in the conversion is gaseous and can easily be removed from the product, and that the precursor is soluble in organic solvents, enabling the material to be cast as a film prior to conversion, giving access to high-quality samples of the polymer. The material produced is more dense than Shirakawa polyacetylene but has a similar degree of crystallinity.

4.2.2 Morphology and Structure

In discussing the morphology and structure of polyacetylene it is important to distinguish between the different types of material investigated, and to recognise that the morphology is fundamentally different from that found in electropolymerised materials. The Shirakawa material, which generally contains traces of the catalyst, consists of an open network of randomly-oriented fibrils giving the material relatively low density and a high surface area (40–$60\,\mathrm{m^2\,g^{-1}}$ [76]). The fibrils have a high degree of crystallinity [28, 29] and have diameters of 20–$50\,\mathrm{nm}$ [76–78] with thin films tending to contain fibrils of smaller diameters than thicker films [79, 80]. The Durham route, produces material which can be seen using SEM to be non-fibrillar and fully dense, and these differences will inevitably affect the electrochemical doping process, particularly with respect to the intercalation of ionic species into the bulk of the material.

Shirakawa polyacetylene has been shown by Raman spectroscopy to have a distribution of chain lengths with a low level (300 ppm) of permanent defects which arise during the isomerisation process [24, 81]. In their initial studies Nigrey et al. reported, that there was no evidence of hydration of the polymer or incorporation of oxygen into the material during electrochemical doping [82] although later work contradicts this [66–68, 83]. X-ray data suggest that polyacetylene electrochemically doped with SbF_6^- has a monoclinic, layer-type structure with stacks of polymer chains separated by layers of dopant ions [84], which allow easy transfer of electrons between chains and implies a maximum doping level of 25% at 100% occupancy of the anion sites, although other possible structures have been identified [85]. Co-insertion of solvent occurs during the doping reaction [86] with the weak interchain bonding allowing structural rearrangement to occur to accommodate dopant ions and associated solvent molecules, and a number of structural phases have been identified for different levels of dopant/solvent incorporation [86–89] with complex sequences of transitions between phases (both with and without clearly-defined stoichiometries) [80, 87, 88].

Durham polyacetylene occurs in a highly disordered state on conversion from the precursor polymer [90], but using stretch orientation techniques during the conversion reaction, a high degree of order with long conjugated sequences can be achieved [91–93].

4.2.3 The Electrochemistry of Polyacetylene

The electrochemistry of polyacetylene has been investigated by a large number of workers since the original synthesis of the material, with particular emphasis on its use in battery systems. It was found at an early stage that both p-doping and n-doping are easily achieved chemically using reagents such as iodine, alkali metal napthalides [94, 95], arsenic pentafluoride or even oxygen, and in 1979 Nigrey et al. [82] reported that both electrochemical oxidation (p-doping) and reduction (n-doping) were possible using Bu_4NClO_4 in methylene chloride, and subsequent work showed that other anions could be used for p-doping [96, 97].

Initial studies using cyclic voltammetry and constant potential charge/discharge experiments in acetonitrile solutions containing Et_4NBF_4 [98], and $LiClO_4$ [76, 99] on both cis- and trans-polyacetylene, showed that the process of p-doping (oxidation) and de-doping (reduction back to the neutral state) is coulombically reversible although the voltammograms were more complex than would be expected for a simple one-electron transfer, with evidence of two reduction peaks (Fig. 8) attributed in part to structural reorganisation [98]. Efficiencies of up to 100% have also been obtained for the charge injection/removal process using potential step experiments [79], although the efficiency falls as the maximum doping level employed is increased

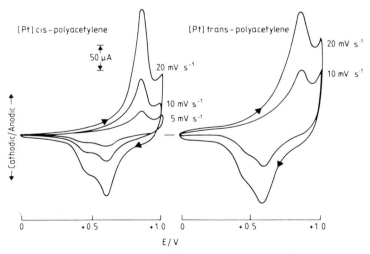

Fig. 8. Cyclic voltammograms of polyacetylene films on a platinum surface measured in acetonitrile containing 0.1 M Et_4NBF_4. Reproduced from [98].

[78, 79]. Billaud et al. [100] identified two separate processes for the p-doping reaction from cyclic voltammetry using $LiAsF_6$ in CH_3NO_2 and concentrated H_2SO_4, one corresponding to the reversible doping process (with associated changes in the conductivity of the material) and another due to irreversible doping occurring in conjunction with degradation of the polymer π-system. Similar results were found by others [63, 101–103] with the number of times that the polymer can be cycled in the p-doping regime being dependent on the potentials involved, with evidence from ESCA [101] of irreversible reactions between the polymer and the solvent, and of crosslinking occurring between chains. Chiang et al. [103] proposed a two-component mechanism involving direct oxidation of the polymer double bonds in addition to a reversible doping reaction. There is also a loss of conductivity on standing in electrolyte for long periods (> 200 hours) due to side reactions causing degradation of the material [80, 104] although careful choice and purification of the electrolyte can improve this [105]. The presence of side reactions will inevitably gradually alter the structure of the polymer even if reversible doping is apparently achieved – the doping may not involve exactly the same structural units even though the same charge is transferred. It has been suggested that the maximum observed limit for *reversible* doping of 6.5–7% is due to the fact that at this point the side reactions become significant and reduce the current efficiency and conductivity of the material with up to 20% of the dopant remaining in the polymer on cycling [106], and there is evidence from photoacoustic FTIR spectroscopy of long aliphatic chains in material which has been doped to much higher levels (up to 13.5%) [107]. At very low doping levels (< 10^{-1}%) there is a logarithmic relationship between the level of doping and the open circuit potential of the material [79, 80], but this becomes sigmoidal at higher levels [80]. From ESR measurements Bernier et al. [108] concluded that trans-segments of the material dope preferentially due to their slightly lower (less positive) oxidation potential (arising from the non-degeneracy of the cis- ground state). It also seems that disordered trans- regions dope more rapidly than either cis- or crystalline trans- but that crystalline trans- material de-dopes first [109], although there is a large body of evidence to suggest that isomerisation occurs on doping [81, 99, 109–114] (and that this also reduces the number of short conjugated segments in the chains [81]).

Although the stability of n-doped polyacetylene is good in organic solvents [115], there is a clear dependence of the behaviour of the material (including the maximum doping level achieved) on the nature of the electrolyte system used, although the relationship between the electrode potential and the degree of reduction is independent of the nature of the dopant, being an intrinsic property of the material [116]. Values reported for the doping levels are usually below 10%, but with the correct choice of salt and solvent reversible doping up to 19% can be achieved [86]. The material has been found to be stable on repeated cycling between the doped and neutral states within a limited potential range [89, 117] using $LiClO_4$ in THF, but over a wider range (going less positive than about + 0.6 V vs. Li/Li$^+$) catalytic reduction of the perchlorate anion occurs at the polyacetylene electrode giving rise to very poor coulombic efficiencies for the cycling process [114] although by changing the anion, both much lower and much higher thresholds for the onset of irreversible processes can be achieved [86]. On cycling, the capacity of a polyacetylene/Li battery is reduced, even if coulombic efficiencies of close to 100% are maintained but there is some evidence [114] to suggest that this is due in part to the lithium counter electrode rather than the polyacetylene, with improvements in capacity being achieved if the lithium electrode is replaced.

Kaufman et al. [23, 24] have used electrochemical voltage spectroscopy (EVS) to investigate the oxidation and reduction of the polymer and thereby directly measure the band gap of the material as discussed in Section 2.3. In this way they have measured the potentials at which charge injection into the conduction band and charge removal from the valence band occur as $+ 3.10$ V and $+ 1.75$ V vs. Li/Li$^+$ respectively $(+/- 0.05$ V), with the neutral material being obtained at $+ 2.4$ V vs. Li/Li$^+$. These potentials agree quite well (related to an appropriate reference point) with those calculated by Bredas et al. (by extrapolation of data from short chain oligomers) of $+ 0.4$ V and $- 1.1$ V vs. SCE), and the values reported in earlier studies [76, 99, 103]. Shacklette et al. have also reported good agreement between the band gap found from electrochemical measurements and that obtained from spectroscopic data [118], although there is some variation in reported values for the oxidation and reduction potentials and this has been attributed to differences in the solvent/electrolyte systems used [99]. The kinetics of the electron-transfer process are evidently slow, with typical separation between cyclic voltammetric peaks in excess of 200 mV [99, 103].

The equilibration of the polyacetylene system after a potential step appears to be controlled by the diffusion of the dopant ion into the polyacetylene fibrils [79, 114, 119] so that the open circuit potential will slowly approach an equilibrium value, with the doping process occurring uniformly throughout the material (rather than in localised metallic areas with high doping levels being created). Potential step techniques similar to EVS provide evidence that initially the doping occurs only at the fibril surfaces (which occurs rapidly) followed by a much slower process where the dopant penetrates and redistributes throughout the fibril (i.e., a solid-state rather than a solution diffusion process) [79, 119, 120]. From cyclic voltammetry, charge/discharge cycles and EPR data Padula et al. [121] also found that the kinetics of the p-doping process are controlled by the diffusion of the counter ion into the material and similar results have been obtained by other workers [79, 80, 109, 122] with apparent phase separation between doped and undoped parts of the polymer [109], and a concentration profile for the dopant into the polymer fibrils to their centres [80]. Electrochemical impedance measurements on polyacetylene electrodes during charge and discharge [122–126] have also provided evidence that supports this model, correlating the charge-transfer resistance with the ion insertion/removal process. The diffusion-limited charge-transfer can thus produce low coulombic efficiencies in some experiments, as long periods of polarisation may be necessary to remove all the charge from the doped polymer [122] and only the surface dopant species are available for rapid charge-transfer [69, 127, 128] (a similar model has been proposed for chemical doping of polyacetylene [129]). It has also been suggested that the kinetics of the doping process are dependent on the state of charge of the material, with faster kinetics occurring at higher doping levels [128]. The diffusion coefficient for ClO_4^- in polyacetylene has been determined by Bernier et al. [78] from constant current charging experiments to be 10^{-17} cm^2 s^{-1}, and by Kaufman et al. to be 4.0×10^{-18} cm^2 s^{-1} from open circuit voltage decay measurements [130]. These low diffusion coefficients for the dopant species [131] would seem to restrict the utility of polyacetylene in applications where rapid charge transfer was necessary, such rechargeable batteries, but it is easily demonstrated that high currents can be obtained from such systems [79, 96]. This has been explained [24, 130, 132] on the basis that high electric fields exist within the fibrillar structure under conditions of high current drain, greatly increasing the rate of ionic transport, and also that the nature of the fibril network is such that the ions need only move very small distances (10 nm) during the doping/de-doping processes. This is supported by experiments on polyacetylene deposited onto electron microscope grids where the high currents obtained were attributed to a low iR drop in the electrolyte and improved access to the polymer fibrils for the dopant [133]. Polyacetylene foam electrodes [134] have also been shown to support higher current densities [121, 122], presumably as a higher surface area is available. It has been noted however [131], that in order to correctly assess the diffusion coefficient for the material from electrochemical measurements, it is necessary to consider the actual solution-wetted surface area of the polymer rather than an apparent surface area, and this may have led to inaccurate estimates for D. If the solution-wetted area is estimated from capacitance measurements, a value of 6.0×10^{-12} cm^2 s^{-1} is obtained for the diffusion coefficient using BF_4^- in sulfolane [131]. A consequence of these relatively slow diffusion rates is that steady state conditions are difficult to achieve unless very small amounts of charge (< 0.01 mA cm^{-2}) are transferred during the doping process or very long equilibration times (> 300 hours) are employed [120], with the effect that many values quoted for steady-state open-circuit potentials related to given doping levels really represent quasi-equilibrium values.

Although it was initially believed that polyacetylene was unstable in contact with water under all conditions, it has been successfully chemically doped in aqueous solutions with no apparent degradation of the material [82] and its electrochemistry has also been investigated [135–137] from which it is clear that no degradation occurs in concentrated aqueous electrolytes. Reaction with water can occur under some circumstances however giving rise to sp^3 carbons and carbonyl-type structures [129, 138–141].

4.2.4 Electronic Structure and Conductivity

Although the conductivity of polyacetylene is generally discussed in terms of solitons, the question of the precise nature of the major charge-carriers continues to be a subject of debate, with conflicting evidence from different experiments. Spectro-electrochemical studies provide evidence that the charge in doped polyacetylene is stored in soliton-like species (although this is not the only possible interpretation [142, 143]), with absorptions in the optical spectra corresponding to transitions to states located at mid-gap [24, 89, 119]. The intensity of the interband transitions becomes weaker as the doping level is increased with a corresponding increase in the transition to the dopant-induced levels (Fig. 9). At high doping levels the interband transition completely disappears giving a spectrum characteristic of free-carrier absorption (i.e., metallic conduction), although retaining some evidence of band structure at long wavelengths [119, 144]. A model can thus be constructed in which the charge carriers at low levels of doping are considered to be solitons [24, 145, 146], with a transition from a soliton lattice to metal-like behaviour occurring on doping to higher levels [76, 147] coincident with structural phase changes (Section 4.2.1). There is considerable evidence however that the measured conductivity is dominated by effects on a larger scale, i.e., charge transfer between chains, fibrils etc. and involves a variable range hopping mechanism, obscuring the mechanism of conduction on the polymer chains themselves [148, 149].

It has been noted by a number of workers [23, 119] that plots of charge vs. potential (such as those obtained from EVS) show hysteresis, and that this is independent of the nature of the counter ion incorporated during the doping process [23, 118]. The origin of this hysteresis apparently does not arise from diffusion limited process failing to attain equilibrium at the point when the measurements were made [23], iR drops (as it is still seen when very low currents are employed), or from irreversible chemical reactions (coulombic efficiencies of up to 100% can be obtained for the charge injection/removal process [23, 78, 79]). It has been suggested that the hysteresis occurs as the result of structural phase changes within the material [86, 89] and also that it is due to the nature of the charge species involved in the doping process. If the injection of charge results in the formation of polarons, the electrochemical potential of the material will be pinned near the band edge, and if the removal of charge occurs via solitons (i.e., from levels located at mid-gap) the potential

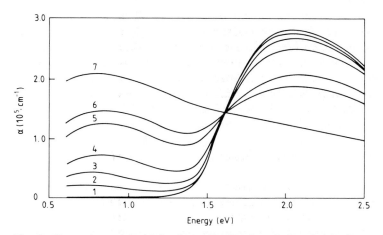

Fig. 9. Absorption spectra (of polyacetylene) taken during the doping cycle at different voltage: Curve 1 – 2.2 V (y = 0). Curve 2 – 3.28 V (y = 0.003). Curve 3 – 3.37 V (y = 0.0065). Curve 4 – 3.46 V (y = 0.012). Curve 5 – 3.57 V (y = 0.027). Curve 6 – 3.64 V (y = 0.047). Curve 7 – 3.73 V (y = 0.078). Reproduced from [119].

will be pinned at the mid-gap energy [23, 119]. Thus initially the doping may involve the creation of polarons which subsequently interact to form solitons, from which subsequent removal of charge occurs.

4.2.5 Electrochemistry at Polyacetylene Electrodes

By far the most frequently proposed application of polyacetylene in electrochemical systems is for use as an electrode material for rechargeable batteries [79, 99, 114, 133, 150–157], giving an open circuit voltage of approximately 3.7 V vs. Li/Li$^+$ [76, 99], and indeed most papers on the electrochemistry of the material make some reference to battery applications. Most systems proposed employ organic solvents, although reports of batteries using aqueous [135, 137] or solid state electrolytes [153, 158] with good long-term stability have appeared. Photovoltaic devices incorporating polyacetylene have also been suggested [159, 160].

There are reports of polymerisation of pyrrole [161, 162] and aniline [163] onto polyacetylene, to give oxygen and water stability [161], although there is some evidence for the polyacetylene acting electrocatalytically, oxidizing the pyrrole with no concomitant polymerisation.

4.3   Polyaniline

Polyaniline has been known in a variety of forms for over a century [164], and as such must be considered the oldest of the conducting polymers. Aniline blacks and other oligometric forms of the material have often been produced on electrode surfaces as non-conducting passivating films (electrochemical oxidation of aniline without polymer formation is also possible [165, 166]), but as with many other examples of the conducting polymers, it is only relatively recently that the optimum conditions for producing high quality films have been investigated. The material is amorphous, stable up to 300 °C and is insoluble in most solvents including water [165, 167–169], (although details of soluble polyanilines have been published [170]), but will dissolve in concentrated H_2SO_4. Polyaniline has a complex chemistry and shows many features which set it apart from other conducting polymers, and is unusual in that it is prepared in aqueous solution in its doped, conducting form [171].

4.3.1   Polymerisation

Aniline black was first prepared in the last century [164] when it was found that anodic oxidation of aniline at a Pt electrode in aqueous H_2SO_4 gave a dark green powdery product. Aniline blacks have subsequently produced under a variety of conditions *via* both chemical and electrochemical routes with a low level of interest being maintained over the years up to the 1960s, but did not arouse wide-spread interest until relatively recently (for a more detailed account of early work on aniline blacks see reference 167 and references therein), when high quality films of polyaniline and substituted polyanilines [172] have been prepared [173–182].

Several variations of a basic reaction mechanism have been proposed [167, 173, 174, 183–186] which differ in some details but all involve the formation of radical cation intermediates from the aniline, and a general reaction scheme is shown in Fig. 10. Substituted anilines have been polymerised [182], and although initially N-substituted derivatives were thought to lead to non-polymeric products via a mechanism which avoids the radical intermediate and results in tail-to-tail coupling [167], successful film formation has been reported [179, 187] albeit with no electroactivity in the latter case [187]. Reducing the concentration of the monomer in the solution reduces the conductivity of the product [188], possibly due to morphological effects, although there is evidence of side reactions occurring with the production of a green soluble product sometimes being seen [173].

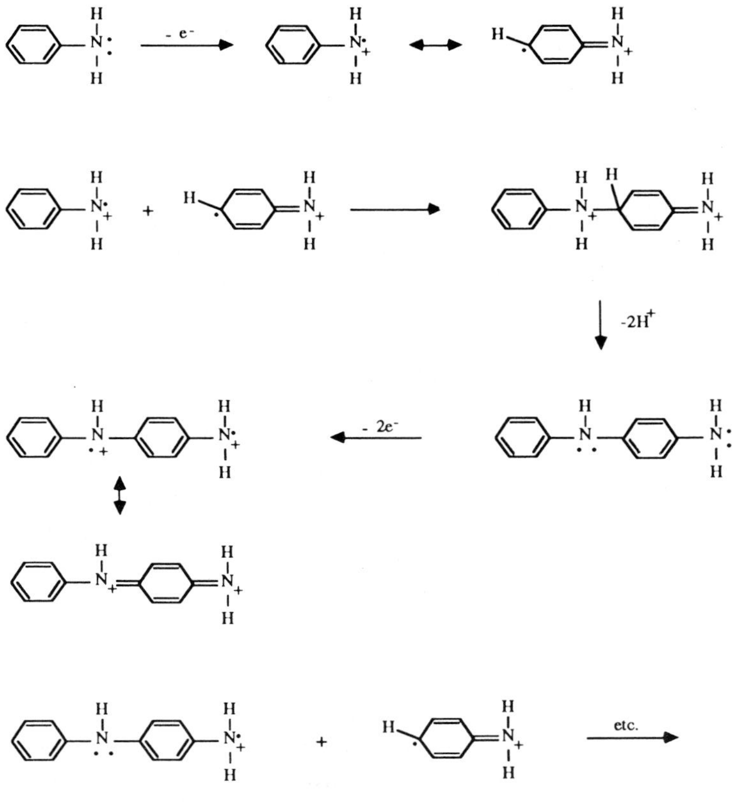

Fig. 10. General reaction scheme for the polymerisation of aniline.

4.3.2 Morphology and Structure

It was concluded at an early stage that the basic structural unit in polyaniline was that of the aniline octamer [189–191], but given this, a number of variations are possible containing different combinations of quinoid and benzenoid sequences, based around aniline dimers which are either fully reduced or fully oxidised (these dimers are generally referred to as the 1A and 2A forms respectively, as shown in Fig. 11). Several octameric compounds (NB, not polymers) containing specific combinations of the 1A and 2A structural units were isolated at an early stage [189], leucoemeraldine (of the form $1A_4$), protoemeraldine ($1A_32A$), emeraldine ($1A_22A_2$), nigraniline ($1A2A_3$) and pernigraniline $2A_4$. The emeraldine octamer has a deep green colour, is a base which will dissolve in some organic solvents and can be oxidized to nigraniline, which has similar properties, but is dark blue [167, 192]. It has been shown from IR spectra that samples of polyaniline contain quinoid rings [193, 194], and initially several forms of the polymer were identified from optical measurements [167, 168, 173, 195, 196], including the fully reduced form $(1A)_x$ (corresponding to the luecoemeraldine octamer), an intermediate radical cation form, and the fully oxidised form $(2A)_x$ (corresponding to the pernigraniline octamer), all of which are reported to be stable in aqueous solution, but it is clear that a given sample may contain various combinations of these [197], and that their relative ratios will depend on the conditions and treatments to which the material is exposed [195, 197] (different structural phases of the material have been identified by a.c. impedance measurements [198]). Given this complexity it is difficult to draw specific conclusions concerning the detailed structure of polyaniline, although a more general picture can be obtained.

Assuming that the polymerisation does occur *via* a radical intermediate, then coupling is possible at all three positions on the ring (albeit with differing probabilities) and a number of alternative products to the emeraldine-based structure can be

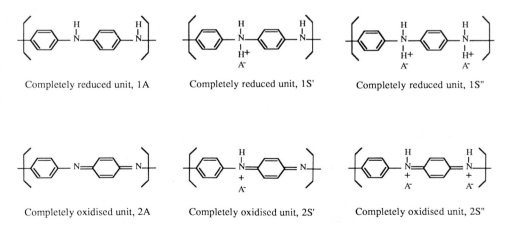

Completely reduced unit, 1A   Completely reduced unit, 1S'   Completely reduced unit, 1S"

Completely oxidised unit, 2A   Completely oxidised unit, 2S'   Completely oxidised unit, 2S"

Fig. 11. Dimeric structural units found in polyaniline.

envisaged [199] including both head-to-head and head-to-tail coupling of the mono-mer units [166]. This is supported by experiments using aqueous sulphate (pH1), phosphate buffer (pH7) or acetonitrile containing $NaClO_4$ and pyridine as the solvent system [200], where it was found that coupling occurs *via* the para- positions of the phenyl ring at pH1, but *via* the meta- positions at pH7 (or using the non-aqueous solvent system), leading to a non-conjugated polymer backbone. Similar results were found by Genies et al. [182] using HF solutions, where ortho- linkages appear to be formed, although Volkov et al. [165] also polymerised various aniline derivatives using a variety of solvent and reaction conditions, and found from IR and XPS measurements that all the monomers gave polymers with essentially the same basic polyemeraldine structure.

Theoretical calculations suggest that the torsion angle between adjacent rings in the chain is close to 90° in the polymer [201–203] although some overlap between the π-systems of adjacent rings is achieved *via* the orbitals of the nitrogens, but these results are not conclusive – the methods used for the calculations are very sensitive to the initial parameters chosen [201, 203]. It would be expected the steric hindrance between hydrogens on adjacent aromatic rings would prevent completely planar configurations [202]. Two alternative configurations (*syn-* and *anti-*) are possible for the regions involving quinoid rings, with slightly different energies, and studies on model compounds suggest that both will occur in polyaniline, with a small energy barrier for interchange between the two [202].

As with other electropolymerised materials, anions are incorporated into the material during the polymerisation reaction [180, 200], with a maximum occupancy of one anion per two aniline units, and up to 30–40 protonated nitrogens in the fully oxidised 2S" form [204]. Small quantities of oxygen are also incorporated into the films (< 1 oxygen per 100 carbon atoms), but the material does not show any air-sensitivity [204].

The polymerisation appears to involve a nucleation process similar to that of the deposition of metals [182], and electrochemically-prepared polyaniline will form dense, non-fibrillar thin films [165, 173], but thicker films (> 150 nm) become less densely packed and more fibrous [176, 182]. This may be due a change in the deposition mechanism when the film becomes sufficiently thick to inhibit direct access to the platinum by unreacted monomer [176].

4.3.3 The Electrochemistry of Polyaniline

The electrochemistry of polyaniline is more complex than that of other conducting polymers and given the large number of possible structures for the material, it is not surprising that many possible reaction schemes have been suggested [181, 182, 195–197, 205]. Many of the properties of the material are pH-dependent [173, 174, 206], including the open circuit potential [207] which is most positive at pH 0, and this is further complicated by the fact that not all the polymer chains are necessarily in exactly the same state at any given time [197]. Above pH 3 polyaniline films do not show any electroactivity, but are not electroactive even at low pH with meta-coupling [200], most probably due to their low conductivity (Section 4.3.4). Polyaniline also exhibits electrochromic behaviour, being yellow below 0 V and passing through green and blue to black if the potential is taken to + 1.0 V [173] (Fig. 12).

Initial reports of cyclic voltammetry of the material stated that the voltammogram contained three peaks [165, 173, 174, 182, 183], the positions and relative intensities of which varied with film thickness, pH and after exposure to air [173] (Fig. 13), and although the work of Kobayashi et al. [196] showed that the cyclic voltammogram is not affected by the nature of the electrolyte, there does appear to be some dependence of cyclic voltammograms on the anion as a result of morphological variations [208]. The peak currents i_{pa} and i_{pc} were found to scale linearly with sweep rate, as would be expected for a surface reaction [173] and performing the voltammetry in acetonitrile gave very slow kinetics with broad peaks [173]. It has been reported that some regions of the material react preferentially during cycling [207], but this may have been related to poor contact between the polymer and the electrode and has not been found subsequently [173]. Further work by Kobayashi et al. [180, 209] using cyclic voltammetry showed a more complex picture than initially supposed. It was found that the CV initially shows two processes with oxidation peaks located at approximately + 0.15 V and + 0.8 V, which correspond to the change of colour of the material from yellow to black, but that on cycling these peaks diminish in intensity

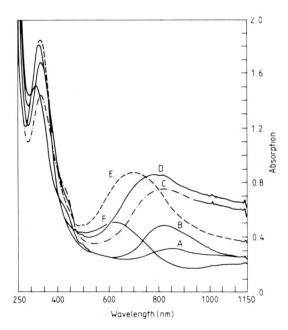

Fig. 12. Absorption spectra of polyaniline film on a semitransparent Pt electrode immersed in 1 M HCl. Spectra (a) to (f) were recorded at potentials of − 0.2, + 0.6, + 0.8, + 1.0, + 1.2 and + 1.4 V relative to a silver reference electrode. Reproduced from [173b].

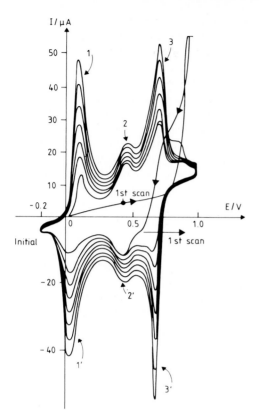

Fig. 13. Electrodeposition of PANi in N-F solution (NH$_4$F, 2.35 HF) 0.02 M aniline, by potential sweeping 20 mV s^{-1} from $-$ 0.2 V to 1.0 V. Reproduced from [182].

and a third process becomes evident with an oxidation peak at approximately + 0.5 V, which increases in intensity as the initial two processes decrease. This third peak then itself diminishes to become almost undetectable after 300 cycles, by which time the material shows little sign of any optical changes on cycling, remaining black. The interpetation of these results was that the most anodic peak corresponds to a degradation of the material and that the peak which appears + 0.5 V is associated with products of the degradation. Provided the potential is not swept beyond about + 0.7 V, the degradation process does not occur and the peak at + 0.5 V does not appear. Under these circumstances the material can be reversibly cycled (i.e., from $-$ 0.2 V to + 0.6 V) with concomitant colour changes (yellow at $-$ 0.2 V and green at + 0.3 V to + 0.6 V), but the degradation process would occur simultaneously with the polymerisation process if potentials in excess of about + 0.7 V are used, with the result that many early reports relate to partly-degraded material [180]. A mechanism for the degradation process was also proposed [209], in which benzoquinone is the main product (the appearance of the peak at approximately + 0.5 V in the cyclic voltammogram is in accord with the known electrochemical behaviour of benzo-quinone). It was found that if the polymer is cycled in the range -0.2 V to $+0.6$ V so that no degradation occurs, the cyclic voltammogram is strongly dependent on the pH of the solution [196], with the peaks becoming broader and being shifted to less positive potentials as the pH is increased. The origin of the peaks and their pH-dependence were believed to be due to a proton exchange reaction (with a correspond-ing colour change from yellow to an intermediate yellow-green state), with a separate anion insertion process (corresponding to a change from the yellow-green state to green) giving rise to the large capacitive currents seen in the cyclic voltammogram [196].

MacDiarmid et al. [181] proposed a model including four general forms of the polymer, two forms incorporating the 1A and 2A dimers (i.e., fully reduced or fully oxidised amine forms), and two incorporating protonated salt forms of these, the 1S and 2S forms, which has since been expanded to include two extra diprotonated salt forms – referred to as the 1S" and 2S" [195] (Fig. 11). Using this picture, the A and S forms can be interconverted by addition or removal of protons (i.e., exposure to acid or base), and the 1 and 2 structures by electrochemical oxidation and reduction, but both this and the similar model of Genies et al. [182, 183] do not consider the degradation process seen by Kobayashi [209], attributing the cyclic voltammetric peaks at $+0.5$ V to an interconversion between different forms of the polymer. Although specific combinations of these repeat units can be identified in principle, it now seems likely that in any material which can actually be prepared, the polymer will contain chains in which these units occur in a variety of combinations, in virtually any ratio depending on the precise conditions of pH and potential employed [203], and in order to fully characterise the behaviour of the material, several workers have constructed three-dimensional plots of potential, pH and conductivity (or charge) e.g. [205].

On the basis of theoretical calculations Chance et al. [203] have interpreted electrochemical measurements using a scheme similar to that of MacDiarmid et al. [181] and Wnek [169] in which the first oxidation peak seen in cyclic voltammetry (at approx. $+0.2$ V vs. SCE) represents the oxidation of the leucoemeraldine $(1A)_x$ form of the polymer to produce an increasing number of quinoid repeat units, with the eventual formation of the $(1A-2S'')_{x/2}$ polyemeraldine form by the end of the first cyclic voltammetric peak. The second peak (attributed by Kobayashi to degradation of the material) is attributed to the conversion of the $(1A-2S'')_{x/2}$ form to the pernigraniline form $(2A)_x$, and the cathodic peaks to the reverse processes. The first process involves only electron transfer, whereas the second also involves the loss of protons and thus might be expected to show pH dependence (whereas the first should not), and this is apparently the case. Thus the second peak would represent the production of the diprotonated $(2S'')_x$ form at low pH and the $(2A)_x$ form at higher pH with these two forms effectively in equilibrium mediated by the H^+ concentration. This model is in conflict with the results of Kobayashi et al. [196] who found pH dependence of the position of the first peak.

To summarise, the electrochemistry of polyaniline shows a complex pattern of behaviour with evidence of both proton addition/elimination and anion insertion/removal (i.e., doping) processes occurring [197, 178, 196, 182] during potential cycling, and an unequivocal assignment of the various different processes has yet to be achieved.

4.3.4   Electronic Structure and Conductivity

The conductivity of polyaniline can be varied over fifteen orders of magnitude under appropriate conditions, up to a maximum of 10^2 S cm^{-1} [168, 207, 210–212], typical values being 10^{-3}–10^{-2} S cm^{-1} when oxidised and 10^{-2}–10^{-1} when reduced, although material with predominantly meta-coupling has a very low conductivity $(10^{-15}$–10^{-14} S cm$^{-1})$ [165, 200]. The conductivity of the material is dependent on the two processes discussed above, (proton exchange and ion insertion/removal) and to fully characterise its behaviour both pH and electrochemical doping must be controlled. Unlike other conducting polymers, a plot of conductivity against potential for the polyemeraldine form of polyaniline shows a maximum, reaching 1 S cm^{-1} between about -0.15 V and $+0.2$ V, falling by several orders of magnitude at both more cathodic and more anodic potentials than this range [195] and the material can be cycled reversibly within the approximate range -0.3 to $+0.4$ V with corresponding reversible changes in conductivity [195].

The model of MacDiarmid et al. [181] can be used to explain some of the processes which can occur in the poly(emeraldine) form (i.e., equal numbers of benzenoid and quinoid repeat units) on changing the pH of the solution with which the material is in contact. In this model, the polymer has the 1A-2S" form at pH 0, with resonance stabilisation and delocalisation occurring, making all the nitrogen atoms equivalent, along with all C–N bonds and all carbon rings, giving rise to high conductivity. A gradual conversion of 2A units to 2S' and 2S" units is envisioned as the pH becomes more acidic, with the conductivity increasing by approximately an order of magnitude per pH unit between pH 6 and pH 0 [181, 213]. Thus protonation of the polyemeraldine form (equal numbers of 1A and 2A repeat units) has the same effect as the doping of materials such as polypyrrole and imparts higher conductivity [181, 195] (the electrochemical doping of polypyrrole is an equivalent process to the protonation of polyaniline in that both result in the removal of electrons from the π-system). It should be emphasised that this relates to the material which remains in the emeraldine structure (equal numbers of benzenoid and quinoid repeat units), and that the fully-reduced $(1A)_x$ leucoemeraldine form of the material shows little change of conductivity on protonation [195]. Raman [214], ESCA [194] and ^{13}C-NMR [202] measurements indicate that during the protonation reaction the number of quinoid rings remains constant, and that the majority of the positive charge is located on the nitrogen atoms.

In view of the complexity of the material, it is difficult to unequivocally assign a particular mechanism for electronic conduction in the polymer, although some evidence exists to suggest that it involves three-dimensional variable-range hopping as found for other polymers [213]. It has also been suggested that the conductivity of polyaniline is a combination of both ionic and electronic conductivity [207], and is affected by the nature of the dopant anion [204] e.g., 0.3 S cm^{-1} for SO_4^{2-} doping [204], 0.5–2 S cm^{-1} for acetate [171] and 10 S cm^{-1} for BF_4^- [204]. A mechanism for inter-chain charge-transfer has been proposed involving a redox reaction between the diamine and diimine forms of the polymer [202].

Considering the electronic structure of the polymer, it is clear that the poly(leucoemeraldine) $(1A)_x$ material is an insulator and would be expected to have a large bandgap, as would the fully oxidised, diprotonated form of the material $(2S'')_x$ which has no electrons in the π-system [175]. The possibility of a degenerate ground state arises in the $(2A)_x$ form (with the interchangeability of the quinoid and benzenoid rings) [181], which should in principle allow bond alternation defects (i.e., solitons) as found for polyacetylene [204]. The 1A-2S'' dication form of polyemeraldine occurring at pH 0 can be considered as supporting bipolarons, although it is possible that the structure may have a lower energy if the charges separate into polarons (i.e., radical cations) [203]. The magnetic susceptibility measured for the diprotonated material is inconsistent with the presence of bipolarons, and strongly suggests that separation into a polaron lattice occurs [215], and this is supported by optical and other data [216, 217], although debate still continues [218]. The proposal has also been made that on protonation localised metal-like areas are created in an insulating matrix [171, 216, 217], rather than producing a uniform material. The conductivity of polyaniline can be cycled over a narrow range by successive exposure to water vapour and subsequent pumping [171] and this is explained on the basis of this metallic island model, with the water vapour decreasing the resistance between these islands and thus increasing the conductivity. Thus the limiting factor for the conductivity of the material is seen in this instance as being the inter-domain resistance rather than intra- or inter-chain electron transfers (a diffusion coefficient for charge carriers within polyaniline of 10^{-10}–10^{-8} cm^2 S^{-1} has been estimated [200]). Monkman et al. [197] however found no evidence of a metal-like state, and conclude that the bipolaron model is the more likely, although noting again that the complexity of the material precludes a completely unequivocal description at this stage.

4.3.5   Electrochemistry at Polyaniline Electrodes

In view of the uncertainty which surrounds the exact nature of the material and the mechanisms involved in its reactions, it is not surprising that relatively little work has been done on applications of the material, and its use as an electrode material, although some references do exist in the literature. It has been reported that the electrochemistry of ferrocene is similar to that at a platinum electrode, although the E^0 for the reaction is shifted slightly cathodic (by about 20 mV) [173], and there is some evidence that the kinetics of the ferrocene couple are faster at polyaniline electrodes than at Pt [174]. Various battery systems have also been proposed incorporating polyaniline [207, 219, 220], based on both the reduced and oxidised forms [220], and with reasonably high coulombic efficiencies (in excess of 90% [220]), and a system has even been proposed incorporating polyaniline as a sensor for redox reagents such as $Fe(CN)_6^{3-}$ [175]. Sulfonated cobalt pthalocyanines have also been incorporated into polyaniline giving a material at which catalytic reduction of oxygen occurs [221].

4.4 Polyazulene (Fig. 14)

The electropolymerisation of azulene was first reported by Tourillon and Garnier [39], and was found to give smooth homogeneous films on Pt and ITO glass, which were insoluble in common organic solvents and were thermally stable up to approximately 400 °C. Bargon et al. [41] attempted to polymerise a variety of azulene

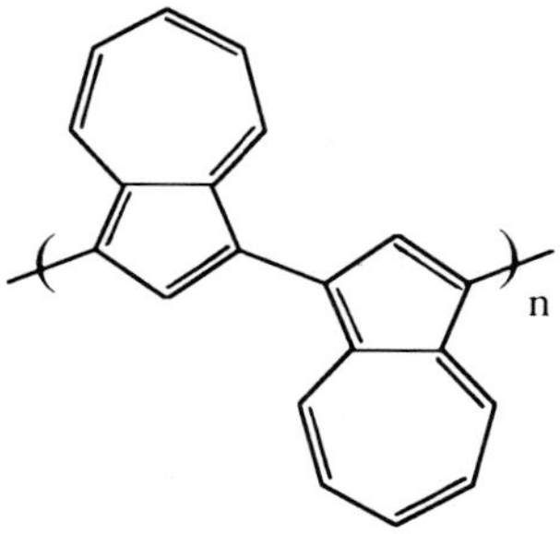

Fig. 14. The structure of polyazulene.

derivatives, finding that only azulene itself and 4,6,8-trimethylazulene would polymerise (Fig. 15), and concluded that the 1- and 3-positions of the monomer were involved in the polymer linkage, and that the seven-membered ring was not involved in the polymer backbone. Rotating disk studies show that current efficiencies of 100% can be achieved for the polymerisation [222], and an ECE mechanism has been suggested [41], where polymerisation occurs *via* radical cations arising from both the monomer and oligomers produced at earlier stages of polymerisation. ESR measurements on polyazulene [40] show interesting features which can be explained by a model in which the neutral material contains a large number of spins due to the presence of inherent defects. The material is stable on cycling [223] and shows a two-stage doping process – the presence of the dopant anion is thought to give rise initially to spinless, charged solitons, causing a reduction in the number of spins seen by ESR whilst at higher doping levels, these solitons would be expected to dimerise to give polarons, with a resulting increase in the number of spins [40, 223]. It is also possible that at very high dopant levels bipolarons are formed, with a subsequent reduction in the number of spins. The optical spectra of polyazulene show complex changes on doping with ClO_4^- [40], from which a unique 'dual band structure' has been proposed, corresponding to a chemical structure where each monomer unit in the polymer backbone carries a single unpaired electron.

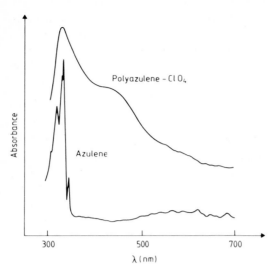

Fig. 15. Visible spectrum of azulene and polyazulene-ClO₄ (on Nesa glass). Reproduced from reference 41, by kind permission of Gordon and Breach Science Publishers S.A.

4.5 Poly(paraphenylene) (Fig. 3)

Like many of the other conjugated polymers poly(paraphenylene) was initially prepared *via* chemical synthesis (in a highly crystalline form) [30, 86, 87, 224–226] and could not easily be produced in forms other than as a powder. Ivory et al. [227] demonstrated that the material could be chemically doped to a highly conducting state (up to 10^4 S cm^{-1}), since when electrochemical synthesis of the polymer has been achieved *via* a number of different monomers [38, 46–49, 228–231], producing material with a range of conductivities (Table 1). By including Friedel–Crafts catalysts AlCl₃ [47] and CuCl₂ [46, 232] in the electrolyte solution, films with conductivities in the region of 10^{-4}–10^{-5} S cm^{-1} [47] and 100 S cm^{-1} [46] have been obtained. The role of the catalyst is not clear, but it has been suggested that with AlCl₃ the reaction may involve the formation of a complex between the catalyst and the benzene, or a species of the form $R_3N^+AlCl_3^-$ in conjunction with the ammonium electrolyte salt (no product is seen in the absence of the electrolyte) [47]. The material produced appears to contain a large number of structural defects, which may explain the low conductivities achieved. In the absence of catalysts, passivating films have been obtained on platinum from polymerisation of benzene in sulfur dioxide [233] although reversible behaviour has also been seen from the same system [49, 230]. Using a two phase system of benzene and HF at 7 °C, Rubenstein [48] produced semi-conducting amorphous films which had conductivities of up to 10^{-2} S cm^{-1} when doped with AsF_6^-. In addition to the desired para-linkages the material showed evidence of ortho- and meta-links and oxygen-containing side groups. Dibromobenzene (both the 1,4- and 4,4'- isomers) can also be cathodically polymerised using a nickel catalyst [50, 234] to give poly(paraphenylene) with up to eighteen benzene rings per chain (from IR data), but no polymer could be obtained using the corresponding trimer or tetramer. The reaction mechanism appears to involve the conversion of the catalyst to an arylnickel complex the formation of which can be seen in cyclic voltammograms of the reaction mixture.

The electrochemistry of the polymer has not been widely studied, although both p- and n-doping have been achieved electrochemically [49, 224, 229, 232, 235, 236] and the band gap of the material measured as 2.8 eV [236]. Reduction of the polymer appears to commence at 1.0 V vs. Li/Li$^+$, and oxidation at approximately 3.8 V [236] (Figs. 16 and 17), the latter making the electrochemistry more difficult than for many other materials as a suitable solvent system is required which is stable in this range and has low nucleophilicity [49]. Substituted polyphenylenes have also been investigated, with similar effects found as for other materials [224] i.e., that electron-donating substituents reduce the oxidation potental of the material and that electron withdrawing groups have the opposite effect. The structure of the poly(paraphenylene) is thought to involve pairs of polymer chains packed in layers with the dopant intercalated, also in layers, although with evidence of several structural phases [86, 87, 89]. The material has been p-doped to levels of up to 10% using ClO_4^- and PF_6^- with coulombic efficiencies of 64–70% [236], although using low charge and discharge currents and keeping the potential of the polymer below 4.2 V, efficiencies of 100% were obtained for PF_6-doping. Optical spectra obtained for the p-doped material [229] are consistent with the formation of bipolarons as discussed in Section 2.2, although there is evidence of hysteresis in plots of doping level vs.

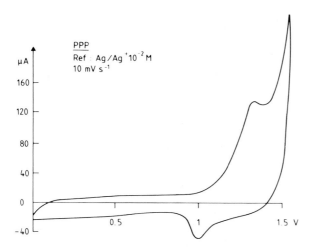

Fig. 16. Cyclic voltammogram of poly(paraphenylene) PPP in CH$_3$CN-0.2 M LiClO$_4$ using PVF + C paste on 0.2 sq. cm Pt electrode. Reproduced from ref. 224, by kind permission of Gordon and Breach Science Publishers S.A.

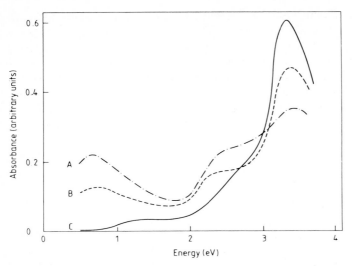

Fig. 17. Absorption spectra of poly(p-phenylene) film at various applied voltages in the electrolysis cell. Voltages of (A) − 3.4, (B) − 3.3 and (C) 1.0 V are applied between poly(p-phenylene) and the counter-electrode (Ni). Reproduced from [232].

potential [86, 89] as found for polyacetylene (Section 4.2) and the low coulombic efficiencies seen at higher charge/discharge rates have been attributed to the oxidation of impurities and production of permanently-doped regions in the material. Similar experiments using Bu_4NPF_6 in SO_2 have yielded an average charge/discharge capacity of 85 A h kg^{-1}, with one charge per four phenylene units [49].

Reversible n-doping has been achieved [229, 232, 236] with coulombic efficiencies of 90% and doping levels of greater than 20% [236], although some degradation of the electrolyte was seen. From the *in situ* optical data a band structure corresponding to the formation of bipolarons was proposed (with a band gap of 3.0 eV) with the unusual feature of the levels within the gap moving towards the band edges on increasing the doping level (attributed to structural changes in the polymer on doping). Electrochemical oxidation and reduction of the related polymer poly(1,4-phenylene-1,2-diphenylvinylene) (PDPV) has also been achieved as films [237] and in solution [238].

5   Five-Membered Heterocycles and Derivatives

5.1   Polypyrrole (Fig. 18)

Pyrrole was first polymerised in 1916 [239, 240] by the oxidation of pyrrole with H_2O_2 to give an amorphous, powdery product known as pyrrole black, which was found to be insoluble in organic solvents. The structure of the material has yet to be elucidated, although elemental analysis reveals that it contains significant quantities of oxygen and also C, H, and N in proportions that suggest linked pyrrole units, with a molecular weight in the region of 800–1000 [241]. Little further interest was shown until pyrrole was electrochemically polymerised in the form of continuous films [242–245], since when polypyrrole and many of its derivatives have received a great deal of attention (perhaps more than any other material with the exception of polyacetylene).

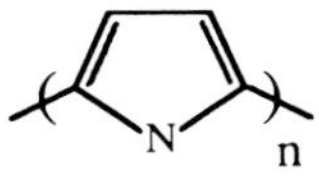

Fig. 18. The structure of polypyrrole.

A large number of different pyrrole-based polymers have now been electrochemically synthesised, using a variety of conditions, and these are summarised in Table 2, although it should be noted that the size of this field and its rate of growth mean that it is impossible to make such a table completely comprehensive, and that reports of related new materials, particularly of copolymers incorporating pyrrole are continually appearing in the literature. Water-soluble polypyrroles have also recently been reported [246].

It is clear from Table 2 that a wide variety of conditions and counterions have been used for the preparation of pyrrole-based polymers, and this makes comparison between different data difficult, although some general observations can be made. The polymer is stable under vacuum [56, 243], but is reactive to O_2 when left in the neutral state if prepared in the absence of oxygen [58, 243, 244, 261, 262]. If electrochemically oxidised the material shows considerably less sensitivity to oxygen (as measured by its conductivity) even at elevated temperatures [57, 245, 262], except when in the form of very thin films [262]. The nature of the reaction with oxygen is not clear but evidence suggests that the neutral material is actually doped on exposure to oxygen to a state similar to that when electrochemically oxidized [58], although ESCA [262, 263] and mass spectrometry [264] show evidence of the formation of carbonyl groups, also with a high degree of OH substitution on the pyrrole nitrogen in the latter case. The electrochemically oxidized material is also decomposed by strong base [58, 265] and halogens [82], and a mechanism has been proposed to account for the effects of base, which involves the gradual formation of carbonyl groups on the pyrrole rings which irreversibly affects their electrochemical behaviour [265]. Reversible water uptake by the polymer has also been reported [266].

5.1.1 Polymerisation

Although pyrrole can be chemically synthesised [239–241, 267, 268], electropolymerisation is easily achieved and is the most common preparative method. This was first reported by Dall'Olio et al. [242] who prepared brittle polymer films on Pt by

Table 2. Electrochemical synthesis of polypyrroles.

Monomer (concn.)	Electrode/solvent[a]	Electrolyte (concn.)	E_{pol}(V)[b]	Counter ion[c]	σ(S cm^{-1})[d]	Ref.
Pyrrole						
pyrrole (10^{-3} M)	Pt/acn	Et$_4$NBF$_4$ (0.1 M)	$+1.200$ cv	BF$_4^-$	30–100 (f)	247
pyrrole (10^{-2} M)	Pt/acn	Bu$_4$NClO$_4$ (0.1 M)	$+0.8$	ClO$_4^-$	30–100	39
pyrrole (0.05 M)	ITO/aq	LiClO$_4$ (0.1 M)	$-$	ClO$_4^-$	50	248
		Et$_4$NBF$_4$ (0.1 M)	$+0.56$ to $+0.73$	BF$_4^-$	100	248
		Na$_2$SO$_4$ (0.1 M)	$+0.59$ to $+0.71$	SO$_4^{2-}$	1	248
pyrrole (0.05 M)		aq, KNO$_3$ (1 M)	$+0.6$ ($+0.9$ cv)	$-$	$-$	249
pyrrole (0.05 M)	aq	H$_2$SO$_4$ (10^{-1} M)	$+0.92$	SO$_4^{2-}$	6.5×10^{-2}	250
pyrrole (0.05 M)	aq	NaOH (10^{-1} M)	$+0.92$	OH$^-$	$<10^{-15}$	250
pyrrole (0.1 M)	ITO, Pt/acn	Et$_4$NBF$_4$ (0.1 M)	$+0.8$	BF$_4^-$	100 (f)	251
pyrrole (0.1 M)	polyacetylene	Bu$_4$NBF$_4$, Bu$_4$NClO$_4$ (0.1 M)	$+1.00$	$-$	$-$	252
pyrrole (0.17 M)	AlCl$_3$ melt/	BuPyCl (1:1 molar ratio)	$+0.7$ to $+0.9$ vs. Al/Al^{3+}	Cl$^-$	1	253
pyrrole (0.2 M)	GC/acn-aq(2)/N$_2$	Et$_4$NOTs	$+1.190$	$-$	$-$	254
pyrrole	Pt on glass, acn	AgClO$_4$ (0.1 M)	$+0.7$	ClO$_4^-$	50	58
pyrrole	GC/acn	Et$_4$NOTs (0.023 M)	$+3.0$ vs. C	OTs	105	255
pyrrole	GC/aq	Et$_4$NOTs (0.2 M)	2.6	OTs	59–83	255
pyrrole	Pt/acn	Et$_4$NBF$_4$	$+1.20$	$-$	$-$	256
bipyrrole	acn, nmp	Et$_4$NBF$_4$	$+0.55$	BF$_4^-$	3–60	257
terpyrrole	Pt/acn	Et$_4$NBF$_4$	$+0.26$	$-$	$-$	256
pyrrollo-3,2b-pyrrole	Pt, Au, ITO/acn	NaClO$_4$, BuNClO$_4$, Bu$_4$NBF$_4$ (0.2 M)	$+0.65$	ClO$_4^-$	5.0×10^{-5}	258
Substituted pyrrole						
3-methyl-	Pt, acn	AgClO$_4$ (0.1 M)	$-$	ClO$_4^-$	4	259
3-ethyl- (0.1 M)	ITO, Pt/acn	Et$_4$NBF$_4$ (0.1 M)	$+0.750$	$-$	$-$	55
3-COCH$_3$- (0.1 M)	ITO, Pt/acn	Et$_4$NBF$_4$ (0.1 M)	$+1.130$	$-$	$-$	55
3-COPh- (0.1 M)	ITO, Pt/acn	Et$_4$NBF$_4$ (0.1 M)	$+1.200$	BF$_4^-$	10^{-3} (f)	55
2,3-dimethyl-	Pt/acn	AgClO$_4$ (0.1 M)	$+1.0$ cv	ClO$_4^-$	10	259
2,3-dimethyl- (0.1 M)	ITO, Pt/acn	Et$_4$NBF$_4$ (0.1 M)	$+0.350$	BF$_4^-$	0.2 (f)	55
2,3-diethyl- (0.1 M)	ITO, Pt/acn	Et$_4$NBF$_4$ (0.1 M)	$+0.650$	BF$_4^-$	10^{-6} (p)	55
2,3-diphenyl-	Pt/acn	AgClO$_4$ (0.1 M)	$+1.200$ cv	ClO$_4^-$	10^{-3}	259

N-substituted – pyrrole

	Electrode/Solvent		Potential	Counter ion	Conductivity	Ref.
-methyl- (10^{-3} M)	Pt/acn	Et$_4$NBF$_4$ (0.1 M)	+ 1.120 cv	BF$_4^-$	10^{-3} (f)	247
-methyl- (0.05 M)	ITO/aq	LiClO$_4$ (0.1 M)	+ 0.67 to + 0.81	ClO$_4^-$	5.0×10^{-3}	248
		Et$_4$NBF$_4$ (0.1 M)	+ 0.65 to + 0.79	BF$_4^-$	2.0×10^{-3}	248
		Na$_2$SO$_4$ (0.1 M)	+ 0.54 to + 0.78	SO$_4^{2-}$	1.0×10^{-7}	248
-methyl- (0.1 M)	ITO, Pt/acn	Et$_4$NBF$_4$ (0.1 M)	+ 0.8	BF$_4^-$	10^{-4} (f)	251
-methyl- (0.2 M)	GC/acn-aq(2%)/N$_2$	Et$_4$NOTs	+ 1.150	–	–	254
-ethyl- (10^{-3} M)	Pt/acn	Et$_4$NBF$_4$ (0.1 M)	+ 1.220 cv	BF$_4^-$	2.0×10^{-3} (f)	247
-n-propyl- (10^{-3} M)	Pt/acn	Et$_4$NBF$_4$ (0.1 M)	+ 1.260 cv	BF$_4^-$	10^{-3} (f)	247
-n-butyl- (10^{-3} M)	Pt/acn	Et$_4$NBF$_4$ (0.1 M)	+ 1.220 cv	BF$_4^-$	10^{-4} (f)	247
-iso-butyl- (10^{-3} M)	Pt/acn	Et$_4$NBF$_4$ (0.1 M)	+ 1.240 cv	BF$_4^-$	2.0×10^{-5} (f)	247
-phenyl- (10^{-3} M)	Pt/acn	Et$_4$NBF$_4$ (0.1 M)	+ 1.800 cv	BF$_4^-$	10^{-3} (f)	247
-phenyl- (0.2 M)	GC/acn-aq(2%)/N$_2$	Et$_4$NOTs	+ 1.410	–	–	254
-CH(CH$_3$)(CH$_2$OAc)	–				10^{-6}	260

[a] Electrode/Solvent:
acn – acetonitrile, aq – water, bzn – benzonitrile, mc – methylene chloride, nbz – nitrobenzene, nmp – N-methyl-2-pyrollidone, odcb – o-dichlorobenzene, pc – propylene carbonate, thf – tetrahydrofuran.

NB. Mixed solvent systems are shown as e.g. acn-aq (0.01 M) where the number in parentheses indicates the concentration of the lesser constituent ITO – Indium/tin oxide-coated glass, Ar – Solutions purged with argon, Ar atm – Experiment performed under an argon atmosphere, N$_2$ atm – Experiment performed under a nitrogen atmosphere.

[b] All potentials are measured vs. SCE unless otherwise stated:
E_{pol} – Potentials are as quoted in the original Ref., cv – The potential given was obtained from cyclic voltammetry.

NB. Where only a current density was given in the original Ref. this is quoted in place of the polymerisation potential.

[c] Counter ions are as incorporated during the electrochemical polymerisation process or by subsequent electrochemical doping unless suffixed with (chem.) which indicates the use of chemical doping.

[d] Conductivities: f – film, p – pressed pellet.

oxidation of the monomer in sulfuric acid solutions. The films were reported to have a room temperature conductivity of 8 $S\,cm^{-1}$, and were found to be ESR-active with elemental analysis corresponding to one positive charge per three pyrrole monomer units.

High quality films were first achieved by Diaz et al. [243–245] using a tetraethyl-ammonium tetrafluoroborate (TEATFB)/acetonitrile electrolyte under galvanostatic conditions. Polymerisation from aqueous solution was successful, but films grown from acetonitrile were of a higher quality, although pure acetonitrile containing no water at all gave non-uniform films which adhered poorly to the electrode surface. Increasing the water content in aprotic solvents gave better adhesion of the film.

Since these initial reports, the electropolymerisation of pyrroles has become a routine matter, and films have been prepared under both potentiostatic and galvano-static conditions on a large number of substrates including semiconductors [243, 269] and using various electrolytes (as illustrated in Table 2). Photoelectrochemical gener-ation of the polymer has also been performed on semiconductor substrates [270, 271], which was found to lower the potential at which the pyrrole polymerised compared to platinum metal. Generally polymerisation onto metals gives smoother, more adherent films than materials such as ITO glass or semiconductors [243], although on some metals (A1, In, Ag, Fe) no polymerisation occurs [269]. Other substrates which have been investigated for the electrochemical polymerisation of polypyrrole include carbon fibres [272], and glassy carbon [273]. Initial attempts to polymerise pyrrole under dry-box conditions were unsuccessful suggesting that either oxygen or water were necessary for the reaction to occur, but it was shown that this was solely due to the absence of a suitable couple for the cathode reaction under these conditions, which can be provided by using $AgClO_4$ as the electrolyte (i.e. $Ag^+ + e^- \rightarrow Ag^0$) [58, 274]. Under these conditions it was also found that the potential for polymerisation of the monomer was reduced by 200–400 mV.

The polymerisation reaction appears to be complex [57, 269] with the con-comitant formation of species such as acetonitrile/water complexes, acetonitrile dimers and hydrogen-bonded pyrrole/acetonitrile complexes [269]. Substitution on the pyrrole rings affects the potential at which the polymerisation occurs, and can move the oxidation potential either more or less anodic [259, 262, 275–282] than for the unsubstituted pyrrole monomer, the polymerisation of which commences at approximately +650 mV vs. SCE. The effect of substituents has been explained partly on the grounds of steric interactions (with the increasing substituent size making the polymerisation more difficult), but also considering the interaction between the substituent and the π system of the monomer [280, 283, 284] (with increasing electro-philic character of the substituent making the oxidation of the monomer more difficult), particularly with reference to phenyl substituted pyrroles which show much more anodic Epa values [275, 278–280, 284] in cyclic voltammetry. It has also been suggested [262] that polymerisation of the unsubstituted pyrrole involves the hy-drogen attached to the nitrogen of the pyrrole.

Cyclic voltammograms of solutions of the pyrrole monomers also show a depend-ence on the anion present in the electrolyte and can show multiple peaks with Epa values ranging from +1.0 to +1.3 V in acetonitrile [57, 279]) (+0.8 to +1.1 V vs.

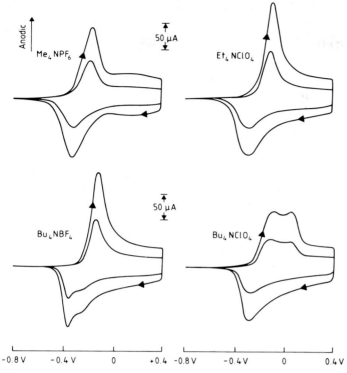

Fig. 19. Cyclic Voltammograms of [Pt]polypyrrole (20 nm thick) in CH_3CN containing various electrolyte salts. Sweep rates: 50 and 100 mV s^{-1}. Reproduced from [262].

SCE in aqueous solutions [249]) (Fig. 19). If the potential sweep is continued too far positive (> + 1.0 V [249, 262, 285]), the film appears to be irreversibly oxidized although there are reports of degradation at lower potentials than this [266], but providing degradation does not occur, the voltammograms show a consistent build up of the polymer on the electrode [249, 262]. The CV also shows a large capacitive current in the region anodic of the redox potential, probably due to the increase in effective surface area of the film [262, 286]. Mirabeau [287] has studied the effects of temperature and current density on the polymerisation of pyrrole on stainless steel from propylene carbonate solution, finding that if the current density is between 1 and 10^5 μA cm^{-2} and the temperature is between $-10\,°C$ and $80\,°C$, a conducting polymer is formed, but if either the current density exceeds 10^5 μA cm^{-2} or the temperature exceeds $80°C$ no product is obtained. If a current density below 1 μA cm^{-2} is used powder is obtained in place of film. Hahn et al. [263] also observed a maximum in plots of conductivity vs. polymerisation current density at 0.75 μA cm^{-2}, and Panero et al. [288] found that a higher maximum doping level is achieved from a higher current density, but make no mention of a maximum occurring. In solutions containing strong nucleophiles such as cyanide or methoxide, attack of the pyrrole ring by the nucleophile to give a derivatised monomer is favoured over polymerisation [289, 290].

The polymerisation reaction appears to involve radicals, although several suggestions have been made as to the initiation and propagation steps involved [259, 269, 291–293] including a free-radical chain polymerisation [269] and a radical-radical coupling mechanism [259, 291, 293] with the coupling being the rate determining step in the film formation (a general reaction scheme is shown in Fig. 20). It has been suggested that the initiator radical could arise from the electrolyte anion rather than the monomer [269], although as noted by Chandler and Pletcher [1] this is unlikely at the potentials normally employed for the reaction, and it is not consistent with the transfer of two electrons per molecule as found by Genies et al. [291]. It is generally assumed that growth of the film occurs *via* a nucleation process similar to that of the deposition of metals (e.g. [266]), but ellipsometric measurements by Jansson et al. [294] suggest that the polymerisation process does not occur *via* localized nucleation, and that the product grows homogeneously on the substrate.

Generally polypyrrole is prepared from electrolyte solutions employing organic solvents, but since the initial work by Diaz et al. [243–245, 295], many workers have

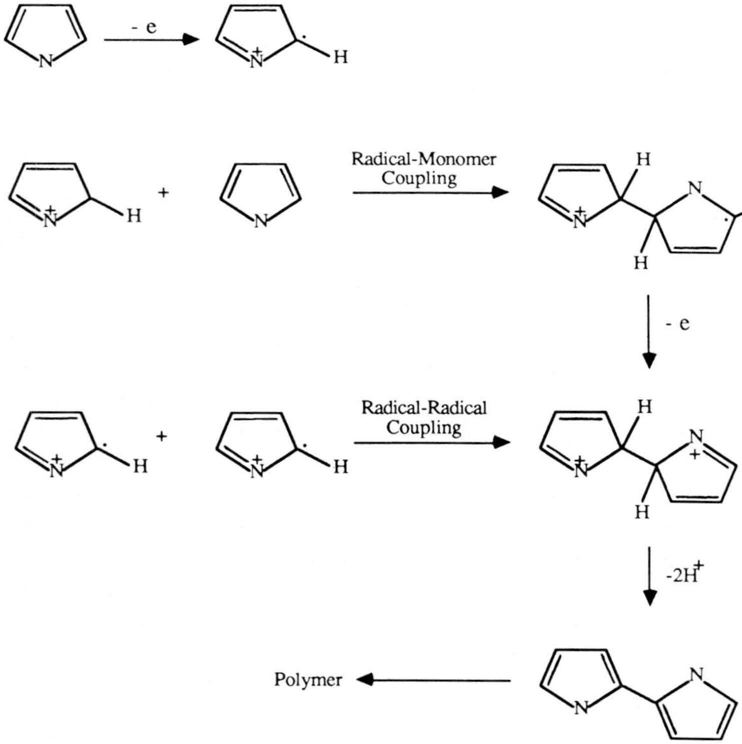

Fig. 20. A possible polymerisation mechanism for five-membered heterocyles showing the two alternative reaction pathways of radical-radical coupling and radical-monomer coupling.

studied the polymerisation reaction in aqueous solution [17, 248–250, 255, 285, 296–301], achieving films with similar properties to those produced from organic solvent solutions. Hyodo and MacDiarmid [248] showed that the presence of a methyl substituent on the pyrrole nitrogen can affect the nature of the polymer formed with certain anions–using BF_4^- or ClO_4^- similar results were obtained as for the corresponding unsubstituted material, but with SO_4^{2-} the polymer has significantly lower conductivity. From spectroscopic measurements, it appears that this is due to the formation of carbonyl groups on the pyrrole ring, as a result of formation of a radical dication during the polymerisation process, promoted by the SO_4^{2-}. Hydrolysis of the cation could then give rise to carbonyl groups, although it is not clear why the SO_4^{2-} ion should promote this when others do not, unless this is due to the stabilising effect of its higher negative charge.

It has been proposed that the mechanism of polymerisation on metals in aqueous solutions is similar to that found in organic solvents, with evidence for instantaneous nucleation and three-dimensional growth [249, 302]. It has been found that a minimum surface coverage of the electrode must be achieved for this to occur [303], which in turn depends on the concentration of the monomer in solution (10^{-3} mol dm^{-3} minimum). The mechanism is thus believed to involve the initial adsorption of the monomer onto the electrode surface, followed by generation of a adsorbed radical cation and subsequent radical coupling [303]. The pH of the solution does not affect the potential at which the polymerisation occurs [249], but at highly alkaline pH the shape of the cyclic voltammogram is significantly altered and irreversible behaviour is seen [304, 285] most probably due to degradation of the film. A different mechanism is proposed for the polymerisation on ITO glass [305], involving initial oxidation of surface Sn groups which then react with a pyrrole monomer to form a bond between an Sn atom and the pyrrole ring. This bonded pyrrole is then re-oxidized and couples with another monomer unit, forming the basis of a radical chain mechanism for the polymer growth.

Bidan et al. [283, 306, 307] have fabricated a number of N-substituted pyrroles with a variety of pendant groups, finding that steric interactions inhibit polymer formation with many substituents having alkyl groups of only two or three carbons, but for larger substituents polymerisation is easily achieved (the optimum being an alkyl chain of six carbons). They have also successfully polymerised compounds consisting of two pyrrole rings linked by a short alkyl chain [283].

A large number of copolymers and composites have been produced incorporating polypyrrole [223, 254, 281, 282, 308–320], including a polypyrrole/polyphenylene oxide [311] copolymer produced by electrochemical polymerisation. Pyrrole has also been polymerised in the presence of latexes [315] (giving a composite which can then be dispersed in organic solvents to enable some degree of processibility), and polymeric electrolytes [312] giving products with good mechanical properties and conductivities of up to 42 S cm^{-1}. Copolymers of pyrrole with substituted pyrroles have been electrochemically synthesised [254, 308, 318] with conductivities ranging over six orders of magnitude depending on the relative ratio of monomers used. Variations are found between the ratio of the different monomer units in the final polymer and the corresponding ratio for the initial electrolyte solution, and these have been

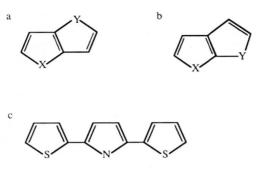

Fig. 21. (a) Pyrollo-3,2b-pyrrole (X = NH, Y = NH); thieno-3,2b-pyrrole (X = NH, Y = S); thieno-3,2b-thiophene (X = S, Y = S). (b) Thieno-2,3b-pyrrole (X = NH, Y = S); thieno-2,3b-thiophene (X = S, Y = S), (c) 2,5-di(2-thienyl)-pyrrole.

explained by the differences in the ease with which radicals are formed by the different substituted monomers [254]. A series of polycyclic monomers containing pyrrole and aliphatic carbon rings have also been electropolymerised [321], as have bicyclic pyrroles (i.e., with two pyrrole rings sharing a common bond as shown in Fig. 21a and b) [258] to give semiconducting materials (with conductivities up to 10^{-5} S cm^{-1}).

Attempts to polymerise pyrrole dimers and trimers were not initially very successful, with evidence for the formation of short chain oligomers rather than continuous films [256], but good quality films have since been obtained using the dimer [257]. The oxidation potential of the oligomer decreases linearly with the number of pyrrole units [256] – probably due to the increased stability of the radical cation, which may also allow the diffusion of the radical further from the electrode surface with subsequent reactions other than polymerisation occurring. Some reports exist in the literature of copolymers of pyrrole with thiophene [83] formed from the two monomers, and also from a trimer (2,5-di(-2-thienyl)-pyrrole) containing pyrrole linked to two thiophene rings *via* the α positions [160] (Fig. 21c), which forms as a powdery film with conductivities of up to 0.1 S cm^{-1} and which is soluble in acetonitrile, retaining its original conductivity on evaporation of the solvent. The polymerisation of bicyclic monomers containing joined pyrrole and thiophene rings [272] sharing a common bond have also been reported (Fig. 21a and b).

5.1.2 Morphology and Structure

Corresponding to the variety of conditions used for the synthesis, the literature contains a large amount of work on the structure and morphology of pyrrole-based polymers. Electrochemically-prepared films are space-filling (i.e, non-fibrillar) [243, 322] and non-crystalline [245, 255, 296, 322, 323], and contain anions from the supporting electrolyte, incorporated into the polymer matrix during polymerisation. Elemental analysis [261, 324] is in accord with a polymer containing primarily monomer units which remain intact after polymerisation, although the precise geometry of the pyrrole rings has been widely discussed (for a comprehensive summary see [325]). The maximum doping level generally corresponds to approximately one fluoroborate anion to four pyrrole rings when BF_4 salts are used as the supporting electrolyte, although this ratio depends on the anion used [58, 259, 323, 326].

It has been found that a wide range of factors affect the morphology and mechanical properties of the material [243, 244, 255, 269, 298, 327, 328]. These include electrode material, solvent and electrolyte salt, oxygen and water content of the system, current density used for the polymerisation reacton, and even the current density history. Although a complete quantitative measure of these effects has not been established, some general observations can be made. Thin films generally appear smooth under an electron microscope, whilst thicker ones have a much more uneven, textured surface [266], and there is some evidence from X-ray data [323, 329] that thin films show preferential alignment of the pyrrole rings parallel to the electrode surface, and that this is enhanced by a lower polymerisation temperature and a higher polymerisation potential. Diaz et al. [243, 244] found that the best results (i.e., smooth adherent films) were achieved by increasing the current density from zero to a final higher value over a period of a few minutes and that low current densities gave rise to smoother films. Smoother morphology also results from the use of lower potentials for the polymerisation [330], possibly because side reactions which could disrupt the conjugation may occur more readily at higher potentials, although the Youngs' modulus decreases with increasing polymerisation potential suggesting that cross-linking does not occur to a large extent [298]. It has been suggested that for polypyrrole tosylate, which is hygroscopic, the polymer is actually plasticised by the solvent, thereby affecting the Youngs' modulus [255], although only relatively small (4–8%) elongations are possible without breaking [331]. Films containing tosylate as the counter ion also show greater long term stability as measured by their conductivity [255, 331].

Although there is some disorder in the structure of the polymer as shown by ^{13}C-NMR [58, 332, 333] and XPS [39, 259, 261, 334, 335], the principle linkage between monomer units is believed to be *via* the α and α' positions on the pyrrole ring [58, 259, 262, 336, 337] and the molecular weight or chain length of the polymer has been estimated by several workers with results ranging from 6–7 monomer units [259] to several thousand [338, 339]. Street et al. [259] used β,β'-substituted monomers to block polymer linkages at these positions and reported films with a higher degree of order and crystallinity, with evidence from X-ray data for localised areas of crystallinity separated by amorphous regions, and this is supported by EELS data [340]. It has also been found [259] that the polymer contains more hydrogen than would be expected (IR [58, 17] and NMR [333] data also show that there are sp^3 carbons in the polymer chain), and it has been suggested [259] that this is consistent with hydrogen addition at β-positions, which would not necessarily disrupt the conjugation of the polymer backbone. A model has been proposed [259] with packing of chains in a monoclinic structure, with anions in the doped polymer intercalated between chains in a plane, rather than between planes. Electron diffraction [341], X-ray diffraction [323] and EELS [340] also suggest that neutral polypyrrole shows some order with a layer structure similar to that of graphite, in regions of approximately 2–50 nm in size

[341] whilst remaining essentially non-crystalline and that substitution disrupts this due to steric effects [259]. The presence of large anions would also be expected to disrupt the structure and reduce the conductivity [259], as would mislinked rings [342], but when the counter ion is an alkylsulfate or alkylsulfonate it appears that the anions form a type of double-layer structure between layers of polymer chains [296, 343] and may act as a template for the molecular organisation [323].

Impedance measurements in aqueous solutions [344] show that coating platinum with polypyrrole results in a forty-fold increase in surface area and Bard et al. [345] and Skotheim [286] found that films are highly porous, as the substrate can still undergo reaction although covered by the film (although at a reduced rate). Cyclic voltammetry and laser interferometry [346] provide further evidence that the polymer is porous in that i_{pa} increases linearly with film thickness, indicating a corresponding increase in the surface area [262], and comparatively large capacitive currents are seen, again suggesting a large effective surface area [286]. Densities of between $0.52 \, \text{g cm}^{-3}$ and $1.58 \, \text{g cm}^{-3}$ have been reported [17, 57], depending on the counter ion incorporated, and decreasing with increasing substituent size for the N-substituted polymers (with the exception of N-phenylpyrrole, which also shows a higher conductivity than would be expected [279]). Electron microscopy shows that the doping/dedoping process can be accompanied by major changes in morphology [17], although it is not clear what the underlying processes might be other than swelling due to intercalation of ions into the polymer matrix.

Films grown in aqueous solutions appear to contain significant quantities of oxygen [242, 301, 304] and it has been proposed that this arises from reactions between the radical cations of the monomer during polymerisation, and water or oxygen [301]. The nature of the oxygen species in the material is not clear, although carbonyl groups have been suggested on the basis of IR spectra [301].

5.1.3 The Electrochemistry of Polypyrrole

The polymer as prepared is formed in the partially oxidised (p-doped) form due to the incorporation of anions from the electrolyte, and a wide range of anions have been used to dope the material (Table 2), including metallic complexes [347–349], but no electrochemical n-doping (with cations) has been reported, possibly because the nitrogen of the pyrrole takes part in other reactions at the potentials required for n-doping [340]. The films can be repeatedly cycled between the neutral and p-doped form in the range $-0.8 \, \text{V}$ to $+0.4 \, \text{V}$ in acetonitrile with no change in the voltammogram [291], but if the sweep is taken positive of $+1.0 \, \text{V}$ vs. SCE a large, poorly-defined irreversible peak occurs which corresponds to a loss of cyclability of the film [262], although the film is stable in the cathodic region down to $-2.3 \, \text{V}$ vs. SCE. Cyclic voltammograms obtained for polypyrrole in both aqueous and non-aqueous systems can be complex and show a dependence on the nature of the dopant anion [57, 249, 262, 322, 350] (Fig. 19), with the appearance of additional peaks in some cases (possibly due to the formation of excited states of ion-pairs) [251, 262, 350], and would be in accord with a multi-step doping process similar to that proposed for chemical doping [56, 335, 351], although Feldberg [352] notes that the high capacitive currents found make it impossible to determine exactly the degree of oxidation (doping) achieved by electrochemical means. Substituted pyrroles also show a variation in the E_{pa} and E_{pc} values for the cycling process [57, 275, 276, 279] with peaks generally occurring at more anodic voltages than for the unsubstituted material [262]. Diaz et al. [279] found that alkyl substituents at the N-position caused the E^0 values for the polymers to be shifted positive relative to pyrrole, and that the magnitude of the shift increases with the size of the substituent, probably due to the steric effects disturbing the planarity of the polymer backbone and destabilising the charged form of the material. This is supported by data from poly(N-phenyl pyrroles) with further substituents attached to the phenyl ring [275, 276, 280] (Table 3). It has also been suggested that hydrogen bonding via the N–H of pyrrole in the unsubstituted polymer causes stabilisation of the polymer cation resulting in its much less anodic E^0 [279].

The actual chemistry of the cycling process appears to be complicated. Diaz et al. [262] found that i_{pa} varies linearly with the sweep rate in the range 10–100 mV s^{-1} indicating that the reaction is occurring at the electrode surface, but elemental analysis [335], XPS [76, 261, 334] and Auger spectroscopy [263] indicate that there is

Table 3. Electrochemical data for N-phenyl-substituted polypyrroles (all data from cyclic voltammetry using 0.1 M Et$_4$NBF$_4$ in acetonitrile except[a, b]).

Monomer (concn.)	E_{pol}	Polymer			Ref.
		E_{pa}	E_{pc}	E^0	
Pyrrole					
-phenyl- (0.2 M)[a]	1.410	623	400	511	254
-2-bromo- (0.2 M)[a]	1.540	–	–	–	284
-3-bromo- (0.2 M)[a]	1.470	–	–	–	254
-4-bromo- (0.2 M)[a]	1.510	–	–	–	280
-phenyl- (10^{-3} M)	1.470	690	595	652	284
-2-methoxy- (10^{-3} M)	1.410	623	400	511	284
-2-chloro- (10^{-3} M)	1.505	770	630	700	284
-2-fluoro- (10^{-3} M)	1.527	850	660	752	284
-2-bromo- (10^{-3} M)	1.540	845	260	552	284
-4-methoxy-	1.310	0.66	0.50	0.58	280
-4-ethoxy-	1.310	0.66	0.55	0.61	280
-phenyl-	1.470	0.69	0.59	0.64	280
-4-methyl- (10^{-3} M)	1.480	0.67	0.54	0.61	280
-4-bromo- (10^{-3} M)	1.510	0.82	0.62	0.72	280
-4-fluoro- (10^{-3} M)	1.520	0.76	0.62	0.69	280
-4-COCH$_3$- (10^{-3} M)	1.520	0.86	0.70	0.78	280
-4-iodo- (10^{-3} M)	1.520	0.72	0.70	0.71	280
-4-chloro- (10^{-3} M)	1.530	0.85	0.69	0.77	280
-phenyl-[b]	–	500	380	430	356
-4-methyl-[b]	–	510	210	340	356
-4-chloro-[b]	–	540	320	430	356

[a] The solvent was Et$_4$NOTs, GC/acn-aq(2%) N$_2$-purged.
[b] The sample was chemically prepared.

a gradual change in the material whereby the original dopant anion is replaced to some extent by oxygen, and there is also evidence to suggest that on cycling oxygen is incorporated as a genuine dopant [56]. Although the source of this oxygen is unclear, XPS data suggest that it may arise from the decomposition of the perchlorate anion [353]. It has been suggested [324, 354] that repeated cycling also reduces the conjugation length of the polymer, increasing the number of short conjugated segments, and these changes also correspond to a reduction in the number of free spins present in the material as measured by ESR [261, 351].

Measurements using the quasi-equilibrium method of gravimetric electrochemical voltage spectroscopy (GEVS) provide evidence that ion pairing can occur within the doped polymer between the dopant ion (e.g., perchlorate) and the positive charges located on the polymer chain [25]. As a result of this, cationic species such as alkali metal ions have a higher mobility within the polymer matrix and can diffuse into and out of the polymer more easily than the actual anionic dopant. Thus on de-doping the material, the cationic species enter the polymer to neutralise the anion charge more quickly than the anionic species can diffuse out, and on subsequent re-oxidation these counter-cations are extracted from the matrix back into the electrolyte solution more quickly than additional anions can migrate in. Electrochemical doping and de-doping are therefore seen as more complex processes, with fluxes of ions of both charges in opposite directions in both cases. Further evidence to support this was obtained by Genies et al. using spectroelectrochemical methods [355] and chronoamperometry [356], who proposed that the perchlorate mobility is the limiting factor during doping, and that the cation mobility is the limiting factor during de-doping [355]. Skotheim et al. [357] also found using XPS measurements that anions appear to be irreversibly incorporated into the polymer matrix on cycling as would be expected from the above discussion, and Tanguy et al. [358, 359] have identified two types of dopant- deeply and weakly trapped. XPS [262, 336, 360] and IR data [17] also suggest that chemical doping/undoping is accompanied by deprotonation/ protonation of the nitrogen in the pyrrole, but it is not clear whether this is applicable to the electrochemical doping process.

It is inevitable that the kinetics of the doping and de-doping processes will also be affected by the size of the ions involved when the steric effects are considered i.e., the movement of the polymer chains, and ions and solvent within the film [262]. Many workers (e.g., [262, 266, 288]) have found that the cyclic voltammetry of polypyrrole films shows non-Nernstian behaviour and this is believed to be due to low ion mobility within the polymer. There is also evidence to suggest that the kinetics of the redox reaction are dependent on the film thickness [327, 288] with Nernstian kinetics being observed with thin films [355]. A number of values have been reported for the diffusion coefficient of dopant ions within the polymer, with values ranging from $5 \times 10^{-12}\,\text{cm}^2\,\text{s}^{-1}$ [287] to $6 \times 10^{-10}\,\text{cm}^2\,\text{s}^{-1}$ [355] for perchlorate (at $25\,°\text{C}$), although Genies and Pernaut [355] note that the measured diffusion coefficient depends to some extent on the bulk solution concentration of the dopant ion and the applied potential. Other more subtle effects have also been seen, with perchlorate having a higher diffusion coefficient in material polymerised from solutions containing PF_6^- than in material which was actually polymerised using ClO_4^-, due to morphological differences between the two materials [361], and fast kinetics have also been reported using molten salt electrolytes [253]. In considering these various results, it should be borne in mind that many of the observed effects will be interdependent. For example the diffusion coefficient will not only depend on film thickness, but also the surface morphology and the bulk structure of the material. These will in turn affect the measured conductivity and doping levels seen for given conditions, but are themselves dependent on the polymerisation conditions used.

Skotheim et al. [286, 357, 362] have performed *in situ* electrochemistry and XPS measurements using a solid polymer electrolyte (based on poly(ethylene oxide) (PEO) [363]), which provides a large window of electrochemical stability and overcomes many of the problems associated with UHV electrochemistrty. The use of PEO as an electrolyte has also been investigated by Prosperi et al. [364] who found slow diffusion of the dopant at room temperature as would be expected, and Watanabe et al. have also produced polypyrrole/solid polymer electrolyte composites [365]. The electrochemistry of chemically prepared polypyrrole powders has also been investigated using carbon paste electrodes [356, 366] with similar results to those found for electrochemically-prepared material.

Cycling experiments coupled with IR measurements suggest that solvent is incorporated into the polymer on cycling, probably in the form of a solvation shell around the dopant ion, and there is also evidence to suggest that the polymer undergoes structural rearrangement on cycling, particularly when the counter ion is a surfactant [265, 340]. In aqueous solutions degradation of the material occurs on cycling [285, 321], and this is pronounced at alkaline pH, possibly due to disruption of the π-conjugation within the chain as a result of nucleophilic attack by OH^- on the pyrrole rings.

5.1.4 Electronic Structure and Conductivity

As with the morphology and structure of the films produced, the conductivity of doped polypyrrole has been found to be highly dependent on the polymerisation conditions used [57, 243, 323] although some conflicting results have been reported. Satoh et al. [298] found that increasing the polymerisation potential in aqueous solutions reduced the conductivity of the product, with maximum conductivity being achieved at a potential of 600 mV. These effects were interpreted as being due to variations in the rate of the polymerisation reaction — at low reaction rates short conjugation lengths dominate, whilst at higher rates the conjugation length is greater, but at too high a potential other reactions could begin to disrupt the conjugation of the chain, reducing the conductivity. Others however [299, 323] have found that a higher potential *increases* the conductivity of the product, and this is attributed to morphological effects rather than changes in the inherent conductivity of the material at a molecular level [323]. Also, increasing the amount of water in the solvent *decreases* the conductivity (going through a minimum at about 3:1 acetonitrile to water) [331] possibly due to differences in solvent polarity affecting the π-system of the polymer, as does increasing the temperature at which the polymerisation is performed both in aqueous [298, 299, 323] and non-aqueous [287, 367] solvents. Data obtained from ESR measurements [367] suggest that the material produced at lower temperature has a higher degree of conjugation.

A number of factors which affect the conductivity of the material can be explained in terms of steric interactions which affect the conjugation of the π-system, i.e., by disrupting the planarity of the polymer backbone and affecting the degree of orbital overlap [57, 259, 308, 368], or due to structural changes involving the co-ordination of the ion [321]. These include the nature of substituents on the pyrrole ring (e.g., the conductivity of poly (β-β diphenyl pyrrole) is smaller by four orders of magnitude than that of β-β dimethyl derivative), the positions at which the monomer units are linked [259, 342] and the nature of the dopant anion [249, 255, 369] (with values ranging from 10^{-3} to 500 S cm^{-1} [17, 34, 57 , 247–249, 279, 296–299, 312, 322, 331, 343, 370]). Materials prepared in aqueous media using alkyl- and aryl-sulfonate salts as the electrolyte show a correlation between the concentration of sulfonate salt in the bathing solution and the conductivity [297, 298], but other factors may be involved as the temperature dependence of the conductivity also varies with the nature of the dopant – the overall dependence being greater for samples with lower conductivities, and with two types of dependence being identified for other anions [57]. Polypyrrole doped with anions containing multiple sulfonate groups has a lower conductivity than with dopants containing a single sulfonate function and this may again be due to steric effects, although the presence of the extra sulfonate groups may result in localisation of the charge on the polymer chains thus reducing the mobility of the charges and the conductivity [371]. It has also been suggested [41, 279] that the conducting form of unsubstituted polypyrrole is stabilised by hydrogen bonding, giving it a higher conductivity than might be expected, although this again may merely be due to a loss of planarity if the hydrogen is replaced by another group [259]. Stretching polypyrrole [367] has produced films with very high conductivities (> 1000 S cm^{-1}) with evidence from X-ray and conductivity data supporting the idea that some alignment of the pyrrole chains had occurred in the stretch direction (there is a degree of anisotropy of the conductivity parallel and perpendicular to the stretch direction).

There is large body of data suggesting that the actual mechanism of conduction involves variable range hopping [243, 255, 324, 326, 341, 370, 372], with the measured conductivity comprising two components, one representing the conductivity along the polymer chain, and the other representing the inter-chain conductivity (i.e., an effective resistance due to the activation energy for the hopping process). The charges in the doped material appear to be localised on the pyrrole rings [350, 373] and although there is considerable evidence from ESR data that initially on doping polypyrrole contains a large number of highly mobile spins [58], EELS measurements [340] show no evidence of a metallic conduction mechanism. The data are however consistent with the formation of polarons, with a concomitant increase in the ESR signal intensity as the dopant level is increased [351], pairing to form bipolarons at higher levels, with an accompanying fall in signal intensity [259, 326, 351, 374, 375]. This is also supported by theoretical calculations [14, 340, 376], and by spectro-electrochemical studies [308, 324, 377], which show the three transitions expected for polarons at low dopant levels, one of which disappears as the dopant level increases as

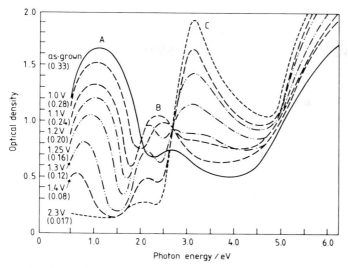

Fig. 22. Dependence of the absorption spectrum of PβDMP-ClO$_4$ (poly-β-dimethyl pyrrole) on the reducing potentials. The voltages written in the figure are the applied voltages between the working and counter electrodes. The numbers in parentheses are the molecular ratios of the perchlorate ion to a pyrrole ring. Each spectrum is obtained using a different free standing film which has almost the same thickness as the others. Reproduced from [377].

would be expected for the pairing to form bipolarons (see Section 2) (Fig. 22), although Nechtstein et al. present a different interpretation of the data [374, 378] concluding that the formation of bipolarons is no more energetically favourable than the formation of two polarons, and expressing the formation of polarons from neutral polypyrrole, and bipolarons from polarons, as processes which each have a characteristic redox potential.

5.1.5 Electrochemistry at Polypyrrole Electrodes

Polypyrrole [244, 262] and its derivatives [379] polymerised onto platinum appear to behave essentially in the same way as bare Pt electrodes with the redox chemistry of a number of species including TCNQ [286]; ferrocene, phenothiazine, chloranil, benzoquinone [337]; ferricyanide, hydroquinone, catechol and p-aminophenol [380] having been investigated, although there is evidence of a catalytic effect with some couples [380, 381]. Santhanam and O'Brien [346] have investigated the mechanism of oxidation of organic species at polypyrrole electrodes using laser interferometry and concluded that the polymer film is porous and that redox reactions involving organic species occur within the polymer matrix. Ferrocene [382], anthraquinone [347], iron pthalocyanine [383] (FePc) and paraquat, phenothiazine, triphenylamine and diphenylene oxide [384] have also been incorporated as dopants during electropolymerisation, retaining their own electroactivity. The pthalocyanine has a catalytic effect, reducing the overpotential for the reduction of oxygen by several hundred mV as compared to a similar electrode containing no added FePc, and this has been further investigated by incorporating cobalt pthalocyanines [385] (CoPc) and sulfonated CoPc [386], giving electrodes which produce water rather than peroxide as the main reduction product of O_2.

Nitration of the surface of polypyrrole and the subsequent reduction of the nitrate groups has been reported [244] and Bidan et al. [306, 307] have investigated the electrochemistry of a number of polymers based on pyrroles with N-substituents which are themselves electrochemically active. Polypyrrole has also been successfully deposited onto polymeric films of ruthenium complexes [387], and has been used as an electrode for the deposition and stripping of mercury [388]. As with most conducting polymers, several papers have also appeared on the use of polypyrrole in battery systems (e.g. [327, 389] and Ref. therein).

5.2   Polyindole (Fig. 23a)

Indole differs from the more simple heterocycles in that when doped, the structure appears to contain one dopant anion to two monomer units, and its conductivity is considerably lower (by four orders of magnitude) when doped to a similar level [39]. From IR spectra, and the fact that N-substituted indoles do not polymerise, the

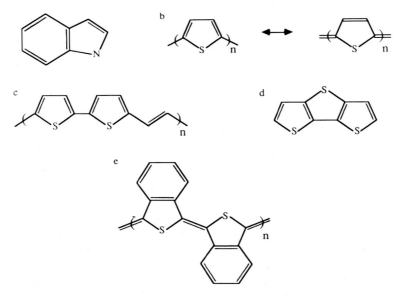

Fig. 23. (a) Indole, (b) polythiophene (the two energetically inequivalent structures are shown), (c) poly(1,2-dithienylethylene), (d) dithienothiophene (dithieno-(3,2b:2′,3′d)thiophene), (e) polyiso-thianapthene.

linkage between monomer units in this case appears to be partly through the nitrogen of the five-membered ring, unlike the simple five-membered heterocycles.

5.3   Polythiophene (Fig. 23b)

As might be expected, the properties of polythiophene show many similarities with those of polypyrrole. As with polypyrrole, polythiophene can be prepared *via* other routes than electrochemical oxidation both as the neutral material [390–392] or in the p-doped form [393]. This material is produced as an infusible black powder which is insoluble in common solvents (and stable in air up to 360 °C), with conductivities ranging from approximately 10^{-11} S cm^{-1} in the neutral form [390] to 10^2 S cm^{-1} when doped [19, 393, 394]. Early work on thiophene polymers showed that the p-doped material is air-sensitive in that the conductivity decreases on exposure to the atmosphere [20, 395] although no evidence of oxygen-containing species was seen in XPS measurements [19].

5.3.1   Polymerisation

The electrochemical synthesis of polythiophene was first reported in 1981 [34, 256], using conditions similar to those employed for the polymerisation of pyrrole, with the conductivity of the product ranging from $10^{-3}–10^{-1}$ S cm^{-1}. Other workers [19, 39, 396] repeated this using a variety of supporting electrolytes, polymerising thiophene onto Pt-coated glass to produce thin films from acetonitrile solutions containing small amounts of water (10^{-2} mol dm^{-3}), and also from THF. Similar results were achieved by Kaneto et al. [20, 397, 398] using monomer solutions in oxygen-free anhydrous solvents (acetonitrile and benzonitrile) and ITO glass substrates. As found by Diaz for pyrrole, no polymerisation occurs in the absence of oxygen and water unless a suitable counter electrode couple is provided (e.g., by using AgClO$_4$ as the electrolyte).

Many substituted thiophenes have also been electrochemically polymerised [19, 54, 399–405] (Table 4) as have thiophene dimers [21, 37, 55, 251, 400, 406], trimers [21, 83, 407], and tetramers [256, 406], with the thiophene dimer giving rise to higher quality films than does the monomer [37, 395, 408]. Several polycyclic monomers including a thiophene ring have also been polymerised [408–416], as have a series of compounds consisting of two thiophene rings linked by a polyene chain (Fig. 23c). The polymerisation of dithieno-thiophene (Fig. 23d) results in a polymer which shows remarkable similarity to polythiophene in its properties [409, 410, 414].

In general the mechanism of polymerisation for thiophene appears to be similar to that of pyrrole (Section 4.11.2), occurring *via* a radical coupling mechanism [423] giving mainly α-α linkages [293, 400, 405], and involves oligomer as well as monomer radicals, with evidence to suggest that the polymerisation reaction occurs at a lower

Table 4. Electrochemical synthesis of polythiophenes.

Monomer (concn.)	Electrode/solvent[a]	Electrolyte (concn.)	E_{pol}(V)[b]	Counter ion[c]	σ(S cm^{-1})[d]	Ref.
Thiophene						
thiophene (0.01–0.02 mM)	Pt/acn	Et$_4$NBF$_4$ (0.1 M)	+ 2.06 vs. Pt	BF$_4^-$	–	400
thiophene (0.01 M)	Pt/acn	Bu$_4$NBF$_4$	+ 1.6	BF$_4^-$	10–100	39
thiophene (0.06 M)	ITO/bzn	LiBF$_4$ (0.1 M)	+ 1.6 vs. Ag/Ag$^+$	iodine	44 (273 K)	394
thiophene (0.06 M)	ITO/acn, pc	LiBF$_4$ (0.1 M)	+ 1.6 vs. Ag/Ag$^+$	iodine	9.4 (273 K)	394
thiophene (0.2 M)	acn (anhyd.)	AgClO$_4$ (0.5 M)	+ 2.4 vs. Pt	ClO$_4^-$	0.6	20
thiophene (0.2 M)	ITO, Pt/pc/5 °C/Ar	Et$_4$NPF$_6$, NaAsF$_6$, Et$_4$NBF$_4$ (0.03 M)	5-10 mA cm^{-2}	PF$_6^-$	190	402
thiophene (0.2 M)	ITO/Et$_2$SO$_4$/25 °C	Bu$_4$NAsF$_6$ (4 mM)	2 mA cm^{-2}	AsF$_6^-$	10	417
thiophene (0.2 M)	ITO/pc	Et$_4$NBF$_4$ (0.03 M)	10 mA cm^{-2}	iodine	50 (278 K)	394
thiophene (0.2 M)	Pt/nbz/5 °C/Ar	Bu$_4$NClO$_4$ (0.02 M)	2 mA cm^{-2}	ClO$_4^-$	40 (f)	21
thiophene (0.2 M)	Pt/acn/25 °C/Ar	Bu$_4$NPF$_6$ (0.02 M)	2 mA cm^{-2} (2.02 V cv)	PF$_6^-$	10 (f)	21
thiophene (0.24 M)	ITO/5 °C, room temp	Bu$_4$NBF$_4$, Et$_4$NClO$_4$ (0.1 M), bzn H$_2$SO$_4$ (3 M), Et$_2$SO$_4$ LiTCNQ (0.1 M), Me$_2$SO$_4$	2 mA cm^{-2}	various	10^{-1}–10^{-4}	418
thiophene (0.4 M)	ITO/bzn/30 °C	LiBF$_4$, (0.5 M)	+ 20	BF$_4^-$	106	397
		LiBF$_4$, (0.5 M)	+ 10 (+ 0.75 cv)	BF$_4^-$	20	397
		NaAsF$_6$ (0.5 M)	+ 20	AsF$_6^-$	97	397
		NaPF$_6$ (0.5 M)	+ 20	PF$_6^-$	55	397
thiophene (0.4 M)	ITO/bzn	LiBF$_4$ (0.5 M)	+ 20 vs. Pt	iodine	170	394
thiophene	acn	Et$_4$NBF$_4$ (0.5 M)	+ 1.6	BF$_4^-$	10 (f, p)	251
Substituted thiophene						
3-methyl-	acn	Et$_4$NBF$_4$ (0.5 M)	+ 1.4	BF$_4^-$	100 (p)	251
3-methyl-	acn	LiClO$_4$ (0.1 M or saturated)	+ 1.5	–	–	54
3-methyl-	Pt/acn	Et$_4$NBF$_4$ (0.5 M)	+ 1.86 cv	–	–	400
3-methyl- (0.2 M)	ITO, Pt/5 °C/pc/Ar	Et$_4$NBF$_4$, Et$_4$NPF$_6$, (0.03 M)	5-10 mA cm^{-2}	various	450–510	402
3-methyl- (0.2 M)	ITO/5 °C/nbz/Ar	Bu$_4$NClO$_4$, (0.02 M)	2 mA cm^{-2}	ClO$_4^-$	120	419
3-bromo-	Pt/acn	Et$_4$NBF$_4$ (0.5 M)	+ 2.10 cv	–	–	400
3-acetonitrile-	Pt/acn	Et$_4$NBF$_4$ (0.5 M)	+ 2.22 cv	–	–	400
3-ethyl- (0.2 M)	ITO, Pt/pc/5 °C/Ar	Et$_4$NBF$_4$,Et$_4$NPF$_6$ (0.03 M)	5-10 mA cm^{-2}	BF$_4^-$	270	402
3-hexyl- (0.09–0.2 M)	ITO/pc/5 °C/Ar	Et$_4$NPF$_6$ (25 mM)	2 mA cm^{-2}	PF$_6^-$	95	404

	Electrode/Solvent[a]	Electrolyte	E_{pol}[b]	Counter ion[c]	Conductivity[d]	Ref.
3-octyl- (0.09–0.2 M)	ITO/pc/5 °C/Ar	Et_4NPF_6 (25 mM)	2 mA cm^{-2}	PF_6^-	78	404
3-dodecyl- (0.09–0.2 M)	ITO/pc/5 °C/Ar	Et_4NPF_6 (25 mM)	2 mA cm^{-2}	PF_6^-	67	404
3-octadecyl- (0.09–0.2 M)	ITO/nbz/5 °C/Ar	Et_4NPF_6 (25 mM)	2 mA cm^{-2}	PF_6^-	17	404
3-icosyl- (0.09–0.2 M)	ITO/nbz/5 °C/Ar	Et_4NPF_6 (25 mM)	2 mA cm^{-2}	PF_6^-	11	404
3-methoxy- (1.34 mM)	GC/acn-aq (3:1)	$NaClO_4$	+ 1.32		–	403
3-methoxy- (1.34 mM)	acn	$LiClO_4$ (0.1 M or saturated)	+ 1.24			54
2,3-dimethyl-	Pt/acn	Et_4NBF_4 (0.5 M)	+ 1.35	BF_4^-	50 (p)	251
3,4-dibromo-	Pt/acn	Et_4NBF_4 (0.1 M)	+ 2.23 cv		–	400

Thiophene oligomers

	Electrode/Solvent[a]	Electrolyte	E_{pol}[b]	Counter ion[c]	Conductivity[d]	Ref.
2,2'-bithiophene (10^{-4}–10^{-5} M)	Pt/acn	Et_4NBF_4 (0.5 M)	+ 1.32		–	395
2,2'-bithiophene (0.01 M)	ITO/acn	$LiClO_4$ (0.1 M)	100 μA cm^{-2}			408
2,2'-bithiophene (0.01 M)	Pt/acn (anhyd.)/Ar	$LiClO_4$ (0.1 M)	+ 0.7 vs. Ag/Ag$^+$	ClO_4^-	0.04	395
2,2'-bithiophene (0.1 M)	acn	$LiClO_4$ (0.5 M)	0.5 mA cm^{-2}			37
2,2'-bithiophene (0.2 M)	Pt/nbz/5 °C/Ar	Bu_4NClO_4 (0.02 M)	2 mA cm^{-2} (1.32 cv)	ClO_4^-	0.02 (p)	21
2,2'-bithiophene (0.2 M)	Pt/acn	Bu_4NPF_6 (0.5 M)	2 mA cm^{-2} (1.32 cv)	PF_6^-	0.6 (p)	21
2,2'-bithiophene	Pt/acn	Et_4NBF_4 (0.5 M)	+ 1.1 cv	BF_4^-	10^{-4} (f)	251
3,2'-bithiophene	Pt/acn	Et_4NBF_4 (0.5 M)	+ 1.1 cv			55
trithiophene (66 mM)	Pt/nbz/5 °C/Ar	Bu_4NClO_4 (0.02 M)	2 mA cm^{-2} (1.05 cv)	ClO_4^-	0.005 (p)	21
trithiophene (0.01 M)	ITO/acn	H_2SO_4 (0.1 M)	1 mA cm^{-2}	SO_4^{2-}	0.02 (gel)	21
		Bu_4NBF_4 (0.1 M)	1 mA cm^{-2}	BF_4^-	0.012 (p)	21

NB. Mixed solvent systems are shown as e.g. acn-aq (0.01 M) where the number in parentheses indicates the concentration of the lesser constituent

[a] Electrode/Solvent:
acn – acetonitrile, aq – water, bzn – benzonitrile, mc – methylene chloride, nbz – nitrobenzene, nmp – N-methyl-2-pyrollidone, odcb – o-dichlorobenzene, pc – propylene carbonate, thf – tetrahydrofuran.
ITO – Indium/tin oxide-coated glass, Ar – Solutions purged with argon, Ar atm – Experiment performed under an argon atmosphere, N$_2$ atm – Experiment performed under a nitrogen atmosphere.

[b] All potentials are measured vs. SCE unless otherwise stated:
E_{pol} – Potentials are as quoted in the original Ref, cv – The potential given was obtained from cyclic voltammetry.
NB. Where only a current density was given in the original Ref. this is quoted in place of the polymerisation potential.

[c] Counter ions are as incorporated during the electrochemical polymerisation process or by subsequent electrochemical doping unless suffixed with (chem.) which indicates the use of chemical doping.

[d] Conductivities: f – film, p – pressed pellet

Table 4. Continued

Monomer (concn.)	Electrode/solvent[a]	Electrolyte (concn.)	E_{pol}(V)[b]	Counter ion[c]	σ(S cm^{-1})[d]	Ref.
trithiophene (0.01 M)	ITO/nmp	Bu₄HSO₄ (0.1 M)	1 mA cm^{-2}	HSO$_4^-$	10^{-8} (f)	21
trithiophene	Pt/nbz/25 °C	LiClO₄ (0.02 M)	2 mA cm^{-2} (1.05 cv)	ClO$_4^-$	0.02 (p)	21
dithienobenzene	acn	Et₄NBF₄ (0.5 M)	+ 1.3	BF$_4^-$	4.4×10^{-4} (p)	55
thieno-3,2b-pyrrole (1 mM)	acn/Ar	Et₄NBF₄, LiClO₄ (0.1 M)	0.6 vs. Ag/Ag$^+$	BF$_4^-$	5.0×10^{-3}	411
thieno-2,3b-pyrrole (1 mM)	Au, ITO/acn	Et₄NBF₄, LiClO₄ (0.1 M)	0.7 vs. Ag/Ag$^+$	–	–	415
thieno-3,2b-thiophene (1 mM)	Pt/CH₂Cl₂(anhyd.)/Ar	Bu₄NClO₄ (0.1 M)	+ 1.4	BF$_4^-$	3.0×10^{-4} (f)	412
thieno-2,3b-thiophene (1 mM)	Au, ITO/acn	Et₄NBF₄, LiClO₄ (0.1 M)	+ 1.10 vs. Ag/Ag$^+$	–	–	415
dithieno- (3,2b:2′3′d)-thiophene	Pt/CH₂Cl₂ (anhyd.)	Bu₄NClO₄, Bu₄NPF₆, (0.1 M)	+ 1.2	ClO$_4^-$ PF$_6^-$	3.0×10^{-3} 1.0×10^{-2}	410, 414 413
2,5-di(2-thienyl)pyrrole	Au/acn/N₂ atm	AgOTs (72–85 mM)	− 0.5 to − 0.2	OTs	10^{-8}-0.1	420
dithieno-(CH=CH)n-compds (0.05–0.1 M)	Pt/Ar n = 1, acn n = 2, thf, nbz	Bu₄NClO₄ (0.05 M)	2 mA cm^{-2} 0.2–4 mA cm^{-2}		3.7×10^{-2} 4.2×10^{-6}-0.5	421 421
1,2-di(2-thienyl)ethylene (0.1 M)	acn/25 °C	Bu₄ClO₄, Bu₄NBF₄ (0.05 M)	0.5–2 mA cm^{-2}	ClO$_4^-$	3.9×10^{-2}	421
1,2-di(2-thienyl)ethylene (0.2 M)	acn	Bu₄NClO₄, Bu₄NBF₄ (0.1 M)	+ 1.100	ClO$_4^-$	40	422

[a] Electrode/Solvent:
acn – acetonitrile, aq – water, bzn – benzonitrile, mc – methylene chloride, nbz – nitrobenzene, nmp – N-methyl-2-pyrrolidone, odcb – o-dichlorobenzene, pc – propylene carbonate, thf – tetrahydrofuran.

NB. Mixed solvent systems are shown as e.g. acn-aq (0.01 M) where the number in parentheses indicates the concentration of the lesser constituent

ITO – Indium/tin oxide-coated glass, Ar – Solutions purged with argon, Ar atm – Experiment performed under an argon atmosphere, N₂ atm – Experiment performed under a nitrogen atmosphere.

[b] All potentials are measured vs. SCE unless otherwise stated:
E_{pol} – Potentials are as quoted in the original Ref., cv – The potential given was obtained from cyclic voltammetry.
NB. Where only a current density was given in the original Ref. this is quoted in place of the polymerisation potential.

[c] Counter ions are as incorporated during the electrochemical polymerisation process or by subsequent electrochemical doping unless suffixed with (chem.) which indicates the use of chemical doping.

[d] Conductivities: f – film, p – pressed pellet

potential on existing polymer than on bare metal [423]. Potential-step experiments [424] have shown that instantaneous nucleation occurs on a bare electrode, followed by rapid formation of a polymer monolayer. Time-resolved spectroscopic measurements [425] have also shown that initially short-chain oligomers are produced, with longer chains appearing subsequently, and that the characteristic electrical properties of the bulk material do not appear until the growth centres have largely overlapped and coalesced, with the optical characteristics typical of the bulk material appearing suddenly (rather than gradually as the result of interactions between growing polymer chains). As with pyrrole, the precise nature of the electrode reaction will depend on the conditions used, with the additional possibility of nucleophilic reactions between the radical cation intermediate and species. In very dry acetonitrile, nucleation is inhibited [426], resulting in unstable films which are not easy to reduce back to the neutral state. As with other materials, substitution on the thiophene ring affects the oxidation potential *via* the radical cation intermediate, with groups such as CN^- and NO_2 affecting the stability of the intermediate to such an extent that polymerisation does not occur, whilst the presence of electron donating groups such as CH_3 have the opposite effect, shifting the oxidation potential of the monomer in a cathodic direction relative to the unsubstituted material [400, 427]. The oxidation potential of the monomer decreases as the chain length of the substituent is increased [21, 83], again implying a greater stability of the intermediate. It has been noted that with the 3-methyl derivative, colouration of the monomer solution occurs in the region of the anode during polymerisation [54], possibly due to the presence of low-molecular weight oligomers which are soluble, or due to a high concentration of more long-lived radicals. Cyclic voltammograms of 3-methyl thiophene also show two reduction peaks during the initial stages of the polymerisation (i.e., when the polymer film is thin), but as the film becomes thicker the second peak disappears, and this has been attributed to structural rearrangement. This is supported to some extent by chrono-coulometric measurements [251, 396] and SEM which shows decreasing homogeneity of the polymer with increasing thickness [428]. Cyclic voltammograms of the methoxy derivative also show an additional peak [54]. With larger β-substituents the polymerisation is less successful, with evidence of large quantities of short-chain oligomers being formed [400] and a linear relationship has been found between the oxidation potential of the monomer and that of the corresponding polymer, indicating that the substituents affect both the monomer and polymer in similar ways, although to a greater extent for the latter [400, 427]. A number of monomers with larger alkyl substituents on the 3-position have been successfully chemically and electrochemically polymerised [399, 429–431], producing polymers which have increasingly high solubilities as the size of the alkyl substituent is increased. Surprisingly, although the doped, conducting form of many of the polymers is soluble in common solvents the conductivity of the solid does not appear to be adversely affected by the presence of the substituent (which might be expected to disrupt the structure of the material due to steric interactions) with all the materials investigated giving conductivities in the region of $1-10\,S\,cm^{-1}$. Water-soluble polythiophenes have also been synthesised, in which the substituent group on the thiophene ring also acts as dopant [246], a development which may have much wider applicability.

The use of a lower current density during the polymerisation results in higher quality films than when high current densities are employed [395] and the use of oxygen and water free conditions gives films with higher conductivities [399]. Kobel et al. [432] found that the use of propylene carbonate as the solvent for the monomer gave the highest quality films of poly(3-methyl thiophene) (the polymerisation of 3-methyl-thiophene has also been achieved in aqueous media [405]), although significant quantities (15–20%) of the solvent were incorporated into the polymer during the polymerisation process.

Several workers have investigated the effect of temperature on the polymerisation of thiophene [418, 433, 434] and derivatives [419] finding clear evidence of a dependence of the properties of the film produced, with the highest quality material being prepared at 5°C. This appears to be the result of longer conjugation lengths and a lower number of non α-α linkages [433] (which disrupt the conjugation [342]) and although the pristine material appears to show little variation, on cycling the higher temperature material shows more evidence of degradation. The rate of the polymerisation reaction increases with temperature, but appears to decrease above about 16°C [434] with evidence of structural changes.

5.3.2 Morphology and Structure

Although there are some differences between the two materials, the basic structure of polythiophene appears to be essentially the same as that of polypyrrole. FTIR [435–437] and ^{13}C-NMR [436] show characteristic absorptions for polythiophene regardless of dopant, indicating that the structure is essentially retained even when heavily doped, and chain lengths of up to 300 have been calculated for poly(3-alkyl thiophenes), although other estimates have been considerably lower than this [429, 438]. X-ray and scanning electron microscopy (SEM) [400] show that the electrochemically-prepared material is amorphous, although it has also been suggested that the polymer is basically granular [20, 400, 407] or fibrillar [428] (there is evidence to suggest that the chemically-prepared material is crystalline [37]). The material is apparently quite porous [405, 409] with coulometric data providing evidence that thiophene polymers have a large internal surface area [439].

As found for polypyrrole, the morphology of the thiophene polymers depends on a number of factors including the nature of the substrate, the growth rate and the current and potential at which the polymerisation proceeds [395, 440]. Thin films are highly homogeneous and show little evidence of solvent incorporation [426]. but as the film thickness is increased this homogeneity is lost, possibly due to the cumulative effect of structural defects such as chain folding, cross linking α-β coupling etc. [428] or as a result of inhomogeneity of the electric field at the polymer surface [395]. The use of high monomer concentrations in the polymerisation solution also tends to produce more powdery films [425].

Tourillon and Garnier [428] found using SEM that the morphology of the material is affected by the nature of the dopant anion and that there is considerable swelling of the polymer on doping (with increases in thickness of the order of 40% for poly(3-methyl thiophene)) noting that these effects are likely to be related to the structure and charge distribution on the ion. XPS and IR measurements have also provided some evidence that the dopant species may themselves be altered during the doping process in poly(3-methyl thiophene) with the possibility of reduction or oxidation depending on the species involved e.g., AsF_6^- being converted to AsF_5, and BF_4^- being converted to BF_3 by the elimination of fluoride [432, 441], which may be involved in the degradation of the polymer on cycling [442]. There is considerable evidence from FTIR and other techniques to suggest that polythiophenes contain predominantly α-α linkages [407, 435–437], and as might be expected, this is particularly true of the polymer derived from terthiophene, which is highly ordered and can be produced with variety of morphologies (by varying the solvent and supporting electrolyte), including a fibrillar gel, granular powders, smooth films, and even as a crystalline product [407]. Tourillon and Garnier also found from IR spectra [19, 396] that electrochemically-prepared poly(3-methyl thiophene) also shows a high degree of order, again with predominantly α-α linking between thiophene rings whilst unsubstituted polythiophene has a much greater degree of disorder due to the possibility of β-linkages. From IR and Raman measurements Akimoto et al. [394] concluded that whilst long chain lengths are readily found in the polymer, these must contain a high degree of conjugation to achieve conductivities as high as those observed. This has been further investigated by Roncali et al. [21] and others [55, 83, 251] who found by polymerising thiophene oligomers, that the conductivity *decreases* as the number of monomer units in the starting material increases, possibly due to a reduction in the degree of conjugation resulting from the lower reactivity of the radical cation involved in the polymerisation reaction. The α- and β-positions on the rings also become increasingly equivalent as the number of monomer units in the radical cation is increased, increasing the likelihood of non-α-α linkages [21, 83, 402, 406, 410, 421].

Elemental analysis shows that the polymer generally contains four monomer units per dopant ion [20], and that there is also more hydrogen than would be expected (cf. polypyrrole) [20, 395], although this may vary depending on the starting material [409, 410, 414]. Even in the neutral form, the polymer contains a small quantity of anions (0.5–1%) [19], although Waltman et al. [400] found that the extent to which the counter ion is incorporated into the polymer on polymerisation depends strongly on the nature of the β-substituent (if present).

5.3.3   The Electrochemistry of Polythiophene

Polythiophene films can be electrochemically cycled from the neutral to the conducting state with coulombic efficiencies in excess of 95% [443], with little evidence of decomposition of the material up to $+1.4$ V vs. SCE in acetonitrile [37, 54, 56, 396, 400] (the 3-methyl derivative being particularly stable [396]), but unlike polypyrrole, polythiophene can be both p- and n-doped, although the n-doped material has a lower maximum conductivity [444]. Cyclic voltammetry shows two sets of peaks corresponding to the p- and n-doping reactions, with E^0 values at approximately $+1.1$ V and -1.4 V respectively (*vs.* an Ag^+/Ag reference electrode) in good agreement with those obtained in theoretical treatments by Bredas et al. [445], and with the peak anodic current scaling linearly with sweep rate as would be expected for a surface-reacting species. The behaviour of polythiophene and derivatives in aqueous solutions is essentially similar to that in organic solvents, except that it is only stable up to $+0.8$ V *vs.* SCE, and the peaks seen in cyclic voltammetry are somewhat broader than would be expected for a Nernstian reaction [54] (possibly due to difference in solvent uptake), and the initial cycle does not appear to be completely reversible [446].

Electrochemical impedance measurements [409] have been used to investigate the behaviour of thiophene-based polymers with evidence to suggest that the doping reaction is limited by diffusion of the dopant into the polymer (as found for other materials), and a correlation has been established between the capacitance of the film and the level of doping. Data from a variety of techniques suggests that for material produced from the dimer and trimer [21], the reversibility of the doping reaction becomes less as the chain length of the starting material is increased, and that this may be connected to the shorter conjugation lengths found for these materials. Cyclic voltammograms of poly(dithieno-thiophene) also show that the doping process is not electrochemically reversible and are complex with the possiblity of some structural reorganisation occurring, although high coulombic efficiencies have been reported [410].

Poly(3-methoxy thiophene) behaves similarly to the other derivatives, and like some of the larger alkyl-substituted thiophene polymers shows some degree of solubility in acetonitrile [54]. Comparatively high doping levels have been achieved with this material (up to 50% using CF_3SO^{-3}) although the doping process may be complex with some regions more heavily doped than others (which may remain completely undoped [428]).

5.3.4 Electronic Structure and Conductivity

From EVS [37] and spectroelectrochemical measurements, it appears that the electronic structure of polythiophene is very similar to that of polypyrrole as might be expected. Figure 24 shows a typical series of absorption spectra for polythiophene at various doping levels achieved by *in situ* electrochemical doping, with the π-π^* interband transition (the intensity of which decreases as the dopant level increases) clearly visible at approximately 2.5 eV. Two peaks which increase in intensity can also be seen at lower energy, which can be assigned to transitions involving defect levels within the band gap [37, 408]. From the presence of only two such transitions, it can be deduced that there are two unoccupied levels within the band gap consistent with the formation of bipolarons (Section 2), and this is supported by ESR data [447], although Kaneto et al. have suggested that at low dopant levels polarons are present, and that at very high doping levels the bipolaron levels merge to form bands [448, 449]. Kaneto et al. [444] also found that changing the nature of the dopant anion has no effect on the changes seen in the optical spectra indicating that they are associated with the polymer rather than any other species present, although steric factors result in a dependence of the

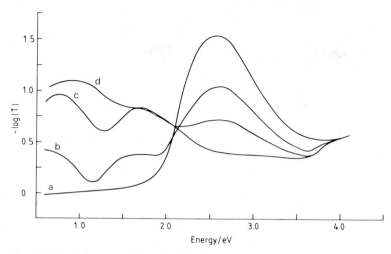

Fig. 24. Optical spectra for polythiophene electrochemically doped with ClO_4^- at potentials of (a) 2.5 V, (b) 3.7 V, (c) 3.9 V, (d) 4.1 V vs. Li/Li^+.

Table 5. Electrochemical doping data for polythiophenes.

Monomer	Counter ion	Degree of oxidation (%)	$\sigma(S\,cm^{-1})$	Ref.
Thiophene	Neutral	1	1×10^{-7} (p)	19
	$CF_3SO_3^-$	30	10–20	19
	ClO_4^-	26	0.6	20
	ClO_4^-	30	10–20	19
	BF_4^-	6	0.02	400
	BF_4^-	30	10–20	19
	BF_4^-	–	20–106	397
	PF_6^-	–	55	397
	PF_6^-	6	0.02	400
	HSO_4^-	22	0.10	400
	AsF_6^-	–	97	397
	$FeCl_4$	26	50	450
3-methyl-	Neutral	0.5	1.5×10^{-7} (p)	19
	$CF_3SO_3^-$	30	30–50 (p)	19
	ClO_4^-	0.027	3.0×10^{-12}	441
	ClO_4^-	0.10	4.3×10^{-9}	441
	ClO_4^-	0.14	3.3×10^{-8}	441
	ClO_4^-	4.2	74	441
	ClO_4^-	4.6	120	441
	ClO_4^-	25	10–30 (p)	19
	PF_6^-	12	1	400
	AsF_6^-	6.0	110	441
	BF_4^-	–	36	441
3,4-dimethyl-	$CF_3SO_3^-$	30	10–50 (p)	19

polymer conductivity on the nature of the anion used (Table 5). From similar data Chung et al. [37] found that the energies of the transitions from the defect states are symmetrical about the gap centre, and concluded that the sulphur of the thiophene ring does not play a large part in determining the electronic structure of the material (if the sulphur carried a significant portion of the charge on the chain, this symmetry would be disrupted), and that it merely stabilises the lower energy form of the non-degenerate polyene chain of the backbone.

Optical and XPS spectra obtained by Tourillon and Garnier [19] for unsubstituted and β-substituted polythiophenes show evidence of highly mobile spins in the polymer with free carriers moving along the polymer chains, and Kaneto et al. [20] proposed that the conduction mechanism involves three-dimensional variable-range hopping. Roncali et al. [21] found some dependence of the measured conductivity of polythiophene on the nature of the solvent/electrolyte system used in doping experiments, most probably due to steric effects, and others [19, 20] have suggested that the macroscopic conductivity is probably controlled by the polymer morphology. The conductivity obeys a $T^{1/4}$ law, although from data obtained for material grown from the trimer and doped with sulphuric acid, it appears that at low temperatures, temperature-independent conductivity occurs [407].

5.3.5 Electrochemistry at Polythiophene Electrodes

Unlike polypyrrole, polythiophene has not been widely studied as an electrode material, although several applications have been proposed for thiophene polymers such as batteries [37, 161], photovoltaic devices [451], microelectronic devices [439, 452] and in electrochromic displays [395, 396, 408] (as they have pronounced colour changes on switching and shorter switching times than other similar materials). It has also been shown that as with polypyrrole, the redox chemistry of ferrocene is essentially the same on thiophene polymers as at platinum [405], and Waltman et al. [400] have reported the oxidation and reduction of triphenylamine at a poly(3-methyl thiophene) electrode. Metal clusters have also been incorporated into poly(3-methyl thiophene) [150, 453] during the polymerisation reaction to give materials which act electrocatalytically for proton reduction [453] with the metal (e.g., Cu), which can be reversibly reduced and re-oxidized [150], included homogeneously into the polymer matrix during the polymerisation.

5.4 Poly(isothianapthene) (Fig. 23e)

The synthesis of poly(isothianapthene) (PITN) is an example of the second generation of conducting polymers, which have been prepared in order to produce a material with specific properties. Given the two inequivalent structures of polythiophene which give rise to its non-degenerate ground state (Section 5.3), stabilisation of the higher energy form would be expected to lower the energy of the π^* antibonding orbitals and thus reduce the bandgap of the material [454, 455]. In order to achieve this, the iso-thianapthene monomer, a bicyclic compound which has the basic structure of thiophene with a benzene ring attached, has been polymerised *via* both chemical [43, 456] and electrochemical routes [43]. The polymer can be reversibly doped (as shown by cyclic voltammetry [44, 454]), and as expected has a lower bandgap (approximately 1 eV from optical spectra) than polythiophene, with the presence of the aromatic ring having the desired stabilising effect on the higher energy configuration of the thiophene moiety. The material is blue as prepared (with a low level of p-doping), becomes yellow/green when more heavily doped, and has a very low absorption in the visible range spectra, characteristic of free-carriers, with a conductivity of approximately 50 S cm^{-1} when very heavily p-doped [44]. The structure of the electrochemically prepared material is of an open network, with some evidence from X-ray data of crystallinity [44]. The p-doping reaction can be followed *via* cyclic voltammetry which shows an anodic peak at approximately $+ 0.6$ V vs. SCE corresponding to the colour change from blue-black to yellow-green and a cathodic peak at approximately 0 V vs. SCE corresponding to the reverse process. The material is very stable under these conditions and can be cycled indefinitely between these two states, but if the positive scan limit is extended to $+ 1.3$ V, a second electrochemically irreversible process is seen with concomitant degradation of the polymer. Extending the negative scan limit to $- 1.9$ V vs. SCE shows no evidence of a reversible n-doping process, but again there is evidence of degradation of the material. The optical spectra are very similar to those of polythiophene and suggest in a similar way that the charge acquired *via* doping is stored in bipolarons [44] (Fig. 25).

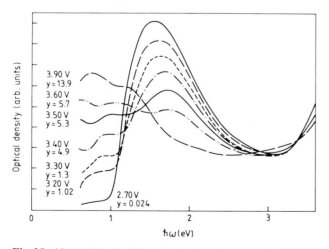

Fig. 25. Absorption coefficient vs. photon frequency for PITN. The data were obtained *in situ* during the injection part of the electrochemical doping cycle with (ClO_4) as the dopant. The cell voltages and corresponding dopant concentrations (in mol%) are indicated. Reproduced from [456b].

5.5 Polyfuran (Fig. 26a)

Little work has been carried out on polyfuran, but it appears similar to its analogues (polypyrrole, polythiophene etc.) in that the material is insoluble and thermally stable, with a structure having four monomer units per dopant ion when doped, with the charge delocalised throughout the polymer chain [39]. The polymer has been produced on ITO and platinum from benzonitrile, THF and *N,N*-dimethylformamide solutions in the absence of water using a number of different electrolyte salts [42], with the highest quality films being obtained with $AgClO_4$ in benzonitrile. The conductivity of the material obtained was found to be $10^{-11}\,S\,cm^{-1}$ when neutral and $10^{-5}\,S\,cm^{-1}$ when doped [42], although higher conductivities of up to $1\,S\,cm^{-1}$ can be reversibly obtained on contact with water. The optical spectra were found to be similar to those obtained for polythiophene, with evidence of mid-gap states tentatively assigned to bipolarons.

5.6 Polyselenophene (Fig. 26b)

Although selenophene is known and can be polymerised in a similar manner to thiophene, pyrrole and furan, it is less easy to synthesise than the other monomers, and there are few reports concerning the polymer. Chemical synthesis of the polymer has been reported [457, 458] and Yoshino et al. first reported the successful electro-polymerisation in 1983 [459] and subsequently published a more detailed investigation [52]. Unlike thiophene and pyrrole, selenophene has only been successfully electro-polymerised in a small number of solvent/electrolyte systems, with the best results being achieved using benzonitrile/$LiBF_4$ from which high quality thin films can be prepared with conductivities ranging from $10^{-10}\,S\,cm^{-1}$ in the neutral form to $10^{-3}\,S\,cm^{-1}$ when heavily (approximately 40%) doped [52]. The *in situ* optical spectra obtained in conjunction with electrochemical doping show close similarities to polythiophene, with absorption peaks which can be attributed to bipolarons and possibly to polarons at very low doping levels [52] (Fig. 27), but the loss of oscillator strength from the interband transition is much less than seen for polythiophene, even at high dopant levels. The ESR data are also similar to those measured for polythiophene, but suggest in conjunction with the less marked changes in the optical spectra that delocalisation of the electrons on the polymer chain with doping is somehow limited, possibly by shorter conjugation length. It is also possible that the larger Se atom precludes coplanarity of the rings.

Methyl- and methoxy-substituted polyselenophenes have also been electropoly-merised on platinum [54], and as found for other substituted monomers, they poly-

Fig. 26. (a) Polyfuran, (b) polyselenophene.

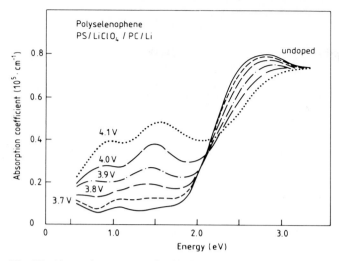

Fig. 27. Absorption spectra of polyselenophene at various doping levels. Reproduced from [52].

merise at less anodic potentials than the parent compound. The materials are stable to high temperatures and do not dissolve in common organic solvents or acids, and cyclic voltammetry of both derivatives in acetonitrile/LiClO$_4$ shows a reversible doping process, with peak height varying linearly with sweep rate. The redox chemistry of ferrocene at electrodes covered with films of either polymer appears to be essentially unchanged from that seen on a bare platinum electrode although the peak separations are too great for a completely Nernstian process. The polymerisation of di-selenophene has also been reported [55] but no details are given other than the conductivity of the product (10^{-5} S cm^{-1}). Polytellurophene has also been chemically-synthesised but remains a poor semiconductor even when doped [458].

6   Other Materials

In addition to the materials discussed above, a number of other conducting polymers have been studied using electrochemistry, albeit to a much lesser degree. Although amorphous films are produced by the electrooxidation of carbazoles [460–462], it seems that these consist largely of low molecular weight oligomers (possibly only as high as the tetramer) and do not have a high electrical conductivity. The films do show some interesting redox properties, but cannot be considered conducting polymers in the sense of the other materials discussed earlier. Reports of conducting polymers from pyrene [460, 463], fluorene [464, 465], quinolines [466, 467], and a fluorinated five-membered ring [468] have also appeared in the literature, but little further work on their characterisation or electrochemical properties has appeared.

7   Conclusion

The investigation and use of conducting polymers in, or using electrochemistry is now evidently well-established, and it has been shown that there are many broad similarities between different materials, in that at low doping levels, energy levels are created within the bandgap, with associated structural changes, and that these levels may further interact when the doping level is increased, although there still remain many questions to be fully resolved concerning the precise details of polymerisation and doping reactions, and the nature of the charge-carriers themselves.

From an initial period of investigation during which many theories and applications have been proposed, the field is now moving to a better understanding of these materials from which a more informed appreciation of their potential can be realised. Of the many proposed applications for conducting polymers in electrochemical systems, perhaps the two most promising are in the fields of battery technology and photovoltaic devices. It should be borne in mind however, that both of these are established industries, and for a new material to replace an existing technology and move from the research laboratory to the world of applications, there must be advantages to the manufacturer either in terms of price, ease of production or performance, and it is fair to say that at present none of these materials can offer significant advantages in any of these areas. Whilst many of the polymers are relatively easy to produce in the laboratory, precisely controlling properties such as morphology and impurity levels has yet to become a matter of routine. Processibility and handling of the materials must also be improved if applications other than small-scale specialist devices are to become practical, and perhaps the biggest driving force to this development would be the discovery of a large-scale application for which these materials were uniquely suited.

This field has encouraged many fruitful collaborations between chemists and physicists and is an excellent example of how the boundaries between subjects are becoming increasingly artificial in modern science. In view of the above considerations, the great challenge to the ingenuity of the electrochemist and physicist alike must surely lie with the production of materials with properties tailored to the requirements of the user, both in physical terms, with more processible and stable materials such as the substituted vinylene polymers and thiophenes (see also Chapter 10 of Ref. [149]), and also in terms of their electrical properties, with smaller band-gaps permitting doping over a narrower, more accessible range of potentials [469].

8   References

1. G.K. Chandler, D. Pletcher, Roy. Soc. Chem. Spec. Period. Rep. Electrochem. 1985, 10, 117.
2. R.E. Peierls, "Quantum Theory of Solids", Oxford University Press, 1955.
3. J.L. Bredas, R.R. Chance, R. Silbey Phys. Rev., 1982, B26. 5843.
4. W. Su, J.R. Schrieffer, A.J. Heeger, Phys. Rev. Lett., 1979, 42, 1698.
5. W. Su, J.R. Schrieffer, A.J. Heeger, Phys. Rev., 1980, B22, 2099.

6. M.J. Rice, Phys. Lett., 1978, 71A, 152.
7. H. Takayama, Y.R. Lin-Lin, K. Maki, Phys. Rev., 1980, B21, 2388.
8. B.R. Weinberger, E. Ehrenfreund, A.J. Heeger, A.G. MacDiarmid, J. Chem. Phys., 1980, 72, 4749.
9. M. Nechtstein, F. Devreaux, R.L. Green, T.C. Clark, G.B. Street, Phys. Rev. Lett., 1980, 44, 356.
10. J.B. Goldberg, H.R. Crowe, P.R. Neuman, A.J. Heeger, A.G. MacDiarmid, Mol. Cryst. Liq. Cryst., 1981, 72, 253.
11. J.L. Bredas, B. Themans, J.M. Andre, R.R. Chance, D.S. Boudreaux, R. Silbey, J. de Physique Colloque C3, 1983, 44, 373.
12. A.R. Bishop, D.K. Campbell, K. Fesser., Mol. Cryst. Liq. Cryst., 1980, 77, 253.
13. J-L. Bredas, G.B. Street, Acc. Chem. Res. 1985, 18, 309.
14. J-L. Bredas, B. Themans, J.M. Andre, Phys. Rev., 1983, B27, 7827.
15. J-L. Bredas, B. Themans, J.G. Fripiat, J.M. Andre, R.R. Chance, Phys. Rev. 1984, B29, 6761.
16. J-L. Bredas, Mol. Cryst. Liq. Cryst. 1985, 118, 49.
17. R. Qian, J. Qiu, Polym. J. 1987, 19, 157.
18. D.T. Glazhofer, J. Ulanski, G. Wegner, Polymer 1987, 28, 449.
19. G. Tourillon, F. Garnier, J. Phys. Chem. 1983, 87, 2289.
20. K. Kaneto, K. Yoshino, Y. Inuishi, Jpn. J. Appl. Phys. 1982, 21, L567.
21. J. Roncali, F. Garnier, M. Lemaire, R. Garreau, Synth. Met., 1986, 15, 323.
22. A.H. Thompson, Physica (Utrecht), 1980, 99B, 100.
23. J.H. Kaufman, J.W. Kaufer, A.J. Heeger, R. Kaner, T-C. Chung, A.G. MacDiarmid, Phys. Rev., 1982, B26, 4.
24. J.H. Kaufman, T.C. Chung, A.J. Heeger, J. Electrochem. Soc., 1984, 131, 2847.
25. J.H. Kaufman, K.K. Kanazawa, G.B. Street, Phys. Rev. Lett. 1984, 53, 2461.
26. M. Stamm, Mol. Cryst. Liq. Cryst., 1984, 105, 259–71.
27. Y.W. Park, M.A. Druy, C.K. Chiang, A.G. MacDiarmid, A.J. Heeger, H. Shirakawa, S. Ikeda, J. Polym. Sci., Polym. Lett. Ed., 1979, 17, 195.
28. Y. Ito, H. Shirakawa, S. Ikeda, J. Polym. Sci., Polym. Chem. Ed., 1974, 12, 11.
29. Y. Ito, H. Shirakawa, S. Ikeda, J. Polym. Sci., 1975, 13, 1943.
30. P. Kovacic, A. Kyriakis, J. Am. Chem. Soc., 1963, 85, 454.
31. J.H. Edwards, W.J. Feast, Polymer, 1980, 21, 595.
32. J.H. Edwards, W.J. Feast, D.C. Bott, Polymer, 1984, 25, 395.
33. S. Antoun, D.R. Gagnon, F.E. Karasz, R.W. Lenz, Polym. Bull., 1986, 15, 181.
34. A.F. Diaz, Chem. Scripta, 1981, 17, 145.
35. M.C. Dos Santos, Solid State Commun., 1984, 50, 389.
36. J-L. Bredas, R.L. Elsenbaumer, R.R. Chance, R. Silbey, J. Chem. Phys., 1983, 78, 5656.
37. T.C. Chung, J.H. Kaufman, A.J. Heeger, F. Wudl, Phys. Rev., 1984, B30, 702.
38. M. Satoh, F. Vesugi, M. Tabata, K. Kaneto, K. Yoshino, JCS Chem. Commun., 1986, 979.
39. J.G. Tourillon, F. Garnier., J. Electroanal. Chem., 1982, 135, 173.
40. S. Hayashi, S. Nakajima, K. Kaneto, K. Yoshino, Solid State Commun., 1986, 60, 545.
41. J. Bargon, S. Mohmand, R. Waltman, Mol. Cryst. Liq. Cryst., 1983, 93, 279.
42. T. Ohsawa, K. Kaneto, K. Yoshino, Jpn. J. Appl. Phys., 1984, 23, L603.
43. F. Wudl, M. Kobayashi, A.J. Heeger, J. Org. Chem., 1984, 49, 3382.
44. J-L. Bredas, G.B. Street, B. Themans, J.M. Andre, J. Chem. Phys., 1985, 83, 1323.
45. M. Satoh, F. Vesugi, M. Tabata, K. Kaneto, K. Yoshino, JCS Chem. Commun., 1986, 550.
46. M. Satoh, K. Kaneto, K. Yoshino, JCS Chem. Commun., 1985, 1629.
47. K. Kaeriyama, M. Sato, K. Samono, S. Tanaka, JCS Chem. Commun., 1984, 1199.
48. I. Rubenstein, J. Polym. Sci., Polym. Chem. Ed., 1983, 21, 3035.
49. M. Dietrich, J. Mortensen, J. Heinze, JCS Chem. Commun., 1986, 1131.
50. G. Froyer, F. Maurice, J.Y. Goblot, J. F. Fauvarque, M. A. Petit, A. Digua, Mol. Cryst. Liq. Cryst., 1985, 118, 267.
51. M. Satoh, K. Kaneto, K. Yoshino, JCS Chem. Commun., 1984, 1627.
52. K. Yoshino, Y. Kohno, T. Shiraishi, K. Kaneto, S. Inoue, K. Tsukagoshi, Synth. Met., 1985, 10, 319.
53. K. Yoshino, K. Kaneto, S. Inoue, K. Tsukagoshi, Jpn. J. Appl. Phys., 1983, 22, L698.
54. G. Dian, G. Barbey, B. Decroix, Synth. Met., 1986, 13, 281.
55. J.M. Bureau, M. Gazard, M. Champagne, J.C. Dubois, G. Tourillon, F. Garnier, Mol. Cryst. Liq. Cryst., 1985, 118, 325.
56. G. Tourillon, F. Garnier., J. Electrochem. Soc., 1983, 130, 2042.

57. M. Salmon, A.F. Diaz, A.J. Logan, M. Krounbi, J. Bargon, Mol. Cryst. Liq. Cryst., 1982, 83, 265.
58. G.B. Street, T.C. Clarke, M. Krounbi, K.K. Kanazawa, V. Lee, D. Pfluger, J.C. Scott, G. Weisser, Mol. Cryst. Liq. Cryst., 1982, 83, 253.
59. M.A. Druy, J. Electrochem, Soc., 1986, 133, 353.
60. G. Tourillon, P.C. Lacaze, J.E. Dubois J. Electroanal. Chem., 1979, 100, 247.
61. M. Armand J. de Physique, Colloque C3, 1983, 44, 551.
62. H. Lund, in "Organic Chemistry – An Introduction and a Guide", 2nd Edn, M.M. Baizer, H. Lund, Eds. N.Y., Marcel Dekker Inc., 1983, pp. 161–219.
63. S.L. Mu, Synth. Met., 1986, 14, 19.
64. C.K. Chiang, C.R. Fincher, Y.W. Park, A.J. Heeger, H.Shirakawa, E.J. Louis, S.C. Gau, A.G. MacDiarmid, Phys. Rev. Lett., 1977, 39, 1098.
65. D.J. Berets, D.S. Smith, JCS Faraday Trans., 1968, 64, 823.
66. R.J. Mammone, A.G. MacDiarmid, JCS Faraday Trans. I, 1985, 81, 105.
67. F. Genoud, F. Devreux, M. Nechtstein, J.P. Travers, J. de Physique Colloque C3, 1983, 44, 291.
68. R. Huq, G.C. Farrington, J. Electrochem. Soc., 1984, 131, 819.
69. G.C. Farrington, B. Scrosati, D. Frydrych, J. DeNuzzio, J. Electrochem. Soc., 1984, 131, 7.
70. M. Zagorska, A. Pron, J. Przyluski, B. Krische, G. Ahlgren, Mol. Cryst. Liq. Cryst., 1985, 121, 225.
71. M.A. Druy, Synth. Met., 1986, 15, 243.
72. M. Faultier, M. Armand, M. Audier, Mol. Cryst. Liq. Cryst., 1985, 121, 333.
73. H.W. Gibson in ref. 149, Ch. 11, pp. 405–440.
74. P. Ehrlich, W.A. Anderson in ref. 149. Ch. 12, pp. 441–488.
75. S. Chen, H. Shy, J. Polym. Sci. Polym. Chem. Ed., 1985, 23, 2441.
76. P.J. Nigrey, D. McInnes, D.P. Nairns, A.G. MacDiarmid, A.J. Heeger, J. Electrochem. Soc., 1981, 128, 1651.
77. A.J. Heeger, A.G. MacDiarmid, Synth. Met., 1979/80, 1, 101.
78. P. Bernier, A. El Khodary, F. Maurice, C. Fabre, P. Mirebeau, A. M. Ledunois, J. de Phys. Colloque. C3, 1983, 44, 583.
79. K. Kaneto, M. Maxfield, D.P. Nairns, A.G. MacDiarmid, A.J. Heeger, JCS Faraday Trans. I, 1982, 78, 3417.
80. J.R. Reynolds, J.B. Schlenoff, J.C.W. Chien, J. Electrochem. Soc., 1985, 132, 1131.
81. H. Eckhardt, S.W. Steinhauser, Mol. Cryst. Liq. Cryst., 1984, 105, 219.
82. P.J. Nigrey, A.G. MacDiarmid, A.J. Heeger, JCS Chem. Commun. 1979, 594.
83. O. Inganas, B. Liedberg, W. Chang-Ru, H. Wynberg, Synth. Met., 1985, 11, 239.
84. S. Flandrois, C. Hauw, B. Francois, J. de Physique Colloque C3, 1983, 44, 523.
85. M. Monkenbusch, G. Wieners, Makromol. Chem. Rapid Commun., 1983, 4, 555.
86. L.W. Shacklette, J.E. Toth, N.S. Murthy, R.H. Baughman, J. Electrochem. Soc., 1985, 132, 1529.
87. L.W. Shacklette, N.S. Murthy, R.H. Baughman, Mol. Cryst. Liq. Cryst., 1985, 121, 201.
88. R.H. Baughman, L.W. Shacklette, N.S. Murthy, G.G. Miller, R.L. Elsenbaumer, Mol. Cryst. Liq. Cryst., 1985, 118, 253.
89. L.W. Shacklette, R.L. Elsenbaumer, R.H. Baughman, J. de Physique Colloque C3, 1983, 44, 599.
90. D.C. Bott, C.K. Chai, J.H. Edwards, W.J. Feast, R.H. Friend, M.E. Horton, J. de Physique Colloque C3, 1983, 44, 143.
91. P.D. Townsend, C.M. Pereira, D.D.C. Bradley, M.E. Horton, R.H. Friend, J. Phys., 1985, C18, L283.
92. G.L. Leising, E. Faulques, S. Lefrant, Synth. Met., 1985, 11, 123.
93. R.H. Friend, D.D.C. Bradley, P.D. Townsend, D.C. Bott, Synth. Met., 1987, 17, 267.
94. C.K. Chiang, M.A. Druy, S.C. Gau, A.J. Heeger, E.J. Lewis, A.J. MacDiarmid, Y.W. Park, H. Shirakawa, J. Am. Chem. Soc., 1978, 100, 1013.
95. T-C. Chung, A. Feldblum, A.J. Heeger, A.G. MacDiarmid, J. Chem. Phys., 1981, 74, 5504.
96. D. MacInnes, M.A. Druy, P.J. Nigrey, D.P. Nairns, A.G. MacDiarmid, A.J. Heeger, JCS Chem. Commun., 1981, 317.
97. Z. Kucharski, M. Lukasiak, J. Suwalski, A. Pron, J. de Physique Colloque C3, 1983, 44, 321.
98. A.F. Diaz, T.C. Clarke, J. Electroanal. Chem., 1984, 111, 115.
99. P. Nigrey, A.G. MacDiarmid, A.J. Heeger, Mol. Cryst. Liq. Cryst., 1982, 83, 309.
100. D. Billaud, D. Begin, P. Mirabeau, Mol. Cryst. Liq. Cryst., 1985, 121, 211.
101. R. Bitthin, B. Nagele, G. Koehler,, R. Denig, Mol. Cryst. Liq. Cryst., 1985, 121, 221.

102. P. Passiniemi, J.E. Osterholm, Mol. Cryst. Liq. Cryst., 1985, 121, 215.

103. C.K. Chiang, E.A. Blubaugh, W.T. Yap, Polymer, 1984, 25, 1112.

104. J.E. Osterholm, H.K. Yasuda, L.L. Levenson, J. Appl. Polym. Sci., 1982, 27, 931.

105. J.B. Schlenoff, J.V. Chien, Synth. Met., 1988, 22, 349.

106. G. Ahlgren, B. Krische, A. Pron, M. Zagorska J. Polym. Sci. Polym. Lett. Ed., 1984, 22, 173.

107. F.G. Will, J. Electrochem. Soc., 1985, 132, 518.

108. P. Bernier, A. El Khodary, F. Rachdi, J. de Phys. Colloque. C3, 1983, 44, 307.

109. P. Meisterle, H. Kuzmany, G. Nauer, Phys. Rev., 1984, B29, 6008.

110. E. Faulques, S. Lefrant, J. de Physique Colloque C3, 1983, 44, 337.

111. F. Mueller, P. Meisterle, M. Kuzmany, Mol. Cryst. Liq. Cryst., 1985, 121, 237.

112. A.G. MacDiarmid, A. J. Heeger, Polym. Prepr., 1982, 23, 241.

113. Y.W. Park, A.J. Heeger, M.A. Druy, A.G. MacDiarmid, J. Chem. Phys., 1980, 72, 946.

114. R.B. Kaner, A.G. MacDiarmid, Synth. Met., 1986, 14, 3.

115. R. Huq, G.C. Farrington, J. Electrochem. Soc., 1985, 132, 1432.

116. R. B. Kaner, S.J. Porter, A.G. MacDiarmid, JCS Faraday I, 1986, 82, 2323.

117. S. Lefrant, P. Bernier, R.B. Kaner, A.G. MacDiarmid, Mol. Cryst, Liq. Cryst., 1985, 121, 233.

118. L.W. Shacklette, R.H. Baughman, N.S. Murthy, Bull. Am. Phys. Soc., 1983, 28, 320.

119. A. Feldblum, J. H. Kaufman, S. Etemad, A.J. Heeger, T-C. Chung, A.G. MacDiarmid, Phys. Rev. Lett., 1982, B26, 815.

120. F.G. Will, J. Electrochem. Soc., 1985, 132, 2351.

121. A. Padula, B. Scrosati, M. Schwarz, U. Pedretti, J. Electrochem. Soc., 1984, 131, 2761.

122. S. Morzilli, M. Patriarcha, B. Scrosati, J. Electroanal. Chem., 1985, 191, 147.

123. T. Osaka, T. Kitai, Bull. Chim. Soc. Jpn., 1984, 57, 3386.

124. T. Ohsaka, T. Kitai, Bull. Chem. Soc. Jpn., 1984, 57, 759.

125. G. Nagele, G.E. Nauer, K. Kuzmany, J. Kurti, Synth. Met., 1987, 21, 293.

126. T.R. Jow, L.W. Shacklette, J. Electrochem. Soc., 1988, 135, 541.

127. F.G. Will, J. Electrochem. Soc., 1985, 132, 2093.

128. B. Scrosati, A. Padula, G.C. Farrington, Solid State Ionics, 1983, 9–10, 447.

129. F. Rachdi, P. Bernier, J. Chem. Phys., 1984, 80, 6285.

130. J.H. Kaufman, E.J. Mele, A. J. Heeger, R. Kaner, A.G. MacDiarmid, J. Electrochem. Soc., 1983, 130, 571.

131. F.G. Will, J. Electrochem. Soc., 1985, 132, 943.

132. J.H. Kaufman, A.J. Heeger, R. Kaner, E.J. Mele, A.G. MacDiarmid, J. de Physique Colloque C3, 1983, 44, 577.

133. J.C.W. Chien, J.B. Schlenoff, Nature, 1984, 311, 362.

134. G.E. Wnek, J.C.W. Chien, F.E. Karasz, M.A. Druy, J.W. Park, A.G. MacDiarmid, A.J. Heeger J. Polym. Sci., Polym. Lett. Ed., 1979, 17, 779.

135. A.G. MacDiarmid, R.B. Kaner, R.J. Mammone, A.J. Heeger, J. de Physique Colloque C3, 1983, 44, 543.

136. B. Kaner, A.G. MacDiarmid, R.J. Mammone, ACS. Symp. Ser. V242 (Polymer Electronics), 1984, 575.

137. W. Wanqun, R.J. Mammone, A.G. MacDiarmid, Synth. Met., 1985, 10, 235.

138. A. Pron, E. Faulques, S. Lefrant, Polym. Commun., 1987, 28, 27.

139. S.I. Yaniger, M.J. Kletter, A.G. MacDiarmid, Polym. Prepr., 1984, 25, 264.

140. Z. Kucharski, A. Pron, M. Josefowicz, J. Suwalski, S. Lefrant, S. Krichene, Solid State Commun., 1984, 51, 853.

141. A. Pron, C. Budrowski, J. Przyluski, Polymer, 1983, 24, 1294.

142. R.H. Friend, Phys. Chem. Ions. Condens. Matter, NATO ASI, Cambridge, England 6–17th Sept, 1983, D. Reidel, Dordrecht, Netherlands, 1984.

143. A.J. Heeger, Polymer J., 1985, 17, 201.

144. T.E. Jones, J.C. Hicks, D.L. Stierwalt, D.M. Cavanaugh, D.M. Gottfriedsen, J. de Physique Colloque C3, 1983, 44, 333.

145. K. Kivelson, Phys. Rev. 1982, B25, 3798.

146. A. Feldblum, R.W. Bigelow, H.W. Gibson, A.J. Epstein, D.B. Tanner, Mol. Cryst. Liq. Cryst., 1984, 105, 191.

147. J. Chen, T.C. Chung, F. Moraes, A.J. Heeger, Solid State Commun., 1985, 53, 757.

148. Springer Ser. Solid State Sci. Vol 63 (Electronic Properties of Polymers and Related Com-
     pounds), Proceedings of an International Winter School, Kirchberg, Tirol, Feb 23–March 1,
     1985, Eds. H. Kuzmany, M. Mehring, S. Roth, Springer Verlag, 1985, pp. 286–9.
149. Handbook of Conducting Polymers, T.J. Skotheim (Ed), Marcel Dekker, N.Y., 1986 ISBN-
     08247-7395-0 (Vol I), 0-8247-7454-X (Vol II)
150. K. Shinozaki, Y. Tomizuka, A. Noguri, Jpn. J. Appl. Phys., 1984, 23, L892.
151. T. Nagamoto, H. Kakahata, C. Ichikawa, O. Omoto, Jpn. J. Appl. Phys., 1985, 24, L397.
152. T. Nagamoto, H. Kakahata, C. Ichikawa, O. Omoto, J. Electrochem. Soc., 1985, 132, 1380.
153. F. Bernier, D. Boils, H. Canepa, J. Franco, A. Lecorre, J.P. Leboutin J. Electrochem. Soc., 1985,
     132, 2100.
154. M.F. Rubner, S.K. Tripathy, J. Georger, P. Cholewa, Macromolecules, 1983, 16, 870.
155. R.B. Kaner, A.G. MacDiarmid, JCS Faraday Trans., 1984, 180, 2109.
156. J. Caja, R.B. Kaner, A.G. MacDiarmid, J. Electrochem. Soc., 1984, 131, 2744.
157. M. Aldissi, A.R. Bishop, Synth. Met., 1986, 14, 13.
158. C.K. Chiang, Solid State Ionics., 1983, 9–10, 445.
159. M. Cadene, M. Rolland, M. Abu Lefdil, J. Bougnet, M.J.M. Abadie, Mol. Cryst. Liq. Cryst.,
     1985, 121, 297.
160. J. Sukamoto, O. Ohigashi, K. Matsumura, A. Takahashi, Synth. Met., 1982, 4, 177.
161. K. Kaneto, K. Yoshino, Y. Inuishi, Jpn. J. Appl. Phys., 1983, 22, L567.
162. W.P. Roberts, L.A. Schultz, Polym. Prepr., 1984, 25, 253.
163. M. Aldissi, Synth. Met., 1986, 15, 141.
164. H. Letheby, J. Chem Soc., 1862, 15, 161.
165. A. Volkov, G. Tourillon, P.C. Lacaze, J.E. Dubois, J. Electroanal. Chem., 1980, 115, 279.
166. J. Bacon, R.N. Adams, J. Am. Chem. Soc., 1968, 90, 6596.
167. D.M. Mohilner, R. N. Adams, W.J. Argersinger, J. Am. Chem. Soc., 1962, 84, 3618.
168. J. Langer, Solid State Commun., 1978, 26, 839.
169. G.E. Wnek, Synth. Met., 1986, 15, 213.
170. R. Jiang, S. Dong Synth. Met., 1988, 24, 255.
171. M. Angelopoulos, A. Roy, A.G. MacDiarmid, A. J. Epstein, Synth. Met., 1987, 21, 21.
172. P.L. Snauwert, R. Lazzaroni, J.J. Verbist, Synth. Met., 1987, 21, 181.
173. A.F. Diaz, J.A. Logan, J. Electroanal. Chem., 1980, 111, 111.
173. (b) D. Bloor, A. Monkman, Synth. Met., 1987, 21, 175.
174. R. Noufi, A.J. Nozik, J. White, L.F. Warren, J. Electrochem. Soc., 1982, 129, 2261.
175. E.W. Paul, A.J. Ricco, M.S. Wrighton, J. Phys. Chem., 1985, 89, 1441.
176. C.M. Carlin, L.J. Kepley, A.J. Bard, J. Electrochem. Soc., 1985, 132, 353.
177. B. Wang, J. Tang, F. Wang, Synth. Met., 1986, 13, 329
178. A. Kitani, J. Izumi, J. Yano, Y. Hiromoto, K. Sasaki, Bull. Chem. Soc. Jpn., 1984, 57, 2254.
179. A. Kitani, J. Yono, K. Susaki Chem. Lett., 1984, 9, 1565.
180. T. Kobayashi, H. Yoneyama, H. Tamura, J. Electroanal. Chem., 1984, 161, 419.
181. A.G. MacDiarmid, J.C. Chiang, M. Halpern, W.S. Huang, S.L. Mu, N.L.D. Somasiri, W. Wu,
     S.I. Yaniger, Mol. Cryst. Liq. Cryst., 1985, 121, 173.
182. E. Genies, C. Tsintavis, J. Electroanal. Chem., 1985, 195, 109.
183. E.M. Genies, A.A. Sayed, C. Tsintavis, Mol. Cryst. Liq. Cryst. 1985, 121, 181.
184. T. Yasui, Bull. Chem. Soc. Jpn., 1935, 10, 306.
185. N.E. Khumatov, J. Gen. Chem. USSR, 1952, 22, 627.
186. M Britenbach, K.H. Heckner, J. Electroanal. Chem., 1971, 29, 309.
187. T. Ohsaka, T. Okajima, N. Oyama J. Electroanal. Chem., 1986, 100, 159.
188. S.P. Armes, J.F. Miller, Synth. Met., 1988, 22, 385.
189. A.G. Green, A. Woodhead, J. Chem. Soc., 1910, 97, 2388.
190. A.G. Green, A. Woodhead, J. Chem. Soc., 1912, 101, 1117.
191. R. Willstatter, G. Moore, Chem. Ber., 1907, 40, 2665.
192. C.B. Duke, E.M. Cornwell, A. Paton, Chem. Phys. Lett., 1986, 131, 82.
193. T.H. Hjertberg, W.R. Salaneck, I. Lundstrom, N.L.D. Somasiri, A.G. MacDiarmid, J. Polym.
     Sci., Polym. Lett. Ed., 1985, 23, 503.
194. W.R. Salaneck, I. Lundstrom, T.H. Hjertberg, C.B. Duke, E. Cornwell, A. Paton, A.G.
     MacDiarmid, N.L.D. Somasiri, W.S. Huang, A.F. Richter, Synth. Met., 1987, 18, 291.

195. P.M. McManus, S.C. Young, R.J. Cushmai, JCS Chem. Commun., 1985, 1556.
196. T. Kobayashi, H. Yoneyama, H. Tamura, J. Electroanal. Chem., 1984, 177, 281.
197. A.P. Monkman, D. Bloor, G.C. Stevens, J.C.H. Stevens, J. Phys., 1987, D20, 1337.
198. I. Rubenstein, E. Sabatini, J. Rishpan, J. Electrochem. Soc., 1987, 134, 3078.
199. F. Wudl, R.O. Angus, F.L. Lu, P.M. Allemand, D.J. Vachon, M. Nowak, Z.X. Liu, A.J. Heeger, J. Am. Chem. Soc., 1987, 109, 3677.
200. T. Ohsaka, Y. Ohnuki, N. Oyama, G. Katagiri, K. Kamisako, J. Electroanal. Chem., 1984, 161, 399.
201. S. Stafstrom, J-L. Bredas, Synth. Met., 1986, 14, 297.
202. T. Hjertberg, M. Sandberg, O. Wennerstrom, I. Lagerstedt, Synth. Met., 1987, 21, 31.
203. R.R. Chance, D.S. Boudreaux, J.F. Wolf, L.W. Shacklette, R. Silbey, B. Themans, J.M. Andre, J.L. Bredas, Synth. Met., 1986, 15, 105.
204. W.R. Salaneck, B. Liedberg, O. Inganas, R. Erlandsson, I. Lundstrom, A.G. MacDiarmid, M. Halpern, N.L.D. Somasiri, Mol. Cryst. Liq. Cryst., 1985, 121, 191.
205. E.M. Genies, E. Vieil, Synth. Met., 1987, 20, 97.
206. N. Oyama, Y. Ohnuki, K. Chiba, T. Ohsaka, Chem. Lett. 1983, 1759.
207. M. Josefowicz, L. T. Yu, J. Perichon, R. Buvet, J. Polym. Sci., 1969, C22, 1187.
208. B. Wang, J. Tang, F. Wang, Synth. Met., 1987, 18, 323.
209. T. Kobayashi, H. Yoneyama, H. Tamura, J. Electroanal. Chem., 1984, 177, 293.
210. H.A. Pohl, E.H. Engelhardt, J. Phys. Chem., 1962, 66, 2085.
211. L.T. Yu, J. Petit, M. Josefowicz, G. Belorney, R. Buvet, C.R. Acad. Sci. France, 1965, 260, 5026.
212. M.F. Combarel, G. Belorney, M. Josefowicz, L.T. Yu, R. Buvet, C.R. Acad. Sci. France. 1966, 262, 459.
213. J.P. Travers, J. Chroboczek, F. Devreux, F. Genoud, M. Nechstein, A. Syed, E.M. Genies, C. Tsintavis, Mol. Cryst. Liq. Cryst., 1985, 121, 195.
214. H. Kuzmany, N.S. Sariciftci, Synth. Met., 1987, 18, 353.
215. A.J. Epstein, J.M. Ginder, F. Zuo, H.S. Wu, D.B. Tanner, A.F. Richter, M. Angelopoulos, W.S. Huang, A.G. MacDiarmid, Synth. Met., 1987, 21, 62.
216. A.J. Epstein, J.M. Ginder, F. Zuo, R.W. Bigelow, H.S. Wu, D.B.. Tanner, A.F. Richter, A.G. MacDiarmid, Synth. Met., 1987, 18, 303.
217. A.G. MacDiarmid, J-C. Chiang, A.F. Richter, A.J. Epstein, Synth. Met., 1987, 18, 285.
218. R.M. Baughman, J.F. Wolf, H. Eckhardt, L.W. Shacklette, Synth Met., 1988, 25, 121.
219. R. De Surville, M. Josefowicz, L.T. Yu, J. Perichon, R. Buvet, Electrochim. Acta. 1968, 13, 1451.
220. A.G. MacDiarmid, S.L Mu, N.L.D. Somasiri, W. Wu, Mol. Cryst. Liq. Cryst., 1985, 121, 187.
221. R. Jiang, S. Dong, J. Electroanal. Chem., 1988, 246, 101.
222. S. Bruckenstein, J.W. Sharkey, J. Electroanal. Chem., 1988, 241, 211.
223. O. Niwa, T. Tamamura, JCS Chem. Commun., 1984, 817.
224. M. Bidar, C. Fabre, E.M. Genies, A.A. Sayed, Mol. Cryst. Liq. Cryst., 1985, 121, 241.
225. P. Kovacic, M.B. Feldman, J.P Kovacic, J.B. Land, J. Appl. Polym. Sci., 1968, 12, 1735.
226. T. Yamamoto, A. Yamamoto, Chem. Lett., 1977, 353.
227. D.M. Ivory, G.G. Miller, J.M. Sowa, L.W. Shacklette, R.R. Chance, R.H. Baughman, J. Chem., Phys., 1979, 71, 1506.
228. F. Beck, A. Pruss, Electrochim. Acta, 1983, 28, 1847–55.
229. M. Tabata, M. Satoh, K. Kaneto, K. Yoshino J. Phys. Soc. Jpn., 1986, 55, 1305.
230. M. Delamar, P-C. Lacaze, J-Y. Dumousseau, J-E. Dubois, Electrochimica Acta, 1982, 27, 61.
231. A.F. Shepard, B.F. Daniels, J. Polym. Sci. A-I, 1966, 4, 511.
232. M. Tabata, M. Satoh, K. Kaneto, K. Yoshino, J. Phys., 1986, C19, L101.
233. L.A. Tinker, A.J. Bard, J. Am Chem. Soc., 1979, 101, 2316.
234. J.F. Fauvarque, A. Digua, M.A. Petit, J. Savard, Makromol. Chem., 1985, 186, 2415.
235. L.W. Shacklette, R.L. Elsenbaumer, R.R. Chance, J.M. Sowa, D.M. Ivory, G.G. Miller, R.H. Baughman, JCS Chem. Commun., 1982, 361.
236. F. Maurice, G. Froyer, Y. Pelous, J. de Physique Colloque C3, 1983, 44, 587.
237. D.D.C. Bradley, G.P Evans, R.H. Friend, Synth. Met., 1987, 17, 651.
238. M. Helbig, H.H. Horhold, A.K. Gyra, Makromol. Chem. Rapid. Commun., 1985, 6, 643.
239. A. Angeli, Gazz. Chim. Ital., 1916, 46, 279.
240. A. Angeli, L. Alessandri, Gazz, Chim. Ital., 1916, 46, 283.

241. G.P. Gardini, Adv. Heterocyclic Chem., 1973, 15, 67.
242. A. Dall 'Olio, G. Dascola, V. Varacca, U. Bocche, C.R. Acad. Sci., 1968, 433, 267C.
243. K.K. Kanazawa, A.F. Diaz, W.D. Gill, P.M. Grant, G.B. Street, G.P. Gardini, J.F. Kwak, Synth. Met., 1980, 1, 329.
244. A.F. Diaz, K.K. Kanazawa, G.P Gardini, JCS Chem. Commun., 1979, 635.
245. K.K. Kanazawa, A.F. Diaz, R.H. Geiss, W.D. Gill, J.F. Kwak, J.A. Logan, J.F. Rabolt, G.B. Street JCS Chem. Commun., 1979, 854.
246. A.O. Patil, Y. Ikenaue, N. Basescu, N. Colaneri, J. Chen, F. Wudl, A.J. Heeger, Synth. Met., 1987, 20, 151.
247. A.F. Diaz, K.K. Kanazawa in: Extended Linear Chain Compounds: G. Miller (Ed), Plenum, 1983; Vol 3, pp. 417–441.
248. K. Hyodo, A.G. MacDiarmid, Synth, Met., 1985, 11, 167.
249. S. Asavapiriyanont, G.K. Chandler, G.A. Gunawardena, D. Pletcher, J. Electroanal. Chem., 1984, 177, 229.
250. A. Murthy, S. Pal, K.S. Reddy, J. Mater. Sci. Lett., 1984, 3, 745.
251. M. Gazard, J.C. Dubois, M. Champagne, F. Garnier, G. Tourillon, J. de Physique, Colloque C3, 1983, 44, 537.
252. B. Krische, G. Ahlgren, Mol. Cryst. Liq. Cryst.., 1985, 121, 325.
253. P.G. Pickup, R.A. Osteryoung, J. Am. Chem. Soc., 1984, 106, 2294.
254. J.R. Reynolds, P.A. Poropatic, R.L. Toyooka, Macromolecules, 1987, 20, 958.
255. K.J. Wynne, G.B. Street, Macromolecules, 1985, 18, 2361.
256. A.F. Diaz, J.I. Crowley, J. Bargon, G.P. Gardini, J.B. Torrance, J. Electroanal. Chem., 1981, 121, 355.
257. H. Lindenberger, D. Schafer-Siebert, S. Roth, M. Hanack, Synth. Met., 1987, 18, 37.
258. N. Oyama, T. Ohsaka, K. Chiba, H. Miyamoto, T. Mukai, S. Tanaka, T. Kumagai, Synth. Met., 1987, 20, 245.
259. G.B. Street, T.C. Clarke, R.H. Geiss, V.Y. Lee, N. Nazzal, P. Pfluger, J.C. Scott, J. de Physique Colloque C3, 1983, 44, 599.
260. R.L. Elsenbaumer, H. Eckhardt, Z. Iqbal, J. Toth, R.H. Baughman, Mol. Cryst. Liq. Cryst., 1985, 118, 111.
261. W.R. Salaneck, R. Erlandsson, J. Prejza, I. Lundstrom, O. Inganas, Synth. Met., 1983, 5, 125.
262. A. F. Diaz, J. Castillo, J.A. Logan, W.Y. Lee, J. Electroanal. Chem., 1981, 129, 115.
263. S.J. Hahn, W.J. Gajda, P.O. Vogelhut, M.V. Zeller, Synth. Met., 1986, 14, 89.
264. O. Nishikama, M. Kato, J. Chem. Phys., 1986, 85, 6758.
265. W. Wernet, G. Wegner, Makromol. Chem., 1987, 188, 1465.
266. F. Beck, M. Oberst, Makromol. Chem., Macromol. Symp., 1987, 8, 97.
267. M. Salmon, K.K. Kanazawa, A.F. Diaz, M. Krounbi, J. Polym. Sci., Polym. Lett. Ed., 1982, 20, 187.
268. E.T. Kang, T.C. Tan, K.G. Neoh, Y. K. Ong, Polymer, 1986, 27, 1958.
269. J. Prejza, I. Lundstrom, T. Skotheim., J. Electrochem. Soc., 1982, 129, 1685.
270. R. Noufi, D. Tench, L.F. Warren, J. Electrochem. Soc., 1980, 127, 2310.
271. T. Skotheim, I. Lundstrom, J. Prejza, J. Electrochem. Soc., 1981, 128, 1625.
272. R. Lazzaroni, S. Dujardin, J.P. Boutique, C. Mousty-Desbuquet, J. Riga, J. Verbist, Mol. Cryst. Liq. Cryst., 1985, 118, 249.
273. A.J. Frank, Mol. Cryst. Liq. Cryst., 1982, 83, 341.
274. G.B. Street, T.C. Clarke, K. Krounbi, P. Pfluger, J.F. Rabolt, R.H. Geiss, Polym. Prepr., 1982, 23, 117.
275. M. Saloma, M. Aguilar, M. Salmon, J. Electrochem. Soc., 1985, 132, 2379.
276. M. Salmon, M.E. Carbajal, M. Aguilar, M. Saloma, J.C. Juarez, JCS Chem. Commun., 1983, 1532.
277. Techniques of Electroorganic Synthesis, Vol. V., Part II, N.L. Weinberg (Ed), Wiley, N.Y. 1975.
278. A.F. Diaz, A. Martinez, K.K. Kanazawa, M. Salmon, J. Electroanal. Chem. 1981, 130, 181.
279. A.F. Diaz, J.I. Castillo, K.K. Kanazawa, J.A. Logan, M. Salmon, O. Fajardo, J. Electroanal. Chem., 1982, 133, 233.
280. M. Salmon, M. Carbajal, J.C. Juarez, A.F. Diaz, M. Rock, J. Electrochem. Soc., 1984, 131, 1802.
281. P. Audebert, G. Bidan, Mol. Cryst. Liq. Cryst., 1985, 118, 187.

282. P. Audebert, G. Bidan, Synth. Met., 1986, 15, 9.
283. G. Bidan, M. Gugliemi, Synth. Met., 1986, 15, 49.
284. M. Salmon, M.E. Carbajal, M. Aguilar, M. Saloma, J.C. Juarez, JCS Chem. Commun., 1983, 24, 1532.
285. R. Qian, J. Qiu, B. Yan, Synth. Met., 1986, 14, 81.
286. T.A. Skotheim, S.W. Feldberg, M. Armand, J. de Physique Colloque C3., 1983, 44, 615.
287. P. Mirebeau, J. de Physique Colloque C3, 1983, 44, 579.
288. S. Panero, P. Prosperi, B. Scrosati, Electrochim. Acta, 1987, 10, 1465.
289. K. Yoshida, J. Am. Chem. Soc. 1979, 101, 2116.
290. J.M. Bobbit, C.L. Kulkarne, J.P. Willis, Heterocycles, 1981, 15, 495.
291. E.M. Genies, G. Bidan, A.F. Diaz, J. Electroanal. Chem., 1983, 149, 101.
292. T. Inoue, T. Yamase, Bull. Chim. Soc. Jpn., 1983, 56, 985.
293. B.L. Funt, S.N. Lowen, Synth. Met., 1983, 11, 129.
294. R. Jansson, M. Armin, R. Bjorklund, I. Lundstrom, Thin Solid Films, 1985, 125, 205.
295. A.F. Diaz, B. Hall, IBM J. Res. Develop., 1983, 27, 342.
296. W. Wernet, M. Monkenbusch, G. Wegner, Makromol. Chem. Rapid. Commun., 1984, 5, 157–64.
297. M. Satoh, K. Kaneto, K. Yoshino, Jpn. J. Appl. Phys., 1985, 24, L423.
298. M. Satoh, K. Kaneto, K. Yoshino, Synth. Met., 1986, 14, 289.
299. M. Takakubo, Synth. Met., 1986, 16, 167.
300. T. Skotheim, M. V. Rosenthal, C.A. Linkous, JCS Chem. Commun., 1985, 612.
301. S. Dong, J. Ding, Synth. Met., 1987, 20, 119.
302. S. Asavapiriyanont, G.K. Chandler, G.A. Gunawardena, D. Pletcher, J. Electroanal. Chem., 1984, 177, 245.
303. M.L. Marcos, I. Rodriguez, J. Gonzalez-Velasco, Electrochim. Acta, 1987, 32, 1453.
304. R. Qian, J. Qiu, D. Shen, Synth. Met., 1987, 18, 13.
305. I. Rodriguez, M.L. Marcos, J. Gonzalez-Velasco, Electrochim. Acta, 1987, 32, 1181.
306. G. Bidan, A. Deronzier, J.C. Moutet, Nouv. J. Chim., 1984, 8, 501.
307. G. Bidan, A. Deronzier, J.C. Moutet, JCS Chem. Commun., 1984, 1185.
308. K.K. Kanazawa, A.F. Diaz, M.T. Krounbi, G.B. Street, Synth. Met., 1981, 4, 119.
309. G. Ahlgren, B. Krische, JCS Chem. Commun., 1984, 14, 946.
310. M.A. De Paoli, R.J. Waltman, A.F. Diaz, J. Bargon, JCS Chem. Commun., 1984, 15, 1015.
311. N. Kumar, J. Polym. Sci., Polym. Lett. Ed., 1985, 32, 57.
312. N. Bates, M. Cross, R. Lines, D. Walton, JCS Chem. Commun., 1985, 871.
313. R.B. Bjorklund, I. Lundstrom, J. Elec. Mater., 1984, 13, 211.
314. S.E. Lindsay, G.B. Street, Synth. Met., 1984/5, 10, 67.
315. S.J. Jasne, C.K. Chiklis, Synth. Met., 1986, 15, 175.
316. S.A. Jenekhe, S.T. Wellinghof, J.F. Reed, Mol. Cryst. Liq. Cryst., 1984, 105, 175.
317. O. Niwa, M. Kakuchi, T. Tamamura, Polym. J., 1987, 19, 1293.
318. H. Koezuka, S. Etoh, J. Appl. Phys., 1983, 54, 2511.
319. T.T. Wang, S. Tasaka, R.S. Hutton, P.Y. Lu, JCS Chem. Commun., 1985, 713.
320. R.B. Bjorklund, B. Liedberg, JCS Chem. Commun., 1986, 1293.
321. G. Wegner, W. Wernet, D. T. Glazhofer, J. Ulanski, Ch. Krohnke, M. Mohammadi, Synth. Met., 1987, 18, 1.
322. L.F. Warren, D.P. Anderson J. Electrochem. Soc., 1987, 134, 101.
323. G.R. Mitchell, A. Geri, J. Phys., 1987, D20, 1346.
324. K, Yakushi, L.J. Lauchlan, T.C. Clarke, G.B. Street, J. Chem. Phys., 1983, 79, 4773.
325. B.J. Orchard, B. Freidenreich, S.K. Tripathy, Polymer, 1985, 27, 1535.
326. F. Devreux, F. Genoud, M. Nechtstein, J.P. Travers, G. Bidan, J. de Physique Colloque C3, 1983, 44, 621.
327. T. Osaka, K. Naoi, S. Ogano, S. Nakamura, Chem. Lett., 1986, 1687.
328. M. Satoh, H. Yamasaki, S. Aoki, K. Yoshino, Synth. Met., 1987, 20, 79.
329. R.H. Geiss, G.B. Street, W. Volksen, J. Economy, IBM J. Res. Dev., 1983, 27, 321.
330. J. H. Kaufman, O.R. Melroy, F.F. Abraham, A.I. Nazzal, A. Kapitulnik, Synth. Met., 1987, 18, 19.
331. A.F. Diaz, B. Hall, IBM J. Res. and Dev., 1983, 27, 342.

332. J.C. Scott, P. Pfluger, T.C. Clarke, G.B. Street, Polym. Prepr., 1982, 23, 119.
333. T.C. Clarke, J.C. Scott, G.B. Street, IBM J. Res. Dev., 1983, 27, 313.
334. W.R. Salaneck, R. Erlandsson, J. Prejza, I. Lundstrom, C.B. Duke, W.K. Ford, Polym. Prepr., 1982, 23, 120.
335. P. Pfluger, M. Krounbi, G.B. Street, G. Weiser, J. Chem. Phys., 1983, 78, 3212.
336. P. Pfluger, G.B. Street, J. de Physique, Colloque C3, 1983, 44, 609.
337. A.F. Diaz., J.M.V. Vallejo, A.M. Duran, IBM J. Res. Dev., 1981, 25, 42.
338. G.B. Street, S.E. Lindsay, A.I. Nazzal, K.J. Wynne, Mol. Cryst. Liq. Cryst., 1985, 118, 137.
339. A. Nazzal, G.B. Street, JCS Chem. Commun., 1984, 83.
340. J. Fink, B. Scheerer, W. Wernet, M. Monkenbusch, G. Wegner, H-J. Freund, H. Gonska, Phys. Rev., 1986, B34, 110.
341. E. Buhks, I.M. Hodge, J. Chem. Phys., 1985, 83, 5976.
342. K. Tanaka, T. Schichiri, T. Yamabe, Synth. Met., 1986, 14, 271.
343. W. Wernet, M. Monkenbusch, G. Wegner, Mol. Cryst. Liq. Cryst., 1985, 118, 193.
344. M.J. Van der Sluijs, A.E. Underhill, B. Zaba, J. Phys., 1987, D20, 1411.
345. R.A. Bull, F. Fan, A.J. Bard, J. Electrochem. Soc., 1982, 129, 1009.
346. K.S.V. Santham, R.N. O'Brien, J. Electroanal. Chem., 1984, 160, 377.
347. M. Zagorska, A. Pron, S. Lefrant, Z. Kucharski, J. Sowalski, P. Bernier, Synth. Met., 1987, 18, 43.
348. M.Takakubo, Synth. Met., 1987, 18, 53.
349. M. Zagorska, H. Wycislik, J. Przyluski, Synth. Met., 1987, 20, 259.
350. P. Pfluger, G.B. Street, J. Chem. Phys., 1984, 80, 544.
351. J.C. Scott, P. Pfluger, M. T. Krounbi, G.B. Street, Phys. Rev., 1983, B28, 2140.
352. S.W. Feldberg, J. Am. Chem. Soc., 1984, 106, 4671.
353. P. Pfluger, G.B. Street, Polym. Prepr., 1982, 23, 122.
354. P. Pfluger, V.M. Gubler, G.B. Street, Solid State Commun., 1984, 49, 911.
355. E.M. Genies, J.M. Pernaut, Synth. Met., 1984/5, 10, 117.
356. E.M. Genies, A.A. Syed, M. Salmon, Synth. Met., 1985, 11, 353.
357. T. Skotheim, M. I. Florit, A. Melo, W.E. O'Grady, Phys. Rev., 1984, B30, 4846.
358. J. Tanguy, N. Mermilliod, M. Hoclet, Synth. Met., 1987, 18, 7.
359. J. Tanguy, N. Mermilliod, Synth. Met., 1987, 21, 129.
360. R. Erlandsson, I. Lundstrom, J. de Phys. Colloque. C3, 1983, 44, 713.
361. T. Osaka, K. Naoi, H. Sakai, S. Ogano, J. Electrochem. Soc., 1987, 134, 285.
362. T.A. Skotheim, Synth. Met., 1986, 14, 31.
363. T.A. Skotheim, O. Inganas, Mol. Cryst. Liq. Cryst., 1985, 121, 285.
364. S. Panero, P. Prosperi, B. Scrosati, Electrochim. Acta, 1987, 10, 1461.
365. M. Watanabe, K. Tadano, K. Sanui, N. Ogata, Chem. Lett., 1987, 1239.
366. P. Audebert, G. Bidan, Synth. Met., 1986, 14, 71.
367. M. Ogasawa, K. Funahashi, T. Demurta, T. Hagiwara, K. Iwata, Synth. Met., 1986, 14, 61.
368. S.K. Tripathy, D. Kitchen, M.A. Druy, Polym. Prepr., 1982, 23, 109.
369. S. Tokito, T. Tsutsui, S. Saito, Chem. Lett, 1985, 4, 531.
370. A. Watanabe, M. Tanaka, J. Tanaka, Bull. Chem. Soc. Jpn., 1981, 54, 2278.
371. S. Kuwabata, K-I. Okamoto, O. Ikeda, H. Yoneyama, Synth. Met., 1987, 18, 101.
372. M. Tanaka, A. Watanabe, M. Fujimoto, J. Tanaka, Mol. Cryst. Liq. Cryst., 1982, 83, 277.
373. T. Inoue, I. Hosoya, T. Yamase, Chem. Lett., 1987, 563.
374. F. Genoud, M. Gugliemi, M. Nechtstein, E. Genies, M. Salmon, Phys. Rev. Lett., 1985, 55, 118.
375. J.C. Scott, J-L. Bredas, J.H. Kaufman, P. Pfluger, G.B. Street, K. Yakushi, Mol. Cryst. Liq. Cryst., 1985, 118, 163.
376. J.L. Bredas, J.C. Scott, K. Yakushi, G.B. Street, Phys. Rev., 1984, B30, 1023.
377. K. Yakushi, L.J. Lauchlan, G.B. Street, J. Chem. Phys., 1984, 81, 4133.
378. M. Nechtstein, F. Devreux, F. Genoud, E. Vieil, J. M. Pernaut, E. Genies, Synth. Met., 1986, 15, 59.
379. M. Salmon, M. Aguilar, M. Saloma, JCS Chem. Commun., 1983, 570.
380. S. Dong, J. Ding, Synth. Met., 1988, 24, 273.
381. M.J. Van der Sluijs, A.E. Underhill, B.N. Zaba, Synth. Met., 1987, 18, 105.
382. B. Zinger, L.L. Miller, J. Am. Chem. Soc., 1984, 106, 6861.

383. R.A. Bull, F. Fan, A.J. Bard, J. Electrochem. Soc., 1983, 130, 1636.
384. A. Deronzier, M. Essakalli, J-C. Moutet, J. Electroanal. Chem., 1988, 244, 163.
385. M.V. Rosenthal, T.A. Skotheim, C.A. Linkous, Synth. Met., 1986, 15, 219.
386. F. Mitzutani, S. Iijima, Y. Tanabe, K. Tsuda, Synth. Met., 1987, 18, 111.
387. K. Murao, K. Suzuki, Polym. Prepr., 1984, 25, 260.
388. M.D. Imsides, G.C. Wallace, J. Electroanal. Chem., 1988, 246, 101.
389. M.V. Rosenthal, T.A. Skotheim, A. Melo, M.I. Florit, J. Electroanal. Chem., 1985, 185, 297.
390. T. Yamamoto, K. Sanechika, A. Yamamoto, J. Polym. Sci. Polym. Lett. Ed., 1980, 18, 9.
391. T. Yamamoto, K. Sanechika, A. Yamamoto, Bull, Chem. Soc. Jpn., 1983, 56, 1503.
392. M. Kobayashi, J. Chen, T-C. Chung, F. Moraes, A.J. Heeger, F. Wudl, Synth. Met., 1984, 9, 77.
393. G. Kossmehl, G. Chalzitheodorou, Makromol. Chem. Rapid Commun., 1982, 2, 551.
394. M. Akimoto, Y. Furukawa, H. Takauchi, I. Harada, Y. Soma, M. Soma, Synth. Met., 1986, 15, 353.
395. M.A. Druy, R.J. Seymour, S.K. Tripathy, ACS. Symp. Ser. 242 (Polymer Electronics), 1984, 473.
396. F. Garnier, G. Tourillon, J. Electroanal. Chem., 1983, 148, 299.
397. K. Kaneto, Y. Kohno, K. Yoshino, Y. Inuishi, JCS Chem. Commun., 1983, 382.
398. K. Kaneto, K. Yoshino, Y. Inuishi, Solid State Commun., 1983, 46, 389.
399. S. Hotta, S.D.D.V. Rughooputh, A.J. Heeger, F. Wudl, Macromolecules, 1987, 20, 212.
400. R.J. Waltman, J. Bargon, A.F. Diaz, J. Phys. Chem., 1983, 87, 1459.
401. S. Tanaka, M. Sato, K. Kaeriyama, Polym. Commun., 1985, 26, 303.
402. M.A. Sato, S. Tanaka, K. Kaeriyama, JCS Chem. Commun., 1985, 713.
403. R.L. Blankespoor, L.L. Miller, JCS Chem. Commun., 1985, 90.
404. M.A. Sato, S. Tanaka, K. Kaeriyama, JCS Chem. Commun., 1986, 873.
405. A. Czerwinski, H. Zimmer, C. Van Pham, H.B. Mark, J. Electrochem. Soc., 1985, 132, 2669.
406. J. Heinze, J. Mortensen, K. Hinckelman, Synth. Met., 1987, 21, 209.
407. Y. Yumoto, S. Yoshimura, Synth. Met., 1986, 13, 185.
408. M.A. Druy, R.J. Seymour, J. de Phys. Colloque. C3, 1983, 44, 595.
409. T.R. Jow, K.Y. Jen, R.L. Elsenbaumer, L.W. Shacklette, M. Angelopoulos, M.P. Cava, Synth. Met., 1986, 14, 53.
410. M. Biserni, A. Marinangeli, M. Mastragostino, J. Electrochem. Soc., 1985, 132, 1597.
411. R. Lazzaroni, J. Riga, J.J. Verbist, L. Christiaens M. Renson, JCS Chem. Commun., 1985, 999.
412. R. Danieli, C. Taliani, R. Zamboni, G. Giro, M. Biserni, M. Mastragostino, A. Testoni, Synth. Met., 1986, 13, 325.
413. P. Dimarco, M. Mastragostino, C. Taliani, Mol. Cryst. Liq. Cryst., 1985, 118, 241.
414. M. Mastragostino, B. Scrosati, J. Electrochem. Soc., 1985, 132, 1259.
415. R. Lazzaroni, A. De Pryck, Ch. Debaisieux, J. Riga, J. Verbist, J-L. Bredas, J. Delhalle, J.M. Andre, Synth. Met., 1987, 21, 189.
416. S. Naitoh, Synth. Met., 1987, 18, 237.
417. S. Hotta, T. Hosaka, W. Shimotsuma, Synth. Met., 1983, 6, 319.
418. S. Hotta, W. Shimotsuma, M. Taketani, Synth. Met., 1983, 6, 69.
419. S. Hotta, W. Shimotsuma, M. Taketani, Synth. Met., 1983, 6, 317.
420. G.G. MacLeod, M.G.B. Mahboubian-Jones, R.A. Pethrick, S.D. Watson, N.D. Truong, J.C. Galin, J. Francois, Polymer, 1986, 27, 455.
421. S. Tanaka, M. Satoh, K. Kaeriyama, Makromol. Chem., 1985, 186, 1685.
422. M. Bragadin, P. Cescon, A. Berlin, G.A. Pagani, F. Sannicolo, Synth. Met., 1987, 18, 241.
423. K. Tanaka, T. Schichiri, S. Wang, T. Yamabe, Synth. Met., 1988, 24, 203.
424. A.R. Hillman, E.F. Mallen, J. Electroanal. Chem., 1987, 220, 351.
425. A.R. Hillman, E.F. Mallen, J. Electroanal. Chem., 1988, 243, 403.
426. P.A. Christensen, A. Hamnett, A.R. Hillman, J. Electroanal. Chem., 1988, 242, 47.
427. R.J. Waltman, A.F. Diaz, J. Bargon, J. Phys. Chem., 1984, 87, 4343.
428. G. Tourillon, F. Garnier, J. Polym. Sci. Polym. Phys. Ed., 1984, 22, 33.
429. R.L. Elsenbaumer, K.Y. Jen, R. Oboodi, Synth. Met., 1986, 15, 169.
430. M.J. Nowak, S.D.D.V. Rughooputh, S. Hotta, A.J. Heeger, Macromolecules, 1987, 20, 965.
431. S.D.D.V. Rughooputh, M. Nowak, S. Hotta, A.J. Heeger, F. Wudl, Synth. Met., 1987, 21, 41.
432. W. Kobel, M. Kiess, M. Egli, R. Keller, Mol. Cryst. Liq. Cryst., 1986, 137, 141.
433. K. Tanaka, T. Schichiri, T. Yamabe, Synth. Met., 1986, 16, 207.

434. T. Otero, E. Laretta-Azelain, Polym. Commun., 1988, 29, 21.
435. S. Hotta, W. Shimotsuma, M. Taketani. Synth. Met., 1984/5, 10, 5.
436. S. Hotta, T. Hosaka, M. Soga, W. Shimotsuma, Synth. Met., 1984/5, 10, 95.
437. C. Yong, Q. Renyuan, Solid State Commun., 1985, 54, 211.
438. K.Y. Jen, R. Oboodi, R.L. Elsenbaumer, Polym. Mat. Sci. Eng., 1985, 53, 79.
439. J.W. Thackeray, H.S. White, M.S. Wrighton, J. Phys. Chem., 1985, 89, 5133.
440. S. Osawa, M. Ito, K. Tanaka, J. Kuwano, Synth. Met., 1987, 18, 145.
441. S. Hotta, W. Shimotsuma, M. Taketani, S. Kohiki, Synth. Met., 1985, 11, 139.
442. Y. Takenaka, T. Koike, T. Oka, M. Tanahashi, K-I. Kaneto, K. Yoshino, Synth. Met., 1987, 18, 207.
443. J.H. Kaufman, T-C. Chung, A.J. Heeger, F. Wudl. J. Electrochem. Soc., 1984, 131, 2092.
444. K. Kaneto, S. Ura, K. Yoshino, Y. Inuishi, Jpn. J. Appl. Phys., 1984, 23, L189.
445. J-L. Bredas, R. Silbey, D.S. Boudreaux, R.R. Chance, J. Am. Chem. Soc., 1983, 105, 6555.
446. G. Tourillon, F. Garnier, J. Electroanal. Chem., 1984, 161, 407.
447. K. Tanaka, T. Schichiri, K. Yoshizawa, T. Yamabe, J. Yamauchi, Y. Deguchi, Solid State Commun., 1984, 51, 565.
448. K. Kaneto, Y. Kohno. K. Yoshino, Mol. Cryst. Liq. Cryst., 1985, 118, 217.
449. K. Kaneto, S. Mayashi, S. Ura, K. Yoshino, J. Phys. Soc. Jpn., 1984, 54, 1146.
450. J.E. Osterholm, P. Passiniemi, H. Isatalo, H. Stubb, Synth. Met., 1987, 18, 213.
451. G. Glenis, G. Tourillon, F. Garnier, Thin Solid Films, 1984, 122, 9.
452. M. Aizawa, T. Yamada, M. Shinchara, K. Akagi, H. Shirakawa, JCS Chem. Commun., 1986, 1315.
453. G. Tourillon, F. Garnier, J. Phys. Chem., 1984, 88, 5281.
454. F. Wudl, M. Kobayashi, N. Colaneri, M. Boysel, A.J. Heeger, Mol. Cryst. Liq. Cryst., 1985, 118, 199.
455. J.L. Bredas, Synth. Met., 1987, 17, 115.
456. K.Y. Jen, R. Elsenbaumer, Synth. Met., 1986, 16, 379.
456. (b) N. Colaneri, M. Kobayashi, A.J. Heeger, F. Wudl, Synth. Met., 1986, 14, 45.
457. M. Bezoari, P. Kovacic, S. Gronowitz, A-B. Hornfeldt, J. Polym. Sci., Polym. Lett. Ed., 1981, 19, 347.
458. R. Sugimoto, K. Yoshino. S. Inoue, K. Tsukagoshi, Jpn. J. Appl. Phys., 1985, 24, L425.
459. K. Yoshino, K. Kaneto, S. Inoue, K. Tsukagoshi, Jpn. J. Appl. Phys., 1983, 22, L701.
460. J. Bargon, S. Mohmand, R.J. Waltman, IBM J. Res. and Dev., 1983, 27, 330.
461. G. Mengoli, M.M. Musiani, B. Schreck, S. Zecchin, J. Electroanal. Chem., 1988, 246, 73. and refs. therein.
462. S. Cattarin, G. Mengoli, M.M. Musiani, B. Schreck, J. Electroanal. Chem., 1988, 246, 87. and refs. therein.
463. A.F. Diaz, J. Bargon, Ch. 3. in ref. 149, pp 81–115.
464. J. Rault-Berthelot, J Simonet, J. Electroanal. Chem., 1985, 182, 187.
465. J.F. Oudard, R.D. Allendoerfer, R.A. Osteryoung, Synth. Met., 1988, 22, 407.
466. P. Denisevich, Y.S. Papir, V.P. Kurkov, S.P. Current, A.H. Schroeder, S. Suzuki, Polym. Prepr., 1983, 24, 330.
467. B. Themans, J.M. Andre, J-L. Bredas, Solid State Commun., 1984, 50, 1047.
468. J.G. Eaves, H.W. Munro, D. Parker, Synth. Met., 1987, 20, 387.
469. R.R. Chance, D.S. Boudreaux, J-L. Bredas, R. Silbey, ACS Symp. Ser. V242 (Polymer Electronics), 1984, 433.

Photoelectron Spectroscopy of Practical Electrode Materials

R. Kötz

Paul Scherrer Institut, CH-5232 Villigen PSI, Switzerland

Contents

1   Introduction

Surface sensitive spectroscopic techniques have found increasing acceptance in electrochemistry during the last ten to fifteen years. This development is nicely reflected by the contributions to a series of conferences devoted to "non traditional" studies of the solid/electrolyte interface held in La Colle-sur-Loup 1977 [1], Snowmass 1979 [2], Logan 1982 [3] and Berlin 1986 [4]. In summarizing the conference on "Electronic and Molecular Structure of Electrode Electrolyte Interface" E. Yeager has given an overview of available *in situ* and *ex situ* techniques [5] adopted by electrochemists from surface scientists. Electron spectroscopy for chemical analysis (ESCA) or X-ray photoelectron spectroscopy (XPS as it will be called throughout this chapter) has to be regarded as an *ex situ* technique for electrochemical interface studies. At least since the contribution of Siegbahn [6] and his coworkers, which were awarded with the Nobel Prize in 1981, XPS was developed into a standard surface analytical technique.

   The development of XPS is tightly bound to the development of photoelectron spectroscopy (PS) in general, for which the theoretical background and the experimental facilities for reliable ultra high vacuum (UHV) control had to be provided. The rapid development of PS and surface science can be visualized when keeping in mind that UHV control was only achieved in the late 1950's and that a major step in creating a theoretical background for PS occurred in the early 1960's by the suggestion of the three step model [7]. Reliable spectra of gases adsorbed on surfaces were only obtained around 1970. Excellent overviews of the history of PS are given by Spicer [8] and by Feuerbacher et al. [9].

   In view of the overwhelming success of PS in surface science, it is not surprising that XPS has been used rather early for the study of electrochemically modified electrode surfaces. Winograd et al. [10–12] were the first to use this spectroscopy for the study of oxide formation on Pt electrodes and also for the investigation of metal underpotential deposition (UPD) on Pt. Although a standard surface analytical tool, XPS has not found a corresponding consideration in electrochemistry.

1.1   Scope

In accord with the fact that XPS has become a standard surface science technique but has not been appreciated adequately in electrochemistry, it is the scope of this review chapter to bring XPS nearer to those who work on electrochemical problems and convince electrochemists to use XPS as a complementary technique. It is not the intention to treat fundamental physical and experimental aspects of photoelectron spectroscopies in detail. There are several review articles in the literature treating the basics and new developments in an extensive and competent way [9, 13]. In this article basic aspects are only addressed in so far as they are necessary to understand and interpret XPS data. Some emphasis will be put on those aspects which are peculiar to electrode problems like electrode emersion and transfer from the liquid environment to the UHV chamber and the effects of the relative complexity of the emersed electrochemical interface on XPS data analysis.

What can be learnt from XPS about electrochemical processes will be demonstrated and discussed in the main part of this chapter by means of specific examples. Thereby a survey of new XPS and UPS results on relevant electrode materials will be given. Those electrode materials, which have some potential for a technical application, are understood as practical and will be discussed with respect to the relevant electrochemical process. The choice of electrode materials discussed is of course limited. Emphasis will be put on those materials which are relevant for technical solid polymer electrolyte cells being developed in the author's laboratory.

1.2 Relevance and Complexity of Photoelectron Spectroscopic Investigations of Electrodes

Electrochemical processes are always heterogeneous and confined to the electrochemical interface between a solid electrode and a liquid electrolyte (in this chapter always aqueous). The knowledge of the actual composition of the electrode surface, of its electronic and geometric structure, is of particular importance when interpreting electrochemical experiments. This information cannot be obtained by classical electrochemical techniques. Monitoring the surface composition before, during and after electrochemical reactions will support the mechanism derived for the process. This is of course true for any surface sensitive spectroscopy. Each technique, however, has its own spectrum of information and only a combination of different surface spectroscopies and electrochemical experiments will come up with an almost complete picture of the electrochemical interface. XPS is just one of these techniques.

XPS provides information about the elemental composition of the electrode surface and about the chemical bonding of the elements, that is the environment of the atom. In combination with a sputtering technique XPS also allows – within limits – depth profiling of the elements. When working on practical surfaces and processes which run at atmospheric pressures, XPS becomes an *ex situ* technique, which is a disadvantage. If there is a choice between an *in situ* and an *ex situ* technique the *in situ* technique is to be preferred. In order to overcome the gap between *ex situ* and *in situ* studies, sophisticated preparation chambers and transfer systems have been developed. These allow the sample to be treated under atmospheric pressure or in contact with a liquid electrolyte and subsequently to be brought into the UHV analysing chamber using a controlled path. The *ex situ* character is indeed the weak part of any XPS investigation on electrode surfaces and the point where criticism usually occurs. Today, however, there is promising experimental evidence that *ex situ* investigations do provide information about the geometric and electronic structure of the electrochemical interface [14].

Electrochemical reactions are driven by the potential difference at the solid liquid interface, which is established by the electrochemical double layer composed, in a simple case, of water and two types of counter ions. Thus, provided the electrochemical interface is preserved upon emersion and transfer, one always has to deal with a complex coadsorption experiment. In contrast to the solid/vacuum interface, where for instance metal adsorption can be studied by evaporating a metal onto the surface, electrochemical metal deposition is always a coadsorption of metal ions, counter ions, and probably water dipols, which together cause the potential difference at the surface. This complex situation has to be taken into account when interpreting XPS data of emersed electrode surfaces in terms of chemical shifts or binding energies.

2 Information Provided by XPS and UPS about Electrode Materials

In the following a brief survey will be given of the fundamentals of the photoemission process in general. Later on (Sections 2.3 and 2.7) specific aspects of XPS data evaluation will be discussed with respect to electrochemical applications. Comprehensive reviews on photoelectron spectroscopies have been published before and have given a detailed discussion of theoretical and experimental aspects [9, 13].

2.1 Fundamental and Experimental Aspects

A major step in understanding photoelectron emission from solids was the development of the three step model [7]. According to this model the photoemission process can be regarded as consisting of three consecutive events: (i) the optical excitation of the electron, (ii) the transport of the electron to the surface of the solid and finally (iii) the escape of the electron overcoming the surface barrier potential. A sketch of these processes, including the detection of the electron in the analyzer, is given in Fig. 1 on an energy scale.

During excitation of the electron by a photon with energy hv, conservation of energy leads to equation (1)

$$E_f = E_i + hv \tag{1}$$

The electron is excited from a filled initial state E_i below the Fermi level E_F to an empty final state E_f above E_F. Momentum conservation will be provided by a lattice vector or in some cases by a surface vector. The transition probability is mainly determined by the optical excitation matrix element containing the joint density of states.

The electron with the energy E_f has a certain probability of reaching the surface without a scattering event according to the mean free path. This is what determines

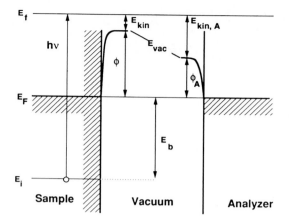

Fig. 1. Sketch of excitation, escape, and detection of photoelectrons on an energy scale.

the surface sensitivity of the photoemission technique and will be discussed in more detail in Section 2.5. In those cases where a scattering event occurs, the electron loses energy and, provided it is emitted, contributes to a background of secondary electrons in the measured spectrum. The intensity of this background increases with decreasing kinetic energy (see Fig. 2). In addition there may be relaxation processes which change the energy of the excited electron which will be discussed in connection with chemical shifts.

At the surface the electron has to overcome the surface barrier which is determined by the work function ϕ. Therefore the kinetic energy of the electron in vacuum can be written as

$$E_{\text{kin}} = E_{\text{f}} - \phi \tag{2}$$

The kinetic energy $E_{\text{kin,A}}$ (see Fig. 1), however, will be measured by use of an analyzer and may differ at the analyzer from the kinetic energy E_{kin} the electron had at the sample. Therefore the work function ϕ_{A} of the analyzer, which can be considered as constant for a given measuring period, has to be used in equation (2). The work function of the sample has no influence in this simple picture on the kinetic energy measured for an electron excited from the bulk of the sample, because E_{kin} is measured with respect to ϕ_{A} of the analyzer.

In Fig. 2 the correlation between the electron distribution curve (EDC) of the solid sample and the distribution of kinetic energies is sketched. The binding energies of the electronic levels can be calculated from the kinetic energy by

$$E_{\text{b}} = h\nu - E_{\text{kin}} - \phi_{\text{A}} \tag{3}$$

In practice the calibration of the analyzer is performed in such a way that ϕ_{A} becomes zero, leading to

$$E_{\text{b}} = h\nu - E_{\text{kin}} \tag{3a}$$

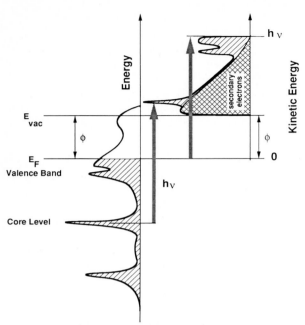

Fig. 2. Projection of electron distribution curve EDC of a sample onto kinetic energy distribution measured by the analyzer.

This calibration procedure has to be repeated once in a while, because the work function of the analyzer changes with time. From Fig. 2 it becomes clear that the work function of the sample can be measured by determining the onset of the secondary electron spectrum.

With respect to the excitation energy $h\nu$, two regimes can be distinguished: ultraviolet photoelectron spectroscopy (UPS) and X-ray photoelectron spectroscopy (XPS). Traditionally the two regimes were connected with two different sources, gas discharge lamps (He I, He II) for UPS and X-ray sources for XPS. Today this separation is rather misleading because of the availability of synchrotron radiation as a continuous light source covering a wide range of energies. In view of the underlying physical processes a distinction into core level and valence band spectroscopy makes more sense. Core level spectroscopy (XPS) provides information about localised core levels and via chemical shifts about the chemical environment. Valence band spectroscopy (UPS) yields data about delocalized valence states and is mainly used for band structure verification.

2.2   Qualitative and Quantitative Analysis of XPS Data

Core level spectroscopy (XPS) provides information about the energetic position of localized core states. Core level energies are characteristic of the element (atom) making XPS a fingerprint technique. For practical use XPS spectra of each element are tabulated [15]. In order to decide whether a specific element is present on a sample or not, the core level with the highest intensity (emission yield) should be investigated. These levels are also tabulated. In those cases where levels of different elements overlap one has to switch to another level. If the sample is unknown an overview spectrum covering the total kinetic energy range has to be taken. These spectra are either limited in resolution or rather time consuming. If the sample composition is known to some extent, only particular energy regions have to be scanned in order to find out the sample's qualitative composition.

In addition to the information on what is in the sample, one usually wants to know how much there is. In order to perform a quantitative analysis on an absolute scale too many parameters regarding source (X-ray flux), sample (roughness scale, distribution of the element in sampled depth, ionization cross section) and analyzer (detection efficiency as a function of energy) have to be known so that this analysis is not feasible. On a relative scale, however, quantitative analysis is straight forward in XPS, provided the amount of one component or the total number of components is known. Still the ionization cross sections have to be known, but on a relative scale only, which is tabulated. The amount of an element in the sample is then directly proportional to the signal amplitude of the core level under investigation. The signal strength is independent of the environment of the atom. For peaks widely separated on the kinetic energy scale, the analyzer transmission as a function of energy has to be considered. There are two main modes in which the analyzer can be run: the fixed analyzer transmission (FAT) and the fixed retardation ratio (FRR) mode. In the FAT mode, peak heights can be taken as they are, while in the FRR mode the analyzer transmission changes in proportion to the electron energy. The advantage of the FRR mode is a constant half width of the measured peaks [16].

For an approximate determination of the sample composition it is often sufficient to measure the peak height of the core level. In general, however, core level structures are asymmetric peaks above a finite background and sometimes accompanied by satellite structures. These structures originate from the many-body character of the emission process. Therefore a peak integration including satellites and asymmetric tails is much more reliable. Due to the above difficulties quantitative analysis of XPS data should be taken as accurate to only within about 5–10%.

2.3 Chemical and Electrochemical Shifts

Beside the elemental composition of the sample surface XPS provides insight into the bonding state of the elements. This information is obtained by analysis of the so-called chemical shift. The binding energy of a given core level is determined by the element only within a limit of roughly ~ 5 eV. This is due to the fact that the same element in different environments exhibits slightly modified core level energies. A change in the environment causes a rearrangement of the valence charges of the atom and a different potential created by all other nuclei and electrons. In addition to this simple ground state approximation, intra- and interatomic relaxation processes have to be taken into account making the calculation of chemical shifts rather difficult.

Chemical shifts are commonly used to identify the chemical environment i.e. the valence state of the atom by comparing the measured binding energy with the binding energies of a set of reference samples. Binding energies of all elements in several different environments are tabulated [15]. In general, the kinetic energy of an electron leaving an atom is reduced with increasing oxidation state of the atom. The higher the oxidation state, the larger the chemical shift (see Section 3.1.2). This simple picture does not always work as is seen for Pb. The binding energies for Pb in the metallic state, in PbO and PbO_2 are almost identical.

The situation becomes even more complex when adsorbates are investigated. Here it has been shown, that the binding energy of a certain adsorbate core level may depend on the substrate and on the coverage. In electrochemistry investigation of adsorbates is of particular interest for electrocatalysis, for the initial stages of metal deposition, oxide formation (OH adsorption) and for the adsorption of counter ions and water for double-layer studies. The difficulties which are encountered in the analysis of chemical shifts in these cases is demonstrated for the O1s level [18] which plays a particular role in aqueous electrochemistry. Binding energies of O1s levels for a number of compounds, molecules and adsorbates as determined by different investigators on various substrates are listed in Fig. 3. It becomes clear that the identification of water for example on an electrode surface using XPS alone will be rather ambiguous. Part of the problem can be avoided by choosing counter ions which do not contain oxygen.

The situation becomes even more complex in the case of electrochemical adsorption. Kolb et al. [17] first demonstrated clearly that the binding energy of adsorbed

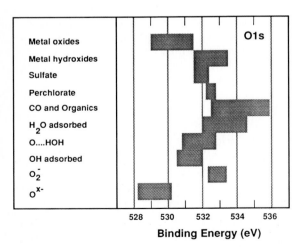

Fig. 3. Binding energy distribution of several oxygen species occurring on emersed electrodes. Binding energies are taken for different substrates from various authors. In part after [15, 18].

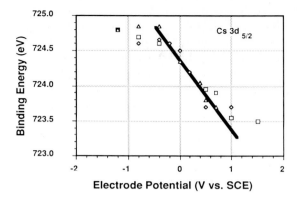

Fig. 4. Binding energy of the Cs2*d* level of Cs counter ions on an emersed Au electrode as a function of emersion potential. After [19].

counter ions (Cs^+) on an emersed Au electrode surface depends on the emersion potential (Fig. 4). This observation was confirmed later for other ions [19, 20]. In all these cases a shift of the binding energy directly proportional to the electrode potential ("electrochemical shift") was observed over a wide potential range. The effect has simply been attributed to the change in the electrostatic potential of the double layer which is felt by the adsorbate [19]. In another possible explanation it is assumed that the non-specifically adsorbed ion has a relatively large distance to the metal surface so that the reference state should be the vacuum (electrolyte) level, rather than the Fermi level. This is the very same argumentation as for the well established Xe surface probe [14, 21].

The origin of the electrochemical shift is not well understood yet. It is very likely that the phenomenon can be explained on the basis of known effects contributing to the chemical shift (see Section 3.2.5). It is clear, however, as long as the effect is not understood, interpretation of binding energies of weakly adsorbed electrochemical species will be difficult.

2.4   Sensitivity

Sensitivity has to be discussed with respect to the two pieces of information provided by XPS: the presence of certain elements and the analysis of their chemical environment.

The amount of a certain element which can be detected by XPS depends mainly on the specific scattering cross section. These cross sections are tabulated [15, 22] relative to either C or Na and vary between 0.5 for B and 22.9 for Cs with C as unity. An easy method for sensitivity determinations of the overall system are adsorption measurements. The amount of the adsorbate can be adjusted either electrochemically (UPD of metals see Section 3.2.5) or in a gas adsorption experiment. In general it has been found that 10% of a monolayer or less can usually be detected on almost any substrate. Sensitivity can in principle be improved by increasing the measuring time and/or the source energy. Both alternatives, however, are limited, not only because of possible beam damage. Finally, it should be mentioned that the absolute XPS signal intensity depends also on the type of surface roughness. Wu et al. [138] have demonstrated that the XPS signal strength of a rough Au surface may be reduced to 1/3 of that of a relatively smooth sample surface.

Sensitivity of chemical shift analysis is determined by the spectral resolution of the XPS system. The resolution of a typical XPS system without a monochromator is ~ 1.0 eV. This corresponds to the intrinsic line width of Al Kα or Mg Kα radiation. The analysing system contributes only little to the overall resolution. This resolution is sufficient to determine the binding energies of most core levels within ~ 0.1 eV. Considerable improvement of the resolution down to 0.3 eV can be achieved by use of a monochromator. The higher resolution has to be paid for by a loss in intensity which, however, is no problem in modern instruments.

The relatively poor resolution of the XPS systems has lead to an extensive use of deconvolution techniques in order to prove the presence of shifted core levels of low intensity in the presence of unshifted levels (thin oxide layers on metal substrates). Deconvolution techniques should be used only in those cases where the presence of multi components is shown up by a shoulder in the intensity distribution. Interpretation of asymmetric peaks in terms of chemical shifts can be misleading in some cases because the asymmetry may change due to a change of the electron population at the Fermi level as was demonstrated for the metallic oxide IrO_2 [23, 24].

2.5   Depth Profiling

As has been mentioned in Section 2.1, the thickness of the surface layer inspected by XPS is determined by the escape depth of the electrons. For the Mg and Al X-ray sources roughly 30 Å were measured. Depth profiling techniques have to be divided into two categories. Depth profiling within the escape depth is provided by angle dependent measurements or variation of the source energy. Monitoring the intensity of different core levels as a function of electron emission angle or source energy provides information about the distribution of elements perpendicular to the sample surface plane. For profiles deeper than the escape depth, ion etching is a well established technique.

In order to change the electron emission angle in commercial XPS systems the sample has to be tilted along an axis normal to the source-detector plane. A change of the electron emission angle from 0 to 60 degrees measured towards the sample normal reduces the effective escape depth to 1/3 according to the cosine law. For a significant increase of the surface sensitivity the escape angle has to be increased beyond 60 degrees. In oxide formation studies this method provides a relative increase of the shifted core level corresponding to the oxidized metal compared to the non-shifted peak of the underlying metal substrate. By monitoring different emission peaks as a function of the angle Pijolat and Hollinger [25] have demonstrated that an accurate picture of the depth profile can be calculated.

A second approach for non-destructive depth profiling is the use of a variety of different excitation energies. As is evident in Fig. 5 the escape depth of the photo-electrons is a function of the kinetic energy i.e. of the excitation energy. There are twin and multi-element anodes commercially available which allow a change of the anode without braking the UHV. Typical elements in addition to Mg and Al are zirconium (ZrLα: 2042.4 eV), silver (AgLα: 2984.3 eV) and titanium (TiKα: 4510 eV). Depth profiling by means of different X-ray sources or a continuous light source (synchrotron radiation) at a given emission angle is expected to be less dependent on the surface roughness than angle dependent measurements with a given source energy [25].

For depth profiling down to several thousand Å below the surface, sputtering techniques are usually applied. Etching of the sample occurs by bombardment of the sample surface with highly energetic ions (mostly Ar). Depending on the acceleration energy of the ions, sputtering rates of ~ 20 Å/min are typical (4 keV). A severe problem is preferential sputtering which may lead to a change in surface composition. In the case of oxides special care has to be taken, because this effect may lead to complete chemical reduction of the oxide presenting a completely altered surface. A list of oxides and their tendency to be reduced during sputtering has been given by Kim et al. [26] and also by Holm and Storp [27].

In order to get an idea about the depth resolution of the system and about sputtering rates, electrochemically prepared Ta_2O_5 serves as a very suitable example. The thickness of the oxide layer can be easily varied by choosing a proper oxidation potential (16 Å/V). Tantalum oxide is not reduced during sputtering according to [26].

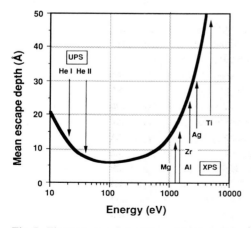

Fig. 5. Electron escape depth as a function of electron energy. The energies of several photon sources are shown.

2.6 Valence Band Spectroscopy

Valence band spectroscopy i.e. monitoring of the delocalized electronic states with binding energies $0 < E_b < 20$ eV can well be performed using XPS. As a consequence of the low energetic resolution of non monochromatized XPS, other light sources are preferable. The most frequently used UPS light source is the He I or II gas discharge lamp which allows a resolution of ~ 0.2 eV. These sources offer the advantage of a very high surface sensitivity (see Fig. 5) and the light can be easily polarized with a gold mirror arrangement.

Valence band spectroscopy (UPS) cannot be regarded as a fingerprint technique like XPS and has been primarily used for verification of band structure calculations. In this context very sophisticated measuring techniques have been developed where almost any accessible parameter (emission angle, polarization, azimuth, excitation energy, etc.) has been varied. These techniques, which require well defined, smooth, single crystal surfaces, have not yet been applied to the investigation of electrodes, although standard procedures for preparation and handling of well defined single crystal electrode surfaces have recently been developed by electrochemists.

Very useful information concerning the surface of emersed electrodes, however, can be deduced from UPS spectra directly, like the electronic density of states at the Fermi level, the position of the valence band with respect to the Fermi level or possible band gap states. The valence band of UPD metals might help to explain the respective optical data (see Sections 3.2.1 and 3.2.5).

In order to obtain meaningful UPS spectra of emersed electrodes, considerable measures have to be taken to prevent surface contamination. The $O2_p$ line around 6 eV binding energy has a large UPS strength and is likely to dominate the valence band spectra of emersed electrodes if a minimal amount of oxygen or CO contaminations is present. On the other hand, this may be an advantage when investigating oxide formation or H_2O adsorption.

2.7 Double Layer Studies, Emersion and Transfer

Although not the subject of this article, double layer studies are briefly discussed in this paragraph in order to demonstrate that *ex situ* XPS studies indeed provide information about the state of the electrode exposed to an electrochemical environment at a defined potential. A crucial step in any *ex situ* experiment is the emersion of the electrode. Here the question arises whether the electrochemical double layer or part of it is preserved at the interface after emersion and transfer. Winograd et al. [10, 11] first demonstrated that the electrode under UHV conditions still "remembers" the electrode potential applied at the time of emersion. These authors investigated oxide formation on Pt and the underpotential deposition of Cu and Ag on Pt by means of XPS and proved that the electrochemically formed oxide layer and the UPD deposited metal stays on the surface of the emersed electrode. The XPS results could be correlated to the electrode potential during oxidation or deposition. Later on, Kolb et al. [17] showed, using XPS, that counter ions building up the double layer can be retained at the electrode surface after emersion and transfer to the UHV system.

The XPS results obtained by Kolb and Hansen are reproduced in Fig. 6 and they clearly demonstrate not only that cations as well as anions stay on the surface but also that the amount of ions exhibits the expected potential dependence even in the case of specific adsorption. The preservation of the double layer charge after emersion was also shown by other techniques like charge monitoring [28] and electroreflectance measurements [29].

In surface science, work function measurements are considered to be rather sensitive towards changes of the sample surface. Work function measurements are used to follow adsorption processes and to determine the dipole established at the surface. During oxygen adsorption and oxide formation the sign of the work function change allows one to distinguish between oxygen atom adsorbed on the surface or sub-surface [30].

In electrochemistry it has been shown that for electron emission into an electrolyte there is a correlation between electrode potential E and the work function of the interface according to [31]

$$\Phi_E = \Phi_M + eg_s (\text{dip}) - eg(\text{ion}) + ed\chi + \mu_{es} \qquad (4)$$

$$E = \Phi_E/e + \text{const.}$$

where Φ_E and Φ_M correspond to the work functions of the electrode in solution and of the bare metal, respectively, eg_s (dip), $eg(\text{ion})$ and $ed\chi$ represent the contribution of dipoles and counter ions at the interface and the change in work function due to the presence of water, respectively and μ_{es} is the chemical potential of electrons in

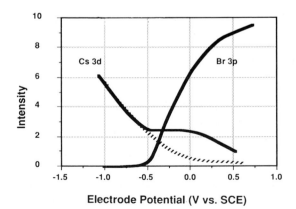

Fig. 6. XPS intensities (arbitary units) of Cs3d and Br3p levels for emersed gold electrodes. Full line corresponds to emersion from 0.1 M KClO$_4$ + 0.01 M CsBr, dashed line to 0.005 M Cs$_2$SO$_4$. After [17].

solution. In the case of the emersed electrode in contact with an inert gas or vacuum eq. (4) should still be valid, provided the electrochemical double layer stays intact and the new reference level, which is now the vacuum level ($\mu_{es} = 0$), is considered [20]. The expected linear correlation between work function and electrode potential was verified by Hansen et al. [32] on a rotating vertical disk electrode measuring the work function just above the electrolyte with a Kelvin probe. In order to be sure that the double layer is not only preserved directly after emersion but also after transfer to the UHV system a similar experiment was performed in the author's laboratory using the photoelectric work function determined from UPS measurements. Again a linear 1:1 correlation was observed for Au [20, 33], Ir [34], Pt [20] and Ag [20] as shown in Fig. 7.

These measurements have verified that the work function of an electrode, emersed with the double layer intact, depends only on the electrode potential and not on the electrode material or the state of the electrode (oxidized or covered with sub-monolayer amounts of a metal) [20]. Work function measurements on emersed electrodes do not serve the same purpose as in surface science investigations of the solid vacuum interface. At the electrochemical interface, any change of the work function by adsorption is compensated by a rearrangement of the electrochemical double layer in order to keep the applied potential i.e. overall work function, constant. Work function measurements, however, could well be used as a probe for the quality of the emersion process. Provided the accuracy of the measurement is good enough, a combination of electrochemical and UPS measurements may lead to a determination of the components of equation (4).

A controversial question, which has been addressed in several publications, is the presence (or absence) of water on the surface of an Au electrode after emersion and

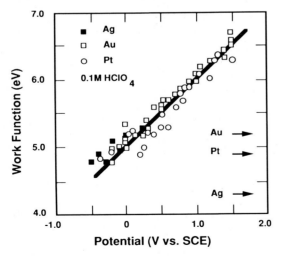

Fig. 7. Photoelectric work function for Au, Pt and Ag electrodes emersed from 0.1 M HClO$_4$ at different potentials. Work function was measured by means of UPS (He I). Arrows indicate work functions of clean polycrystalline surfaces. After [20].

transfer. Water is a major component of the double layer and should be still on the surface after the transfer, as indicated by the observed preservation of the work function, unless its contribution is assumed to be negligible. Several authors have come to different conclusions on this matter, which may be due to different experimental conditions. Peuckert [35], Streblow [36] and D'Agostino [19] stated that water is present on the surface while Kötz et al. [20, 33] and Aberdam et al. [37] think that the amount is rather low, at least below the detection limit of their analysis. It is indeed very unlikely that a complete monolayer as proposed in double layer models stays on the surface. H_2O has been shown to desorb from almost any noble metal surface at room temperature under UHV conditions. H_2O bound to counter ions in the solvation shell, however, may well be stabilized on the surface after emersion and transfer to the UHV chamber [139].

Although the actual composition of the double layer present on an emersed electrode surface is still a matter of debate and experimental investigation, it is clear that the emersed electrode, exposed to the UHV of the XPS system, "remembers" its electrode potential in the electrolyte. Drawing conclusions from these *ex situ* measurements about the electrode surface in contact with the electrolyte is thus sensible. Due to the contributions made by Kolb et al. [14] (and references therein) it is now accepted that the electrochemical interface may be investigated *ex situ*. In addition to photoelectron spectroscopies, which probe the elemental and electronic structure of the double layer, electron diffraction studies (LEED, RHEED) on single crystal electrodes probing the geometric structure give further evidence for the significance of *ex situ* studies on electrode surfaces.

While the presence and composition of the electrical double layer is of vital importance for the study of adsorbates and the top most atomic layer of the electrode, changes in the bulk of the electrode are expected to be affected only to a minor extent. Nevertheless, for XPS *ex situ* studies a transfer system which guarantees as much control as possible should be used in any case, also for the investigation of thick layers or strongly bound adsorbates. For technical electrodes this is, however, not possible in general. A minimum of control is achieved by the use of a glove box which is constantly purged with an inert gas. After an electrochemical experiment in the glove box the electrode is sealed in a transportable chamber, which fits to the insertion lock of the UHV system.

This technique has been improved by minimizing the "glove box" and directly attaching it to the fast insertion lock of the electron spectrometer [38]. This electrochemical chamber is purged with purified argon. Such a system is sketched in Fig. 8 for a Kratos ES300 electron spectrometer equipped with a fast insertion lock. The horizontally mounted sample is exposed to the electrolyte via the meniscus building up on the vertical tube filled with electrolyte. The electrolyte can be moved up and down with the aid of a syringe like arrangement. Counter and reference electrode are connected to this tube. After emersion of the electrode, transfer through the fast insertion lock into the preparation chamber with a pressure in the order of 133.3×10^{-7} pa takes only 5 to 10 minutes. This set-up gives satisfactory results as is indicated by the comparison of two UPS spectra of gold which were recorded with the clean electrode and after exposure to the electrochemical preparation chamber [38].

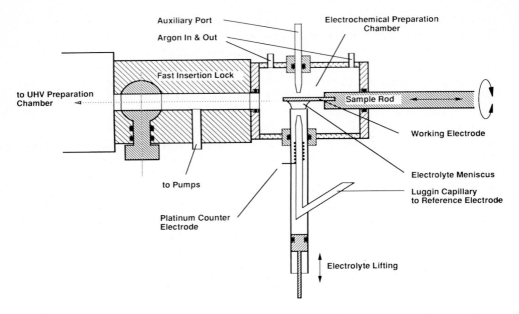

Fig. 8. Electrochemical preparation chamber for a Kratos ES 300 photoelectron spectrometer with fast insertion lock. After [38].

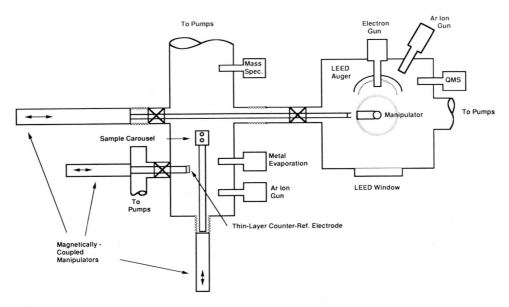

Fig. 9. LEED Auger electrochemical UHV system with electrochemical thin layer cell. After [39].

The best approach is 'of course' a fully UHV compatible set-up with an electrochemical preparation chamber which fulfills UHV standards with respect to cleanliness and pressure control. The electrochemical chamber is filled with argon at atmospheric pressure only during the electrochemical treatment. Afterwards the chamber can be pumped down directly. Such a system was first described by O'Grady [39] and was also used by Ross [40], Hubbard [41], Gerischer [42], and Streblow [36].

The system developed by O'Grady is reproduced in Fig. 9. A key element of this arrangement is the electrochemical thin layer cell, using a combined Pd-hydrogen reference and counter electrode, thus minimizing the amount of electrolyte necessary for the electrochemical treatment. This type of cell is particularly useful for double layer studies but cannot be used for gas evolution or corrosion experiments at higher current densities. For a collection and discussion of other transfer systems the reader is referred to the review article by Sherwood [43].

3 Characterization of Electrodes by XPS

In the following chapter examples of XPS investigations of practical electrode materials will be presented. Most of these examples originate from research on advanced solid polymer electrolyte cells performed in the author's laboratory concerning the performance of Ru/Ir mixed oxide anode and cathode catalysts for O_2 and H_2 evolution. In addition the application of XPS investigations in other important fields of electrochemistry like metal underpotential deposition on Pt and oxide formation on noble metals will be discussed.

The chapter is divided into two subsections the first of which deals with the characterization of electrodes as prepared prior to any electrochemical treatment. The knowledge of the actual surface composition of the fresh electrodes is needed to optimize preparation conditions and to be able to correlate electrochemical performance with surface properties. In Section 3.2 the application of XPS to the elucidation of electrochemical reaction mechanisms will be demonstrated. Here XPS monitors possible changes after controlled electrochemical treatment.

3.1 Characterization of Electrodes as Prepared

3.1.1 Surface Composition and Depth Profiling of Ruthenium–Iridium Based Electrodes

The invention of the ruthenium based DSA® electrode by H.B. Beer has had a major impact on the chlorine industry and has stimulated a number of investigations aimed at the understandng of the electrocatalytic processes occurring during Cl_2 and O_2 evolution at these electrodes. Today 43% of the ruthenium and 25% of the iridium in the western world is used for electrochemical processes. An extensive review on these metallic oxides has been recently published by Trasatti [44], which covers fundamental as well as applied and technical aspects of these materials and their use. General aspects of the Cl_2 evolution process, including ruthenium based anodes, are reviewed by Novak et al. [131].

For technical purposes (as well as in the laboratory) RuO_2 and Ru based thin film electrodes are prepared by thermal decomposition techniques. Chlorides or other salts of the respective metals are dissolved in an aqueous or alcoholic solution, painted onto a valve metal substrate, dried and fired in the presence of air or oxygen. In order to achieve reasonable thicknesses the procedure has to be applied repetitively with a final firing for usually 1 hour at temperatures of around 450°C. A survey of the various processes can be found in Trasatti's book [44]. For such thermal decomposition processes it is dangerous to assume that the bulk composition of the final sample is the same as the composition of the starting products. This is especially true for the surface composition. The knowledge of these parameters, however, is of vital importance for a better understanding of the electrochemical performance including stability of the electrode material.

XPS has been used by several authors to identify the surface and bulk composition of ruthenium based, valve metal stabilized DSA electrodes for chlorine evolution. Augustynski et al. [45] investigated the composition of RuO_2-TiO_2 electrodes before and after electrochemical treatment. They found a surface composition which deviated significantly from that of the bulk. The Ru/Ti ratio corresponded to 0.15 at the surface while in the bulk 0.28 was measured. Similar results were obtained by De Battisti et al. [46] for RuO_2-TiO_2 electrodes with different composition ratios on Ti substrates. Fig. 10 shows the Ru/Ti ratio for different solution compositions as a function of the depth.

The depth profile was obtained using Ar^+ sputtering. It is evident, that the surface deficiency in Ru does not depend much on the bulk composition. Even 50 nm below the surface the nominal composition is not reached. In an earlier investigation by De Battisti et al. [47] on RuO_2-Ta_2O_5 electrodes on Ti substrates the authors looked somewhat deeper beneath the surface (500 nm) and again found notable deviations of the actual from the nominal composition. As is indicated in Fig. 11 a drastic deficiency of Ru at the surface was observed independent of the bulk Ru/Ta ratio. For a nominal composition of 75% (Ru/Ta ratio 3.0) a Ru/Ta ratio of 0.25 was observed at the surface by XPS. Using AES Gorodetskii et al. [48] found similar depth profiles as shown in Fig. 11. These authors ascribed the Ru deficiency to preferential sputtering which is sometimes a problem in Ar^+ etching. In order to check this argument the depth profiles shown in Fig. 10 were reinvestigated by De Battisti et al. using Rutherford Backscattering Spectroscopy (RBS) [46]. The results of RBS were identical to those of the XPS analysis ruling out a significant influence of preferential sputtering.

The origin of the surface segregation during thermal decomposition of these DSA type electrodes is not clear yet. When comparing the ionic radii of Ti^{4+} (0.068 nm) and

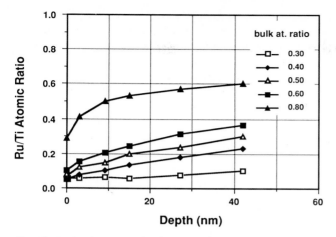

Fig. 10. Ruthenium over titanium ratio for mixed (Ru, Ti)O$_2$ based electrodes prepared by thermal decomposition. Nominal compositions are given in the figure. After [46].

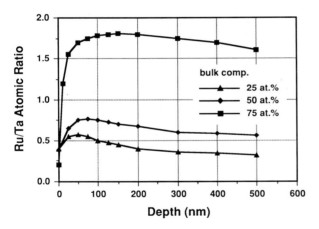

Fig. 11. Ruthenium over tantalum atomic ratio for mixed Ru/Ta based electrodes prepared by thermal decomposition. Nominal compositions are given in the figure. After [47].

Ta^{5+} (0.073 nm) with that of Ru^{4+} (0.067 nm) one might expect the valve metal to diffuse to the surface. At least for RuO$_2$-TiO$_2$ films, however, this argument appears to be rather weak in view of the small difference. Another reason could be the higher reactivity of the valve metals with oxygen. Triggs et al. [49] have demonstrated that two phases are observed in the (Ru$_x$Ti$_{1-x}$) O$_2$ system, one which is Ti rich ($x < 0.02$) and one which is Ru rich ($x > 0.98$). For compositions inbetween, a two phase system is developed. For a better understanding of the formation of Ti- and Ta-rich surface layers, investigations are needed where a systematic variation of the decomposition temperature is performed.

Such temperature variations were performed on mixed oxide powder catalysts of the type $(Ru_{0.25}Ir_{0.25} Sn_{0.5}) O_2$ for O_2 evolution in solid polymer electrolyte electrolyzers [50]. These catalysts are prepared according to a procedure described by Adams [51] which is a precipitation of the oxide from a salt melt of the respective chlorides. The Adams preparation procedure runs at rather low temperatures (350 °C) so that a post treatment at various higher temperatures is possible. The resulting surface composition of such a catalyst as determined by XPS analysis is shown in Fig. 12 and exhibits a significant surface segregation with temperature. Two effects are to be mentioned. First a deficiency in Sn at the surface is observed, which vanishes with increasing temperature and secondly there is an increase in Ir concentration at the surface with respect to Ru independent of the temperature. At relatively high temperatures (800 °C) the nominal composition is approached at the surface.

Surface enrichment of Ir has been observed by means of XPS for Ru/Ir mixed oxides [50, 50a], which are known to form solid solutions over the complete composition range. The effect of Ir enrichment at the surface with respect to Ru, can be rationalized in two ways. The ionic radius of Ir is slightly higher than that of Ru which would favor a segregation of Ir to the surface where the necessary expansion is easily accommodated. On the other hand, the known higher oxidation potential of Ru as compared to Ir would lead to an enrichment in Ru during the early stages of the crystal growth and consequently to a deficiency in Ru during the later stages [50, 50a]. It is not possible to distinguish between the two effects on the basis of available data.

This argument with respect to the ionic radii would predict a surface enrichment of Sn^{4+} (0.083 nm) for the ternary oxide, which is not observed. Hutchings et al. [50]

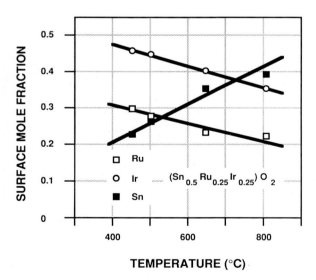

Fig. 12. Surface composition of $(Sn, Ru, Ir)O_2$ catalyst as a function of annealing temperature. After [50].

concluded that Sn forms a metastable phase with the Ru/Ir oxide leading to a separation of Sn at higher temperatures. The reason for the metastable phase was thought to be the difference in electronic properties, that is the difference in bonding, which prevents the formation of a stable phase. This argument would also hold for the above cited valve metals Ti and Ta forming wide-gap semiconductor oxides like Sn. In addition the fortuitously controlled precipitation kinetics were thought to be responsible for producing the mixed $(Sn, Ru, Ir)O_2$ catalyst.

Physical properties of binary or ternary Ru/Ir based mixed oxides with valve metal additions is still a field which deserves further research. The complexity of this matter has been demonstrated by Triggs [49] on $(Ru,Ti)O_x$ who has shown, using XPS and other techniques (UPS, Mössbauer, Absorption, Conductivity), that Ru in TiO_2 (Ti rich phase) adopts different valence states depending on the environment. Possible donors or acceptors are compensated by Ru in the respective valence state. Trivalent donors are compensated by Ru^{5+}, pentavalent acceptors will be compensated by Ru^{3+} or even Ru^{2+}. In pure TiO_2 ruthenium adopts the tetravalent state. The surface composition of the titanium rich phase (2% Ru) was found to be identical to the nominal composition.

3.1.2   Oxidation States of Ruthenium Oxide and Ru-Based Electrodes

Ruthenium like iridium is known for its ability of adopting various valence states which make these elements rather attractive in catalysis. Kim and Winograd [52] were the first who studied the chemical in XPS of different Ru compounds. The results of their extensive work still serve as reference for today's investigators. Kim and Winograd have identified binding energies of Ru-oxygen species (Table 1).

With the help of these reference levels it was possible to determine the catalytic sites on Ru and RuO_2 anodes during oxygen evolution (see Section 3.2.2). The presence of RuO_3 on RuO_2 based anodes for O_2 evolution [53, 54] and Cl_2 evolution [45] was realized rather early and has lead to the proposal of respective reaction mechanisms. As can be seen from Table 1, RuO_3 is not only characterized by a shift in Ru binding energy but also by a shift of the O1s level. Fig. 13 displays the Ru3d level of a RuO_2 electrode on a Ti substrate which was prepared by reactive sputtering of Ru in the presence of O_2. Reactive sputtering can be regarded as a cold preparation because the temperature of the substrate does not exceed 100°C. After annealing at 450°C for

Table 1. Binding energies in eV of the Ru-oxygen system according to Kim and Winograd [52].

Compound	Ru $3d_{5/2}$	O1s
Ru	280.0	—
RuO_2	280.7	529.4
$RuO_2 \times H_2O$	281.4	529.3 (oxide), 530.5 (H_2O)
RuO_3	282.5	530.7
RuO_4	283.3	—

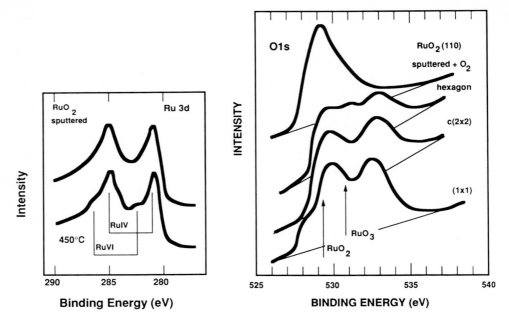

Fig. 13. XPS spectra of the Ru3*d* level of a reactively sputtered RuO$_2$ electrode after preparation (top) and after firing at 450 °C in air for 1 h. After [55].

Fig. 14. O1*s* level of the RuO$_2$(110) surface of single crystalline RuO$_2$ with various surface structures. After [56].

1 hour in air the XPS spectrum exhibits an additional structure indicating the formation of a RuO$_3$ phase probably at the surface [55].

Similar observations were made by Anatasoska et al. [56] on RuO$_2$ single crystals where the RuO$_3$ surface structure shows up most prominently in the O1*s* level. After sputtering and O$_2$ exposure without thermal treatment only one O1*s* peak is observed in Fig. 14. The surfaces of reconstructed and unreconstructed RuO$_2$ (110) faces exhibit two or more peaks. These additional features are assigned to RuO$_3$ and molecular adsorbed oxygen.

The authors also investigated the O/Ru ratios for the different surface structures and found significant changes ranging from 0.37 for the sputtered sample to 2.9 for the unreconstructed (1 × 1) surface. It was proposed that the presence of RuO$_3$ has a major impact on the stability of RuO$_2$ surfaces. The low O/Ru ratio for the sputtered sample indicates preferential sputtering of oxygen (see Section 2.5 and [26]).

3.1.3   Doping Levels and Contaminations of RuO$_2$ and SnO$_2$ Electrodes

In order to detect possible impurities or doping agents at rather low concentrations the SIMS method should be chosen. Compared to XPS the detection limit of SIMS is much lower by orders of magnitude. XPS, however, provides the same information within its sensitivity range, that is in the range of percent.

As a consequence of the preparation procedure of DSA type electrodes chloride may be built into the RuO_2 lattice. The presence of Cl in the electrode and at the surface may change the chemical and electrochemical properties considerably. During the course of their investigations of Cl_2 evolution on RuO_2 and RuO_2-TiO_2 electrodes Augustynski et al. [45] have monitored the XPS $Cl2p$ level of different electrodes. For the electrodes as prepared they found Cl levels of about 4 atom% for the RuO_2 as well as the RuO_2-TiO_2 electrodes. These electrodes were prepared with a final firing treatment at 450 °C for 1 hour. Chloride content of DSA electrodes prepared by thermal decomposition was monitored by EDX [57] by Lodi et al. as a function of temperaure. These authors found a straight temperature dependence of the Cl content, decreasing with increasing firing temperatures.

Application of XPS for the study of donor concentrations in semiconductors is possible only in those cases where the concentrations are high. Because of its relevance to practical applications concerning transparent electrodes and anodic oxidation of refractory organic impurities [58], Sb doped SnO_2 is chosen as an example. This type of electrode is usually prepared by spray hydrolysis, where an alcoholic solution of the chlorides is sprayed onto a heated substrate in air. Doping is achieved by adding $SbCl_3$ to the spray solution. It is known from literature that a doping level of 5% Sb gives highest electrical conductivity [59]. Using XPS it is possible in this case to determine the Sb content of SnO_2 as a function of the solution composition, for example. It was found that a nearly 1:1 correlation exists between the atom% of Sb in the film and in the spray solution. In addition it can be confirmed by XPS that Sb in SnO_2 adopts the $+5$ valence state. In Fig. 15 the XP spectrum of the $Sb3d$ level is reproduced for an SnO_2 sample with 5% Sb demonstrating the overlap of the Sb levels with the $O1s$ level.

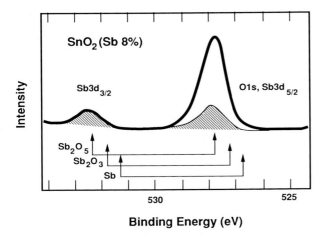

Fig. 15. $Sb3d$ and $O1s$ levels of a SnO_2 film electrode doped with $\sim 8\%$ Sb.

3.2   Electrode Surface Analysis for the Elucidation of Electrochemical Reactions

While characterization of the electrode prior to use is a prerequisite for a reliable correlation between electrochemical behaviour and material properties, the understanding of electrochemical reaction mechanisms requires the analysis of the electrode surface during or after a controlled electrochemical experiment. Due to the *ex situ* character of photoelectron spectroscopy, this technique can only be applied to the emersed electrode, after the electrochemical experiment. The fact that *ex situ* measurements after emersion of the electrode are meaningful and still reflect the situation at the solid liquid interface has been discussed in Section 2.7.

3.2.1   Anodic Oxide Layers

X-ray photoelectron spectroscopy (XPS) of electrodes was first applied to the oxidation of noble metal electrodes. Kim and Winograd investigated in 1971 the electrochemical formation of anodic oxides on Pt [10] and later on Au electrodes [60]. The electrochemical parameters of oxide formation on these noble metal electrodes were well characterized and enabled a direct correlation between *ex situ* XPS and *in situ* electrochemical analysis.

The principle of XPS analysis of such oxides is demonstrated by means of Fig. 16 where the oxidation of a polycrystalline Au electrode was monitored by XPS [61]. The formation of the oxide is well reflected by the additional peak occurring in the Au 4*f* spectrum at 85.7 eV for potentials of 1.5 V and 1.8 V. (Eb for metallic Au: 84 eV).

From the shift in binding energy one can conclude with the help of published reference data [15] that Au in the oxide has the valence state III. For further conclusions with respect to the stoichiometry of the oxide the O1*s* level of the oxide has to be inspected (Fig. 16). The growth of the oxide is again reflected by the increase in O1*s* signal strength for anodic potentials. For a potential of 1.5 V a clear splitting of the O1*s* level arises which can be interpreted in terms of an OH and an O^{2-} species, leading to the conclusion that the composition of the anodic oxide is AuO(OH). The angular dependence of the O1*s* signal (see Fig. 16) indicates that the OH species are at the outer surface. The analysis is supported by the fact that the O/Au ratio for the oxide was found to be 1.8 ± 10%.

Uncertainties in such analysis are buried in the deconvolution process necessary for the quantitative determination of the O/Au ratio for the thin oxide layer, and in the assumption that the total O1*s* signal originates from oxygen of the oxide. The latter problem has been demonstrated in Section 2.3 where O1*s* binding energies were listed for a variety of oxygen species. The high binding energy peak of the O1*s* spectrum in Fig. 16 can be interpreted not only as OH but also as adsorbed oxycounter ion, some organic contamination layer or even as H_2O species. The possible contribution of these species is usually not easy to clarify.

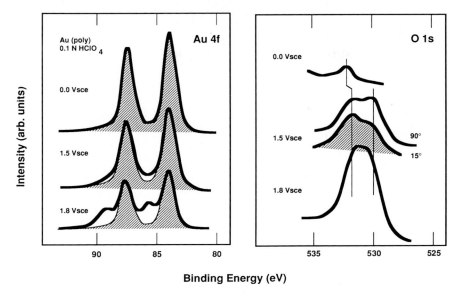

Fig. 16. XPS Au4f and O1s levels for a polycrystalline gold electrode emersed from 0.1 mol L^{-1} HClO$_4$ at different potentials. Shaded Au4f spectra represent the contribution of clean Au. The O1s spectra were recorded with twenty-fold sensitivity at 90° (15° for shaded O1s spectrum).

In view of the above uncertainties it is not too surprising that the results of XPS analysis performed by different authors on anodic oxides on Pt and Au differ significantly. For gold the stoichiometry of the oxide was determined to be Au$_2$O$_3$ [62], Au(OH)$_3$ [60], and AuO(OH) [35]. For Pt the different results were PtO$_2$ on top of PtO [63], PtO or PtO$_2$ at different potentials [10], and Pt(OH)$_4$ [35, 64, 65]. The assignment of Pt(OH)$_4$ by Peukert has been substantiated by the same author using ELS as a complementary technique [64].

In addition to the stoichiometry of the anodic oxide the knowledge about electronic and band structure properties is of importance for the understanding of electrochemical reactions and *in situ* optical data. As has been described above, valence band spectroscopy, preferably performed using UPS, provides information about the distribution of the density of electronic states close to the Fermi level and about the position of the valence band with respect to the Fermi level in the case of semiconductors. The UPS data for an anodic oxide film on a gold electrode in Fig. 17 clearly proves the semiconducting properties of the oxide with a band gap of roughly 1.6 eV (assuming *n*-type behaviour).

The maximum in optical absorption for anodic oxides on Au as determined by McIntyre and Kolb [66] at a photon energy of 4.0 eV may be attributed on the basis of the UPS data to a transition from the maximum of the valence band to the conduction band. The differences between the spectra of the anodized and rinsed sample in Fig. 17 is due to counter ion removal after rinsing the sample with H$_2$O [67].

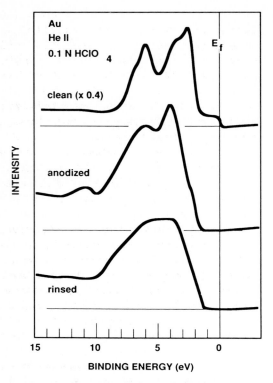

Fig. 17. UP spectra for (top) a clean Au electrode, (middle) after anodization with a current density of 0.1 mA/cm². and (bottom) after rinsing the oxidized sample with ultra pure water.

3.2.2 Oxide Formation and O_2 Evolution on Ruthenium and Iridium

Ruthenium and iridium are known to be the best catalysts for O_2 evolution in acid electrolytes and also for Cl_2 evolution. The invention of the Ru-based DSA electrode for the chlorine evolution process had a major impact on the chlor-alkali industry [44]. In advanced solid polymer electrolyte electrolyzers mixed Ru-Ir oxide catalysts are used at the O_2 evolving anode [68]. Recently it was demonstrated that RuO_2 is also a promising candidate for the cathode catalyst in hydrogen evolving cells. During H_2 evolution, RuO_2 was found to be practically insensitive towards poisoning by metal deposition [69]. As a consequence of the importance of Ru and Ir in electro-chemical technology, the electrochemical behaviour has been studied intensively and the results are published in several review articles [44 and references therein]. In the following the contribution of *ex situ* XPS and UPS investigations to the understanding of these materials will be described.

Ruthenium. In contrast to Au and Pt, the monitoring of oxide formation on Ru and Ir by cyclic voltammetry is not straightforward. There is no sharp edge in the anodic scan, which in the case of Au or Pt indicates the onset of oxide formation, and similarly the well known oxide reduction peak on the cathodic scan is also missing. The formation of the anodic oxide as a function of the electrode potential was monitored by means of XPS on the emersed electrode. The potential dependence of the binding energy of the Ru3d and the O1s level of an Ru electrode in 1N H_2SO_4 is plotted in Fig. 18.

Growth of a thick oxide layer is shown by the step-like shift of both binding energies at a potential of 1.15 Vsce. Taking the data obtained by Kim and Winograd [52] for the different ruthenium oxides as references (see Section 3.1.2) the anodic oxide can be identified as hydrated RuO_2 [54]. The onset of oxide growth on ruthenium coincides with the onset of oxygen evolution. Using the rotating ring-disk electrode Kötz et al. [70] have shown that O_2 evolution in acid occurs on ruthenium at electrode potentials above 1.15 Vsce, and coincides with corrosion of Ru as RuO_4. In contrast to the observations made on RuO_2 electrodes, prepared by thermal decomposition of the chlorides or by reactive sputtering and subsequent heat treatment [55], no clear indication of RuO_3 was found for the anodic oxide on Ru [54]. Assuming that there is a correlation between the stability of ruthenium based electrodes and the stability of the Ru(VI) species, the absence of RuO_3 in the XPS data of the anodic oxide was used to explain the massive corrosion of the metallic electrode during O_2 evolution. Based on XPS investigations, on the rotating disk-ring data and on Tafel measurements a mechanism for the O_2 evolution reaction on Ru-based electrodes was suggested by Kötz and Stucki [54] where the relative stability of Ru(VI) plays a fundamental role (Fig. 19).

As can be seen in Fig. 19, the surface site from which O_2 evolution starts is a Ru^{VI} species like $RuO_3 \times H_2O$ or $RuO_2(OH)_2$. The relative stability of this species is

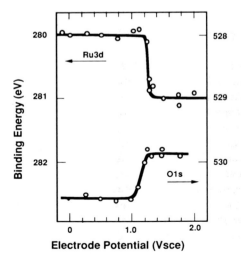

Fig. 18. Dependence of binding energies of the Ru3d and O1s levels on emersion potential for a bulk Ru electrode in 0.5 mol L^{-1} H_2SO_4. After [54].

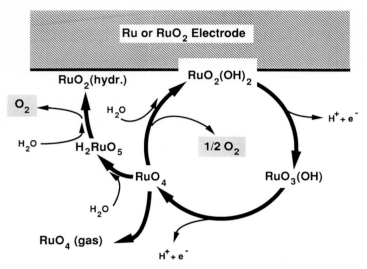

Fig. 19. Oxygen evolution and corrosion model for ruthenium based electrodes. After [54].

decisive for the subsequent reaction of $Ru^{VIII}O_4$, which gives off the oxygen either chemically via corrosion or electrochemically. The former pathway leads to the hydrous oxide layer, the latter pathway leads back to Ru^{VI}, closing the O_2 evolution circle. The importance of Ru^{VI} for the stability of a RuO_2 electrode during O_2 evolution has been demonstrated for reactively sputtered RuO_2 (see Fig. 13) [55]. After thermal treatment, which leads to the formation of RuO_3, the stability of the electrode is improved. The RuO_3 species can be detected on the surface even after O_2 or Cl_2 evolution.

The key role of RuO_3 for the electrochemical stability of RuO_2 based DSA electrodes for Cl_2 evolution was also pointed out by Augustynski et al. [45] on the basis of XPS data. After electrolysis in NaCl solution Augustynski found Cl species on the surface of the DSA electrode and a higher Ru oxide species, most probably RuO_3.

Recently the proposed O_2 evolution mechanism was supported by the results of a DEMS (Differential Electrochemical Mass Spectrometry) study performed by Wohlfahrt-Mehrens and Heitbaum [71] on Ru electrodes. Using this mass spectroscopic technique and ^{18}O labeling for the determination of reaction products during O_2 evolution, it could be verified that the oxygen of the oxide formed on Ru takes part in the O_2 evolution process. The same observation was made for RuO_2 electrodes when using labeled $H_2^{18}O$.

While thick oxide formation on Ru occurs together with O_2 evolution at a potential of 1.15 Vsce, the initial steps of oxide formation are expected to occur at more cathodic potentials of roughly 0.3–0.8 V [72]. Structures in the cyclic voltammogram in this potential region were attributed by Vucovik et al. [73] to hydroxide or oxide adsorption. These oxides are reversibly reduced at a potential of 0.1 V. The presence of a thin oxide layer on Ru at potentials cathodic of 1.15 V was demonstrated by

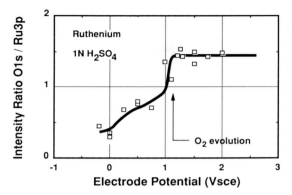

Fig. 20. Ratio of oxygen and ruthenium XPS intensities as a function of electrode potential in $0.5 \, \text{mol} \, \text{L}^{-1} \, \text{H}_2\text{SO}_4$. for bulk Ru.

means of angle dependent XPS measurements of the Ru3d binding energies [54]. In addition the XPS intensity ratio O1s/Ru3p increases monotonously with more anodic potentials. This ratio is plotted in Fig. 20 for ruthenium in 1N H_2SO_4.

The increase in intensity ratio indicates the accumulation of oxygen species on the surface of the electrode. After formation of the thick oxide layer the ratio O/Ru becomes constant. The fact that the adsorbed species do form bonds with the Ru of the electrode surface is clearly shown in the respective UPS spectra for different electrode potentials [74].

Formation of the t_{2g} band with a binding energy of 0.8 eV below the Fermi level in Fig. 21 indicates a change in Ru coordination for potentials above 0.4 V. The increased O/Ru ratio in Fig. 20 is therefore not only a consequence of enhanced counter ion adsorption, but rather a result of oxide/hydroxide formation.

Iridium. Although iridium exhibits solid state properties and electrochemical behaviour similar to those of ruthenium, oxide formation during oxygen evolution is different on iridium electrodes. In contrast to Ru no thick oxide is formed on Ir during O_2 evolution on the clean metal. Monitoring of the oxide thickness by means of XPS after electrochemical treatment reveals a decreasing thickness as soon as O_2 evolution sets in [55]. The maximum thickness of about one monolayer in Fig. 22 is achieved at a potential of 1.3 Vsce in acidic electrolyte. These results are in good agreement with conclusions drawn by Shaidullin et al. [75] on the basis of voltammetric data.

Deconvolution of the XPS spectra for the Ir4f levels reveals a chemical shift of 1.2 eV for the oxidized Ir species at 1.3 Vsce, indicating that Ir occurs in the valence state IV. Kim et al. [60] and also Hall et al. [76] assigned the binding energy of 62 eV with a chemical shift of 1.1–1.2 eV to IrO_2. Work performed by Augustynski et al. [77] lead to the conclusion that the anodic film on Ir is $Ir(OH)_4$, while Peuckert determined the film composition to be $IrO(OH)_2$ [78].

The initial steps of oxide formation on Ir were studied by Mozota and Conway [79] in different electrolytes. It was found that similar to Ru the oxidation of Ir starts

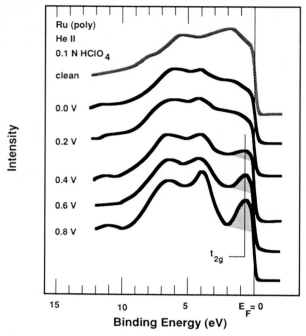

Fig. 21. UPS spectra of a clean (top) polycrystalline Ru electrode and after emersion at different potentials from 0.1 mol L^{-1} HClO$_4$. Shaded peak corresponds to t_{2g} band of the oxide.

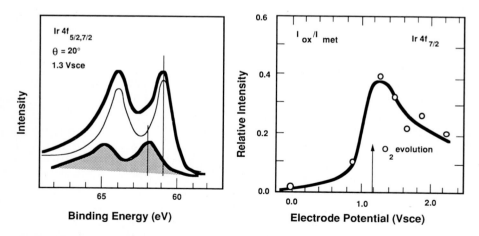

Fig. 22. XPS spectrum of the Ir 4f level of a Ir electrode after polarization at 1.3 Vsce in 0.5 mol L^{-1} H$_2$SO$_4$ for 15 min. The shaded area represents the contribution of the oxide layer (left). Ratio of oxidizer Ir and metallic Ir contributing to the Ir4f level as a function of electrode potential (right). The maxium corresponds to ~1 monolayer (3Å). After [55].

at rather low potentials. OH adsorption was observed at potentials of 0.5 Vrhe for Ir [79] and at 0.25 Vrhe for Ru [73]. While the actual onset of the oxide formation is difficult to deduce from the XPS data in Fig. 22, the application of UPS to emersed iridium electrodes [67] allowed the determination of the onset of surface oxide/hydroxide formation. Taking the occurrence of the t_{2g} band 1.5 eV below E_F in the UPS spectra of electrochemically treated Ir electrodes as an indication for surface oxide formation (see also Fig. 21 for Ru), it was concluded that oxide formation starts below 0.5 Vsce. The irreversibility of the oxide formed at 1.25 V could also be demonstrated by means of UPS. The t_{2g} band developed during the first anodic scan, does not disappear on the consecutive cathodic scan [67].

3.2.3 Stabilization of Ru/Ir Alloys and Mixed Oxides

For Cl_2 or O_2 evolution the stability of ruthenium based electrodes is not sufficient on a technical scale. Therefore the possibility of stabilizing the ruthenium oxide without losing too much of its outstanding catalytic performance was investigated by many groups. For the Cl_2 process, mixed oxides with valve metals like Ti or Ta were found to exhibit enhanced stability (see Section 3.1), while in the case of the O_2 evolution process in solid polymer electrolyte cells for H_2 production a mixed Ru/Ir oxide proved to be the best candidate [68, 80].

Corrosion of Ru/Ir alloys can easily be monitored by following the amount of the corrosion product RuO_4 as a function of alloy composition for otherwise constant parameters. RuO_4 can easily be monitored electrochemically [70]: the cyclic voltammogram of Ru exhibits a pronounced reduction peak at 0.8 Vsce, corresponding to the reduction of RuO_4, provided the electrolyte is not stirred and the anodic scan limit is above the O_2 evolution potential. Upon alloying Ru with Ir this peak is significantly reduced. As a consequence of the much lower corrosion rate of RuO_2 electrodes, electrochemical detection of RuO_4 is not possible. Time until failure of RuO_2 electrodes deposited on Ti substrates are usually determined under constant current conditions by measuring the time until a certain electrode potential is exceeded.

The stabilizing influence of IrO_2 admixtures to RuO_2 has been demonstrated by several groups [80–83]. For electrodes prepared by reactive sputtering of the mixed oxide on Ti substrates, a major reduction of the corrosion rate could be observed for a 20% addition of IrO_2 [83]. In order to find an optimum catalyst composition not only with respect to performance but also with respect to costs (price of Ir is four times that of Ru) the underlying mechanism of stabilization should be understood. Detailed investigations of this aspect were performed for Ru/Ir alloys and for the mixed oxides prepared by reactive sputtering [82, 83] by means of electrochemical techniques and photoelectron spectroscopy. In accord with the scope of this chapter the results of the former techniques will be mentioned only briefly, while results of the latter will be discussed in detail.

Admixtures of Ir to Ru or RuO_2 resulted in an increased Tafel slope for the O_2 evolution reaction. The Tafel slope for the pure Ir compound was achieved for an Ir concentration below 50%. This observation can be taken as an indication that the mechanism of O_2 evolution is governed by the Ir species already at rather low concentrations. Assuming a linear superposition of the contributions of both components would lead to the opposite result, i.e. a Tafel slope corresponding to that of Ru over a wide composition range. In addition to the change in Tafel slope, a linear increase of various redox potentials was observed for the alloys as well as for the mixed oxides. The O_2 evolution potential, the reversible potential for the valence state change from III to IV and the potential for the reduction of the initially formed oxide all increase from the value known for pure Ru or RuO_2 to that of Ir or IrO_2 proportional to the Ir content [82, 83]. As a consequence a $Ru_{0.5}Ir_{0.5}O_2$ mixed oxide exhibits an O_2 evolution mechanism like IrO_2 (Tafel slope) but with reduced O_2 evolution overpotential.

Systematic changes in the cyclic voltammograms of mixed Ru/Ir oxide electrodes with composition were also observed by Angelinetta et al. [50a] and were used for *in situ* determination of the electrode's surface composition. In a recent study performed by Angelinetta et al. [84] on the electrocatalytic properties of Ru/Ir mixed oxides, prepared by thermal decomposition of the chlorides, different results for the dependence of Tafel slopes and overpotentials on the oxide composition were obtained. It was concluded, however, that the results depend very much on the homogeneity of the oxide. While electrodes prepared by reactive sputtering or by precipitation from the salt melt in case of powders closely represent the ideal homogeneous mixed oxide, electrodes prepared by thermal decomposition exhibit some inhomogeneity and are probably better represented by a physical mixture [84]. The contribution of photoelectron spectroscopy to the understanding of the stabilization of Ru based electrodes by Ir will be reported in the following. First of all XPS allows the determination of the electrode's surface composition after O_2 evolution. Fig. 23 shows the surface composition of Ru_xIr_{1-x} alloys as a function of the electrode potential. As soon as O_2 evolution sets in, the Ru component corrodes preferentially giving rise to a significant

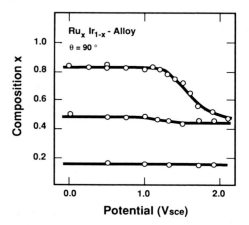

Fig. 23. Surface composition x of a Ru_xIr_{1-x} alloy electrode after electrochemical treatment at different potentials for 5 min in 0.5 mol L^{-1} H_2SO_4. Electron emission angle 90°. After [82].

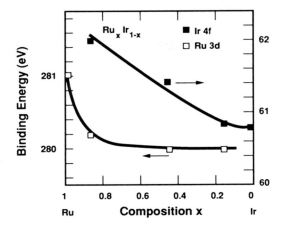

Fig. 24. Ru3d and Ir4f binding energies for a Ru_xIr_{1-x} alloy electrode after polarizaition at 1.7 V for 5 min in 0.5 mol L^{-1} H$_2$SO$_4$ as a function of composition x. Electron emission angle was 90° for Ru3d and 20 for Ir4f.

change in surface composition for the $Ru_{0.85}Ir_{0.15}$ alloy. For a higher Ir content preferential corrosion is inhibited. The reason for this behaviour becomes clear upon inspection of the Ir4f and Ru3d binding energies after O$_2$ evolution at 1.7 V for 5 minutes (see Fig. 24).

While the positive chemical shift for Ru, indicating thick oxide formation, is virtually absent for $x < 0.8$ in Ru_xIr_{1-x}, a positive chemical shift is observed for Ir4f for $x > 0.2$. Obviously the oxidation and corrosion of the ruthenium component is inhibited at the expense of increased oxidation of the iridium sites. Similarly, no preferential corrosion was observed for a $Ru_{0.8}Ir_{0.2}$ mixed oxide during O$_2$ evolution at 1.5 Vsce for 8 hours [83].

In order to understand the observed shift in oxidation potentials and the stabilization mechanism two possible explanations were forwarded by Kötz and Stucki [83]. Either a direct electronic interaction of the two oxide components via formation of a common d-band, involving possible charge transfer, gives rise to an electrode with new homogeneous properties or an indirect interaction between Ru and Ir sites and the electrolyte phase *via* surface dipoles creates improved surface properties. These two models will certainly be difficult to distinguish. As is demonstrated in Fig. 25, XPS valence band spectroscopy could give some evidence for the formation of a common d-band in the mixed oxides prepared by reactive sputtering [83].

The position of the d-derived t_{2g} band in the mixed oxides shifts from 0.8 eV for RuO$_2$ to 1.5 eV for IrO$_2$ proportional to the composition of the oxide. As a consequence of common d-band formation the delocalized electrons are shared between Ir and Ru sites. In chemical terms, Ir sites are oxidized and Ru sites are reduced and electrochemical oxidation potentials are shifted. Oxidation of Ru sites to the VIII valence state is now prohibited. Thus corrosion as well as O$_2$ evolution on Ru sites is reduced which explains the Tafel slope and overpotential behaviour. Most probably Ru sites function as Ir activators [83].

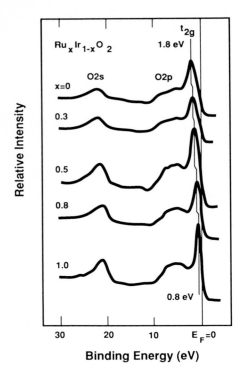

Fig. 25. XPS valence band spectra for reactively sputtered $Ru_xIr_{1-x}O_2$ electrodes on a Ti substrate after preparation for different compositions x. Note the shift in t_{2g} band position. After [83].

The results of the above mentioned study on mixed oxides prepared by thermal decomposition [84] are not in contradiction to the results obtained on reactively sputtered electrodes. A premise for common d-band formation is the formation of a solid solution with homogeneous properties which is probably not obtained during thermal decomposition. Indeed the authors find a trend towards the behaviour of the sputtered electrodes when homogeneity is improved by changing the solvent for the starting compounds.

Stabilization of Ru based oxides by valve metal oxides has not been studied in such detail using photoelectron spectroscopy. The most common compositions, however, with relatively high valve metal content, are not in favor of formation of a solid solution. Studies of the phase formation in Ru/Ti mixed oxides has shown [49] that homogeneous solutions are formed for compositions with Ru < 2% or Ru > 98% (see Section 3.1.1). Therefore electrodes with other compositions are better described as physical mixtures and the electrochemical behaviour is most likely that of a linear superposition of the single components. It has to be considered, however, that the investigations performed by Triggs [49] concern thermodynamic equilibrium conditions. If, by means of the preparation procedure, thermodynamic equilibrium is avoided homogeneous mixed oxides with other compositions may well be produced (see [50]).

3.2.4 Electrochromism of Iridium

Iridium as an electrode material has received considerable attention in the last decade not only because of its excellent catalytic properties but also in relation to the electrochromic effect observed for anodic iridium oxide films (AIROF). Electrochromism of iridium was thought to be of technical relevance for display applications and triggered several studies of the electrochemical and optical properties of AIROFs [67, 85–88].

Thick anodic iridium oxide films are formed by repetitive potential cycling between properly chosen anodic and cathodic limits [89]. The coloration (bleaching) transition is reflected in the cyclic voltammogram by a significant increase (decrease) of the electrode pseudo-capacity at a potential around 0.7 Vsce in acid electrolytes. At potentials above 0.7 V the thick film appears dark blue, while below 0.7 V the film is almost clear.

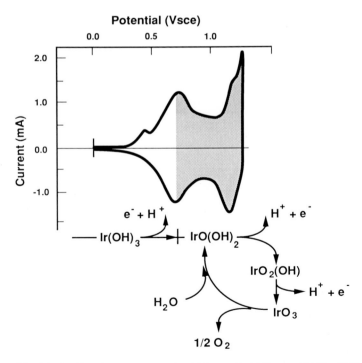

Fig. 26. Cyclic voltammogram of a thick anodic iridium oxide film (AIROF) in 0.5 mol L^{-1} H$_2$SO$_4$. The reaction mechanism for coloration and O$_2$ evolution is indicated.

On the basis of electrochemical, optical and RBS [90] investigations the underlying mechanism was found to be proton insertion during bleaching, which can be written as [91]

$$\text{Ir(OH)}_n \rightleftharpoons \text{IrO}_x\text{(OH)}_{n-x} + x\text{H}^+ + xe^- \tag{5}$$

The number of protons extracted from the film during coloration depends on the width of the potential step under consideration. As can be seen in the formulation of Fig. 26 an additional valence state change occurs at 1.25 Vsce giving rise to another proton extraction. The second proton exchange may explain the observation by Michell et al. [91] who determined a transfer of two electrons (protons) during coloration. Equation (5) is well supported by XPS measurements of the Ir4f and O1s levels of thick anodic iridium oxide films emersed at different electrode potentials in the bleached and coloured state. Deconvolution of the O1s level of an AIROF into the contribution of oxide (O^{2-}, 529.6 eV) hydroxide, (OH, 531.2 eV) and probably water (533.1 eV) indicates that oxide species are formed during anodization (coloration) on the expense of hydroxide species. The bleached film appears to be pure hydroxide (Fig. 27).

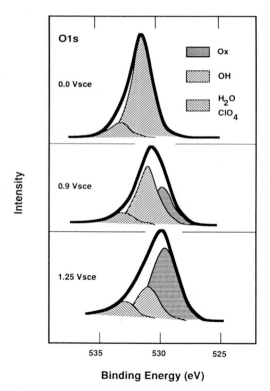

Fig. 27. Deconvoluted O1s levels of a thick anodic iridium oxide film at different potentials. After [34].

Quantitative analysis of the XPS data indicates a constant O/Ir ratio of close to 3 [34] and an OH/O ratio of 2 for a potential of 0.9 V and of 1/2 at 1.25 Vsce. These results, together with electrochemical data substantiate the reaction mechanism given in Fig. 26 for the electrochromic effect and for O_2 evolution.

Interpretation of the Ir4f level is less informative. Independent of the electrode potential a constant binding energy of 62.4 eV was found [34], indicating a hydrous oxide with Ir in the tetra-valent state. The presence of Ir^{VI} has been postulated on the basis of deconvoluted spectra for the AIROF by Angelinetta et al. [83] and also by Kim et al. [60]. Interpretation of Ir4f levels of IrO_2, however, is not straightforward, because of an intrinsic asymmetry of these levels as has been experimentally and theoretically demonstrated by Wertheim and Guggenheim [24].

The question as to why the electrical and optical properties of the anodic film change drastically upon proton insertion could be answered in correlation with UPS investigations (Fig. 28).

The distribution of electronic states of the valence band for the colored film at 1.25 Vsce resembles very much the valence band of pure IrO_2 as reported by Mattheiss [93]. The maximum of the t_{2g} band occurs at 1.6 eV below E_F, the O2$_p$ region extends from roughly 4 eV to 10 eV and a finite density of electronic states at the Fermi level. Upon proton (and electron) insertion the t_{2g} band, which can host 6 electrons, is completely filled and moves to a binding energy of 2.5 eV. Simultaneously, the density of states at E_F is reduced to zero and an additional structure, indicating OH bond formation, occurs in the O2$_p$ band. The changing density of states

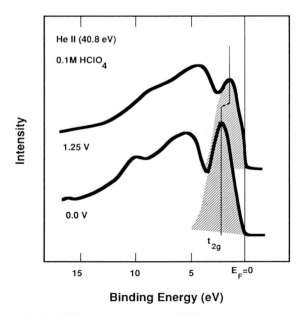

Fig. 28. Valence band spectra (UPS) of an AIROF electrode in the coloured (1.25 V) and in the bleached (0.0 V) state. The electrode was rinsed after emersion with ultra pure H_2O. After [67].

at the Fermi level is the key to the understanding of the electronic and optical propertis of the AIROF. Optical transitions in the visible from filled states below E_F to E_F [93] are no longer possible for the film in the reduced (bleached) state and the film conductivity is significantly reduced.

In order to explain the changing optical properties of AIROFs several models were proposed. The UPS investigations of the valence band of the emersed film support band theory models by Gottesfeld [94] and by Mozota and Conway [79, 88]. The assumption of nonstoichiometry and electron hopping in the model proposed by Burke et al. [87] is not necessary. Recent electroreflectance measurements on anodic iridium oxide films performed by Gutierrez et al. [95] showed a shift of optical absorption bands to lower photon energies with increasing anodic electrode potentials, which is probably due to a shift of the Fermi level with respect to the t_{2g} band [67].

3.2.5   Underpotential Deposition of Metals (UPD)

More than a decade ago, Hamond and Winograd used XPS for the study of UPD Ag and Cu on polycrystalline platinum electrodes [11, 12]. This study revealed a clear correlation between the amount of UPD metal on the electrode surface after emersion and in the electrolyte under controlled potential before emersion. Thereby, it was demonstrated that *ex situ* measurements on electrode surfaces provide relevant information about the electrochemical interface. (see Section 2.7). In view of the importance of UPD for electrocatalysis and metal deposition [132, 133], knowledge of the oxidation state of the adatom in terms of chemical shifts, of the influence of the adatom on local work functions and knowledge of the distribution of electronic states in the valence band is highly desirable. The results of XPS and UPS studies on UPD metal layers will be discussed in the following chapter. Finally the poisoning effect of UPD on the H_2 evolution reaction will be briefly mentioned.

Chemical Shifts and Work Function. The XPS investigations by Hamond and Winograd [11, 12] gave very surprising results. The binding energy of UPD Cu and Ag on Pt was found to be significantly smaller than the bulk binding energy. This negative chemical shift of about -1 eV is in total contradiction to what one would expect on the basis of electrochemically determined electrosorption valencies, which in general indicate a small positive charge on the adatom. The results of Hamond and Winograd were reproduced by Kolb et al. for the system Cu on Au(111) and in the author's laboratory (Figs. 29 and 30) for Cu and Ag on Pt(poly). The Ag3*d* level of UPD silver on Pt shifts from 366.7 eV at $+1.0$ Vsce, corresponding to a coverage of 0.1, to 367.8 eV at -0.1 V. At 0.1 V the coverage is 2 monolayers and the binding energy corresponds to that of bulk Ag. With respect to the substrate/adsorbate combinations investigated up to now, the negative chemical shift of UPD deposited metals appears to be independent of the substrate (Pt, Au, Ru [74]) as well as independent of the adsorbate (Ag, Cu).

Hamond and Winograd who were the first to discover this phenomenon speculated on several explanations but did not come to a final conclusion about the actual

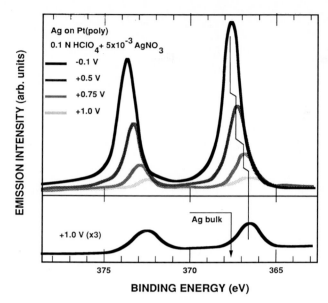

Fig. 29. Ag3d core level spectra of Ag underpotential deposited onto polycrystalline Pt at various electrode potentials. 0.1 mol L^{-1} HClO$_4$ + 0.001 mol L^{-1} AgNO$_3$.

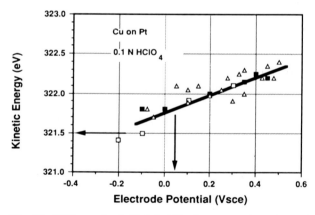

Fig. 30. XPS core level (Cu2p) shifts for copper deposited on Pt. Different symbols correspond to separate experiments (new electrolyte, new electrode). Arrows indicate kinetic energy of bulk Cu and reversible potential for Cu deposition respectively. MgKα source.

origin of the negative chemical shift [12]. While Hamond and Winograd and later Kolb et al. [17] observed a constant negative shift, independent of the electrode potential within the UPD regime, Figs. 29 and 30 clearly exhibit an almost linear shift of the Ag3d and Cu2p levels with electrode potential or in other terms with coverage.

Results similar to those observed for UPD copper on platinum electrodes were recently obtained by Shek et al. [96] in an extensive XPS, UPS and work function investigation of vacuum deposited copper on Pt(111) substrates. In qualitative agreement with the observations at the electrochemical interface, the Cu3d level of vacuum deposited Cu exhibits a negative chemical shift as a function of coverage. For a submonolayer coverage of 0.3 the shift was -0.6 eV. The bulk value for the Cu3d binding energy was reached at a coverage of 2 to 3 monolayers. Possible reasons for this negative chemical shift given by Shek et al. were electron rearrangement within the Cu, a change in final state relaxation energy and/or surface versus bulk effects. Electron transfer between the copper and the platinum substrate was excluded as a reason for the negative chemical shift.

Calculation of core-electron binding energy shifts on the basis of initial and final state effects is extremely difficult and requires a major computational effort. Instead of using these microscopic parameters an approach on the basis of mainly thermodynamic quantities, the Born–Haber cycle, has been proposed recently for the estimation of chemical shifts of rare gas atoms or metal atoms on metal substrates, making use of the so-called equivalent core approximation [97, and references therein]. Application of this approach to Au monolayer and sub-monolayer on Pt(111) indicated that the negative chemical shift observed (1.0 eV for low coverage, 0.4 eV for the monolayer) is mainly due to the cohesive interaction in the monolayer. The dependence of the chemical shift on coverage can be explained in terms of changing coordination numbers.

As long as no other experimental or theoretical evidence is given, one has to assume that the negative chemical shift of UPD metal adsorbates is not a phenomenon confined to the electrochemical interface, but rather a general effect, which can be observed at the metal vacuum interface as well. Unless a consistent model for chemical shifts of metal adsorbates on metal substrates at the metal vacuum interface is developed, deduction of relevant data for charge distribution in the electrochemical double layer from measured core-electron binding energies on emersed electrodes will be extremely difficult. This argumentation does not only hold for investigation of UPD metals, but also for the study of adsorbed counter ions as components of the double layer. Further XPS studies should be performed on single crystalline electrodes. By choosing an UPD system with a rather sharp UPD desorption/adsorption spike, like Pb on Ag, one should be able to separate electrode potential effects from coverage effects, because for such a system the coverage can be changed from 0 to 1 with only a minor variation in potential.

While the above XPS results give the impression, that the electrochemical interface and the metal vacuum interface behave similarly, fundamental differences become evident when work function changes during metal deposition are considered. During metal deposition at the metal vacuum interface the work function of the sample surface usually shifts from that of the bare substrate to that of the bulk deposit. In the case of Cu deposition onto Pt(111) a work function reduction from 5.5 eV to 4.3 eV is observed during deposition of one monolayer of copper [96]. Although a reduction of work function with UPD metal coverage is also observed at the electrochemical interface, the absolute values are totally different. For Ag deposition on Pt (see Fig. 31)

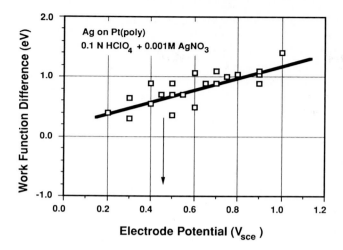

Fig. 31. Work function change, referred to the clean Pt electrode surface, as a function of electrode potential for Ag underpotential deposition onto Pt. The work function of bulk Cu would correspond to -1.

the work function is higher than that of the Pt substrate during deposition of the first monolayer of Ag. At a potential of 0.0 Vsce the work function corresponds to that of clean Pt(poly). The work function of Ag, which is about 1 eV lower than that of Pt, would be reached at a potential of -0.9 Vsce as can be seen by extrapolation of the data in Fig. 31. A comparison of Fig. 7 with Fig. 31 confirms again the conclusion drawn from work function measurements on noble metal electrodes like Au, Pt and Ir that the work function of emersed electrodes is primarily determined by the emersion potential and only to a lesser extent – if at all – by the electrode material or surface modifications as a consequence of electrochemical oxide formation or metal deposition. For Au electrodes, where work function measurements were performed by different authors using different techniques [20, 32], the linear correlation between work function after emersion and electrode potential during emersion was maintained even at electrode potentials anodic of 1.0 Vsce where electrochemical oxide formation occurs (see Fig. 7). At the metal vacuum interface, on the other hand, oxidation would cause a drastic change of the work function in most cases. The most probable explanation for this behaviour is a compensation of the changed surface dipole (electrochemically formed oxide layer or UPD metal adsorbate) by a rearrangement of other dipole contributions like water or counter ions (see Equation 4). The sum of these contributions make the total work functions which, however, is a direct function of the electrode potential.

Valence Band Spectroscopy. Optical and electronic properties of UPD metal flms on metal electrodes have been studied *in situ* by means of differential- and electro-reflectance spectroscopy [98]. Optical absorption bands, however, reflect a combined density of electronic states at a photon energy which is the energetic difference of initial and final state. In order to position any optical transition on the absolute energy scale, the energetic position of at least one of the states involved has to be known. In the case of Cu deposition on Pt [99] this was achieved by assigning the low energetic absorption band to a transition from the filled Cu3d state below E_F to the Fermi level. The analysis of the optical spectra lead to the conclusion, that the Cu3d level of UPD copper on Pt is situated 2.8 eV below E_F, while on the basis of Cu_6 cluster calculations Cu3d should appear at 2.0 eV below E_F.

The actual position of the Cu3d level for UPD Cu on Pt can be determined by means of UPS valence band spectroscopy on emersed electrodes. Fig. 32 displays the spectra obtained at different potentials (Cu coverages) in comparison to the spectra of bulk Pt and bulk Cu. The formation of the Cu3d level is clearly shown for potentials of 0.3, 0.1, and − 0.2 Vsce corresponding to coverages of 0.6, 1.0 and 2.3 monolayers. Additional structures visible for the spectrum at − 0.2 V are due to ClO_4^- adsorption. In agreement with the observations made at the metal vacuum interface by Shek et al. [96] the maximum of the Cu3d level for the Cu adatom is shifted to lower binding energies (2.2 eV at 0.3 Vsce) as compared to the bulk value (2.8 eV). With the help of UPS spectra it is now possible, for the first time, to assign optical transitions measured *in situ* for Cu deposition on Pt. The low energy optical transition occurring at a photon energy of 2.3 eV for low and at 2.5 eV for high coverages [99] is most probably

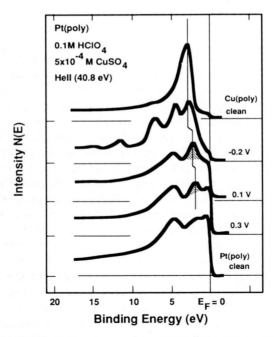

Fig. 32. Valence band spectra (UPS) of a polycrystalline Pt electrode emersed at a different potential where underpotential deposition of Cu (0.3V, 0.1 V) and bulk deposition (−0.2 V) of Cu occurs. Clean Pt and Cu surfaces are shown for comparison.

due to a transition from the Cu3d level to the Fermi level. The shift in photon energy is a consequence of a shift of the Cu3d level with coverage.

UPS valence band spectra were also obtained for the UPD system Ag on Pt. These spectra exhibit again a shift of the Ag4d level of the adatom to lower binding energies when compared to the bulk value. For both Cu and Ag adatoms on Pt the UPS spectra clearly show that bulk properties of the adsorbate layer are achieved for coverages of about 3 monolayers.

H_2 Evolution on RuO_2.   Optimization of technical electrolysis processes in terms of energy consumption and investment cost concerns the reduction of ohmic voltage drops as well as the cathodic and anodic overpotentials. As a consequence of the relatively high overpotentials for the O_2 and Cl_2 evolution reaction most of the research work was addressed to the anode reaction because possible voltage savings appeared to be larger on the anode side. Accordingly, H_2 evolution on Pt or Ni electrodes contributes only little to the overall cell voltage. On long term operation, however, poisoning effects have to be taken into account and occasionally it may be worthwhile to lose some mV in overpotential for the merit of stable long term performance.

Underpotential deposition of heavy metals on H_2 evolving electrodes is a well known problem [133]. The existence of a direct correlation between H_2 evolution activity and metal work function, makes UPD very likely on high work function electrodes like Pt or Ni. Cathode poisoning for H_2 evolution is aggravated by UPD for two reasons. First, deposition potentials of UPD metals are shifted to more anodic values (by definition), and second, UPD favors a monolayer by monolayer growth causing a complete coverage of the cathode [100]. Thus H_2 evolution may be poisoned by one monolayer of cadmium for example, the reversible bulk deposition potential of which is cathodic to the H_2 evolution potential.

A way to get around UPD is the use of pure oxide electrodes or oxide admixtures to metal electrodes. This possibility was suggested for alkaline electrolysis by Nidola et al. [101] as well as for acid electrolytes by Kötz and Stucki [69]. The fact that UPD does not occur on semiconductors and oxides has been demonstrated by Bindra et al. [102] and was also demonstrated for the metallic oxide RuO_2 [69]. On first sight, use of an oxide as a cathode for hydrogen evolution appears not to be sensible because reduction of the oxide to the metal is very likely. XPS investigations on RuO_2 electrodes after H_2 evolution have shown, however, that only a partial reduction to some oxi-hydroxide occurs. The resulting Ru3d and O1s spectra of a RuO_2 electrode before and after H_2 evolution are given in Fig. 33. The stability of the electrode is indicated in terms of a constant binding energy of the Ru3d level at 280.7 eV, indicating that ruthenium is in the valence state IV before and after H_2 evolution. The shift in O1s binding energy can be interpreted as hydroxide formation or hydration. A recent study of the electrochemical properties of IrO_2 electrodes during hydrogen evolution by Boodts and Trasatti [103] also came to the conclusion that only a partial reduction of the oxide occurs. Application of RuO_2 cathodes in solid polymer electrolyte cells for H_2 production has lead to a significant improvement of long term stability [69].

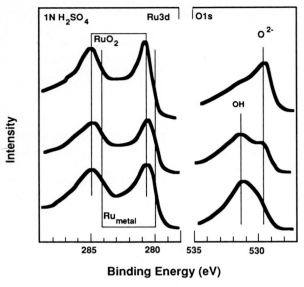

Fig. 33. Ru3d and O1s core levels for a RuO$_2$ electrode before (top) and after H$_2$ evolution in 0.5 mol L^{-1} H$_2$SO$_4$ (middle: 50 mA/Cm^2, 15 min., 25°C; bottom: 50 mA/cm^2, 15 min., 80°C). After [69].

3.2.6   Other Selected Electrode Materials

Iron and Stainless Steel.   The purpose of XPS investigations on typical corrosion systems like iron or stainless steel, is the determination of the composition of the passive surface layer, if possible, as a function of depth. As a consequence of the technical and economic relevance of corrosion reactions, XPS investigations on corrosion systems are numerous. With respect to the application of XPS, there is no difference between corrosion systems and any other electrochemical surface reaction like oxide formation on noble metals. Therefore, in this paragraph only a few recent typical results of such studies, using XPS, will be mentioned. For a detailed collection of XPS corrosion studies the reader is referred to references [43, 104]. A review of aqueous corrosion studies, using XPS, was given by McIntyre for the elements O, Cr, Mn, Fe, Co, Ni, Cu and Mo [105]. The book edited by M. Froment [111] gives an impression of the research achieved on passivity of metals up to 1983.

The corrosion of stainless steel in 0.1 mol^{-1} NaCl solutions at open circuit potential was studied in detail by Brüesch et al. [106] using XPS in combination with a controlled sample transfer system [38]. It was verified by XPS analysis that the passivating film contains chromium oxide. The position and the height of the Cr concentration maximum depends critically on the bulk chromium content of the steel. Significant variations in the electrode passivation properties were observed at a Cr concentration of 12%, while the film behaviour was found to be rather independent of the other components like Mo, Ni, Cu. From the fact that the film structures and compositions determined in a variety of studies are very similar, it was concluded that the nature of the passive film is rather independent of the alloy composition and of environmental exposure. On the basis of the electrochemical and XPS investigations a phenomenological model for the passivation/corrosion transition in binary alloys was developed [107].

The presence of water in terms of OH species in the passive film has been studied by several authors using XPS (and SIMS) in order to distinguish between the hydrated amorphous [134] and the unhydrated polycrystalline [135] models for the passive layer. The presence of OH species concentrated at the surface of the film was confirmed by several investigators for iron [108–110] as well as for iron–chromium alloys [111, 112] or stainless steel [106, 110 111]. The O1s level of a passive film grown on iron in borate buffer is reproduced in Fig. 34. The superposition of two different oxygen species is clearly shown, one at a binding energy of 529.4 eV corresponding to O^{2-} and another at a binding energy of 532 eV corresponding to OH species. After sputtering, the peak with the higher binding energy disappears, indicating that the main portion of the film is free of OH.

XPS was also used for the determination of chlorine in the passive film grown in chlorine containing electrolytes. While chlorine was found in the passive film on pure iron, it was absent for chromium rich stainless steel samples. Chloride content of the passive film is substantially time dependent, increasing with time until film breakdown occurs, and decreasing subsequently [109].

Especially in conjunction with the detection of water or OH species, *ex situ* XPS measurements have been critizised because of possible changes occurring during transfer and exposure of the sample to UHV. Kuroda et al. have demonstrated that structural changes of the passive film indeed occur when electron diffraction studies are performed in a hydrated and 'subsequently' in a dehydrated environment. Structural changes, however, do not necessarily cause changes in elemental composition as determined by XPS.

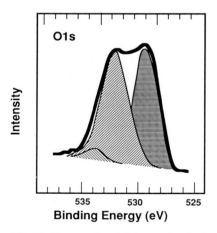

Fig. 34. Deconvoluted O1s core level of a passive layer on iron in borate buffer at 0.65 Vrhe. After [108].

Carbon.   Carbon or graphite electrodes and electrode supports are of considerable interest from a technical point of view. The most prominent application is that for air or oxygen fuel cell electrodes [113]. As is generally accepted, the electrochemical properties of carbon or graphite electrodes are mainly determined by functional groups established on the electrode surface by different treatment processes. XPS allows the determination of the nature of these surface functional groups *via* O1*s* and C1*s* binding energies. As a consequence of the huge number of surface groups possible [114] this approach is not unequivocal and complementary techniques are necessary for a complete analysis.

Different surface functional groups on carbon fibres were analyzed by means of XPS by Sherwood et al. [115]. The different surface groups were generated by chemical and electrochemical oxidation processes and clear changes of the C1*s* emission level were detected as a function of the fibre treatment. The performance of glassy carbon electrodes during O_2 reduction was correlated by Sundberg et al. [116] to surface functional groups detected on the electrode surface. Surface functional groups were characterized by means of XPS. Both investigators were able to distinguish between three different types of surface functional groups. The observed shifts in C1*s* binding energies were interpreted as follows, (i) carbon in phenol, alcohol or ether (C–OH, $\Delta E = 1.2$ eV), (ii) carbonyl or quinon type (C=O, $\Delta E = 2.6$ eV), (iii) carboxylic type groups (O=C–OH, $\Delta E = 4.2$ eV). Electrochemical reduction of the electrode surface causes the C1*s* level assigned to carbonyl groups to disappear.

Graphite electrodes, electrooxidized in sulfuric acid, were studied by Wandrass et al. [117]. Various surface functional groups were detected in terms of chemical shifts of the C1*s* level and the extent of oxidation was monitored in terms of the O/C ratio. Chemical shifts of the C1*s* level of 0.8 eV, 2.1 eV, and 3.5 eV were assigned to CH_x, C–O, and C=O respectively. C–O groups predominate for mild (below 8 C/cm^2) oxidation, while C=O groups were observed for higher oxidation states. Oxidation was found to an extent of 30–70 Å below the surface. These XPS results support the observation by Hine et al. of a "lower" and a "higher oxide" on graphite [131, 118].

The chemical shift of the carbon C1*s* level was used by Rach and Heitbaum [136] for the detection of adsorbed CO or HCO species on platinum electrodes during formic acid oxidation. Although considerable difficulties were encountered as a consequence of carbon contamination of the platinum, the authors were able to detect an HCO species. Methanol oxidation was investigated by Goodenough et al. [137] on Pt/Ru alloy electrodes using XPS. Retardation of the poisoning effect usually occurring on Pt electrodes was demonstrated for the alloy by means of XPS. While some format (Pt–CH$_2$–OH) could be detected on the electrode surface after prolonged methanol oxidation, nearly no Pt–CO groups were found.

Titanium Carbide.   Carbides of transition metals are known for their hardness, wear resistance and also for their high electrical conductivity, which makes them attractive as a refractory coating material for cutting tools or bearings. Only little work has been done on the electrochemical stability of transition metal carbides with the exception of TiC, where a corrosion and passivation mechanism was suggested by Hintermann et al. [119, 120]. This mechanism was confirmed on amorphous TiC produced by metal-organic CVD [121]. In acid electrolyte, TiC starts to dissolve at about 0.6 Vsce, as shown by a current peak in the voltammogram. At more anodic potentials, between 0.8 and 1.5 Vsce, a passive layer is formed, which corrodes at potentials above 1.5 V. Corrosion of TiC is described as follows [119, 120]:

$$TiC + 2H_2O \rightarrow TiO^{2+} + CO_{ads} + 4H^+ + 6e^- \tag{6}$$

The passive layer is subsequently formed from TiO^{2+} and was described as $TiO_2 \times H_2O$. During corrosion CO or CO_2 is formed from the carbon of the carbide. The electrochemical behaviour of TiC in acid electrolytes was reinvestigated with respect to the depassivation of titanium substrates for anodic PbO_2 deposition by M. Cappadonia et al. [122] using XPS.

In addition to the positive chemical shifts mentioned above for different surface functional groups resulting from carbon oxidation, carbide can be distinguished from other carbon species by means of a negative chemical shift between -1.5 eV (WC) and -3.5 eV (HfC). For TiC a binding energy of the C1s level 281.3 eV was found. Fig. 35 shows the C1s level of reactively sputtered TiC on a glass substrate as prepared in comparison to a sample electrochemically treated at 1.25 Vsce (passive region) for 5 minutes in 1N H_2SO_4. The three different carbon peaks nicely indicate the transition from carbon in the form of carbide to carbon in the form of graphite and carboxylic surface groups (CO, CO_2). Simultaneously, part of the TiC is converted to TiO_2 as indicated by the shift in binding energy of the Ti2p level. While the reduction of the C1s level assigned to TiC and the occurrence of a level corresponding to CO or CO_2

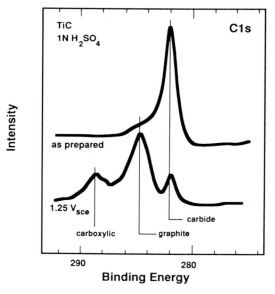

Fig. 35. C1s core level of a TiC electrode as prepared and after electrochemical nolarization at 1.25 Vsce in 0.5 mol L^{-1} H_2SO_4. After [122].

(288.4 eV) is in good agreement with the mechanism suggested by Hintermann et al., the formation of a graphite like layer on the electrode has not been considered before. The detailed study performed by Cappadonia et al. [122] has revealed that TiC films on Ti substrates are well suited for the depassivation of the Ti substrate during anodic PbO_2 deposition, which occurs at a potential of roughly 1.5 Vsce. Corrosion of TiC at potentials more anodic than 1.5 V is indicated by a total disappearance of the C1s level assigned to TiC and the growth of the Ti2p level assigned to TiO_2.

XPS investigations of the composition of the anodically grown passive layer on Ti electrodes were performed by Armstrong and Quinn [123, 124]. The formation of a suboxide layer between the underlying Ti metal substrate and the stoichiometric TiO_2 on top was demonstrated using XPS, AES and electrochemical techniques.

GaAs, CuInS$_2$, CuInSe$_2$. Semiconductor electrodes have received increasing attention as a consequence of their potential application in photoelectrochemical energy conversion devices. In order to achieve optimum efficiency, the knowledge of the surface composition plays a crucial role. Surface modifications may occur during operation of the photo electrode, or may be the result of a chemical or electrochemical treatment process prior to operation.

In place of the many investigations performed in this growing field, only a few examples for XPS investigations performed on semiconductor electrodes will be mentioned. Up to now, a controlled transfer system has been used only for the investigation of $MoSe_2$ [125] and GaAs [126]. Solomun et al. [126] have investigated the surface composition of GaAs electrodes before and after anodic electrochemical treatment in NH_4Cl (pH 7.6) and HCl (pH 1.3). After electrochemical treatment anodic of the flat band potential a significant increase in As and depletion in Ga was found at the electrode surface. Arsenic as well as gallium exhibit positive chemical shifts indicating oxidation of both species to a valence state of III (As_2O_3, Ga_2O_3). The oxide layer is stable towards cathodic corrosion reactions but does not prevent further photooxidation at higher potentials.

Improvement of photoelectrochemical performance of $CuInS_2$ electrodes after surface modification by etching was observed by Goslowsky et al. [127]. XPS analysis showed that etching resulted in modified surface composition with decreased Cu and In and increased sulfur and oxygen content. Further improvement of electrode performance could be achieved by a ruthenium treatment. Ruthenium fixed to the surface after exposure of the etched electrode to a $RuCl_3$ solution forms a Ru–S–O structure with a Ru binding energy of 281.7 eV. A decrease in Cu content on the surface of $CuInS_2$ photoelectrodes was also found by Mirovsky et al. [128] for those samples with higher efficiencies. For $CuInSe_2$ electrodes in polysulfide solution possible S/Se exchange was excluded by means of XPS [128].

Significant depletion of In, enrichment in iodine and simultaneous reduction of selenium to Se^0 was observed by XPS for n-CuInSe$_2$ photoelectrode surfaces in an I_2–I^-–Cu^+–HI electrolyte [129]. Over a distance of 100 Å the composition of the surface layer gradually changes to that of the substrate. With the help of UPS analysis the surface film, assumed to be $CuISe_3$, was characterized as *p*-type semiconductor [130].

4 Concluding Remarks

The various examples discussed above and several others, not mentioned in this necessarily incomplete chapter, demonstrate that XPS and also UPS assist and improve our understanding of the electrode/electrolyte interface and of electrochemical reactions. XPS, UPS and other *ex situ* techniques will continue to play a key role in providing information about the structure and composition of the electrochemical interface on a microscopic scale.

In order to put the various *ex situ* techniques on a safer basis, the effects of electrode emersion and transfer deserve further clarification. Preferably those techniques which can be applied in both environments, *in situ* as well *ex situ*, will serve this purpose.

The interpretation of XPS data is not always straightforward as is exemplified by different conclusions drawn by different investigators for the same electrode reaction. These discrepancies can be overcome if certain standards for electrode preparation, emersion and transfer processes are developed. The effects of the relative complexity of the emersed electrochemical interface on XPS and UPS data analysis in terms of (electro)chemical shifts and work function changes have to be considered.

XPS, an exotic technique for electrochemists 10 years ago, has become a standard working tool which, if the experimental and theoretical boundary conditions are observed, yields reliable data on electrode surface composition and hence electrode processess.

Acknowledgement. I would like to thank Dr. S. Stucki for numerous inspiring discussions, during which some of the model concepts presented were brought forth and for critical reading of the manuscript.

5 References

1. J. Phys., Vol 38, Suppl. 5 (1977).
2. Surf. Sci., Vol. 101 (1980).
3. J. Electroanal. Chem., Vol. 150 (1983).
4. Ber. Bunsenges. Phys. Chem., Vol. 91, 4 (1987).
5. E.B. Yeager, J. Electroanal. Chem. 150 (1983) 679.
6. K. Siegbahn, C. Nordling, A. Fahlman, R. Nordberg, K. Hamrin, J. Hedman, G. Johansson, T. Bergmark, S. E. Karlsson, I. Lindgren, and B. Lindberg, ESCA - Atomic, Molecular and Solid State Structure Studied by Means of Electrone Spectroscopy, Almquist and Wiksells, Uppsala, 1967.
7. C.N. Berglund and W.E. Spicer, Phys. Rev. 136, 1030A (1964).
8. W.E. Spicer, in: Chemistry and Physics of Solid Sufaces IV, R. Vanselow and R. Howe, eds. Springer Verlag Berlin, 1982, p. 1.
9. Photoemission and the Electronic Properties of Surfaces, B. Feuerbacher, B. Fitton, R.R. Willis eds. John Wiley & Sons, 1978.
10. K.S. Kim, N. Winograd and R.E. Davis, J. Am. Chem. Soc. 93 (1971) 6269.
11. J.S. Hammond and N. Winograd, J. Electroanal. Chem. 80 (1977) 123.

12. J.S. Hammond and N. Winograd, J. Electrochem. Soc., 124 (1977) 826.
13. Photoemission in Solids I & II, M. Cardona and L. Ley (eds.) in: Topics in Applied Physics, Vols. 26 & 27, Springer Verlag Berlin, 1978.
14. D.M. Kolb, Z. Phys. Chem. Neue Folge, 154 (1987) 179.
15. Perkin-Elmer Cooperation, Handbook of X-Ray Photoelectron Spectroscopy, Eden Prairie, Minnesota, 1977.
16. Y.M. Cross and J.E. Castle, J. Electron. Spec. Rel. Phen. 22 (1981) 53.
17. D.M. Kolb, D.L. Rath, R. Wille, and W.N Hansen, Ber. Bunsenges. Phys. Chem. 87 (1983) 1108.
18. J.C. Fuggle, in: Handbook of X-ray and Ultraviolet Photoelectron Spectroscopy, D. Briggs, (ed, Heyden & Son Ltd., London, 1977, p. 295.
19. A.T. D'Agostino and W.N. Hansen, Surf. Sci., 165 (1986) 268.
20. R. Kötz, H. Neff, and K. Müller, J. Electroanal. Chem., 215 (1986) 331.
21. K. Wandelt, J. Vacuum Sci. Technol. A2 (1984) 802.
22. M. Cardona and L. Ley, in ref. [13], pp. 82, 83.
23. G.K. Wertheim and P.H. Citrin in ref. [9], p. 197.
24. G.K. Wertheim and H.J. Guggenheim, Phys. Rev. B22 (1980) 4680.
25. M. Pijolat and G. Hollinger, Surf. Sci. 105 (1981) 114.
26. K.S. Kim, W.E. Baitinger, J.W. Army, and N. Winograd, J. Electron. Spec. Rel. Phen. 5 (1974) 351.
27. R. Holm and S. Storp, Appl. Phys. 12 (1977) 101.
28. W.N. Hansen, C.L. Wang, and T.W. Humphreys, J. Electroanal. Chem. 93 (1978) 87.
29. D.M. Kolb and W.N. Hansen, Surf. Sci. 79 (1979) 205.
30. B.E. Hayden, E. Schweizer, R. Kötz, and A.M. Bradshaw, Surf. Sci. 111 (1981) 26.
31. S. Trasatti, in: Comprehensive Treaties of Electrochemistry, Vol. 1, J. O'M Bockris, B.E. Conway and E. Yeager (eds.), Plenum Press, New York, 1980, Ch. 2, p. 81.
32. W.N. Hansen and D.M. Kolb, J. Electroanal. Chem. 100 (1979) 493.
33. H. Neff and R. Kötz, J. Electroanal. Chem. 151 (1983) 305.
34. R. Kötz, H. Neff, and S. Stucki, J. Electrochem. Soc. 131 (1984) 72.
35. M. Peuckert, F.P. Coenen, and H.P. Bonzel, Surf. Sci. 141 (1984) 515.
36. S. Haupt, U. Collisi, H.D. Speckmann, and H.H. Strehblow, J. Electroanal. Chem. 194 (1985) 179.
37. D. Aberdam, S. Traore, R.T. Durand, and R. Faure, Surf. Sci. 180 (1987) 319.
38. H. Neff, W. Foditsch, and R. Kötz, J. Electron. Spec. Rel. Phen. 33 (1984) 171.
39. E. Yeager, W.E. O'Grady, M.Y.C. Woo, and P. Hagans, J. Electrochem. Soc. 125 (1978) 346.
40. R.N. Ross, J. Electroanal. Chem. 76 (1977) 139.
41. A.T. Hubbard, R.M. Ishikawa, and J. Kalekaau, J. Electroanal. Chem. 86 (1978) 271.
42. T. Solomun, W. Richtering, and H. Gerischer, Ber. Bunsenges. Phys. Chem. 91 (1987) 412.
43. P.M.A. Sherwood, Chem. Soc. Rev. 1 (1985) 1.
44. Electrodes of Conductive Metallic Oxides, S. Trasatti (ed.), Part A and B, Elsevier Scientific Publishing Company, Amsterdam, 1980.
45. J. Augustynski, L. Balsenc, and J. Hinden, J. Electrochem. Soc. 125 (1978) 1093.
46. A. DeBattisti, G. Lodi, M. Cappadonia, C. Battaglin, and R. Kötz, J. Electrochem. Soc. 136 (1989) 2596.
47. A. DeBattisti, R. Brina, G. Gavelli, A. Benedetti, and G. Fagherazzi, J. Electroanal. Chem. 200 (1985) 93.
48. V.V. Gorodetskii, Y.Y. Tomashpol'ski, L.B.Gorbacheva, N.V. Sadovskaya, M.M. Pecherskii, S.V. Evdokimov, V.L. Kubasov, and V.V. Losev, Elektrokhimiya 20 (1984) 1045.
49. P. Triggs, Helvetica Physica Acta 58 (1985) 657.
50. R. Hutchings, K. Müller, R. Kötz, and S. Stucki, J. Materials Sci. 19 (1984) 3987.
50a. C. Angelinetta, S. Trasatti, L.D. Atanasoska, and R.T. Atanasoski, J. Electroanal. Chem. 214 (1986) 535.
51. R. Adams and R.L. Shriner, J. Amer. Chem. Soc. 45 (1923) 2171.
52. K.S. Kim and N. Winograd, J. Catalysis 35 (1974) 66.
53. L.D. Burke and J.F. Healy, J. Electroanal. Chem. 124 (1981) 327.
54. R. Kötz, H.J. Lewerenz, and S. Stucki, J. Electrochem. Soc. 130 (1983) 825.
55. R. Kötz, H.J. Lewerenz, P. Brüesch, and S. Stucki, J. Electroanal. Chem. 150 (1983) 209.
56. Lj. Atanasoska, W.E. O'Grady, R.T. Atanasoski, and F.H. Pollak, Surf. Sci. 202 (1988) 142.

57. G. Lodi, C. Bighi, and C. de Asmindis, Mater. Chem. 1 (1976) 177.
58. R. Kötz and S. Stucki, unpublished results.
59. Z.M. Jarzebski and J.P Marton, J. Electrochem. Soc. 123 (1976) 299C.
60. K.S. Kim, C.D. Sell, and N. Winograd in Proc. Electrocatalysis, Electrochem. Soc., M.W. Breiter (ed.), (1974) 242.
61. R. Kötz and H. Neff, unpublished results.
62. T. Dickinson, A.F. Povey, and P.M.A. Sherwood, J. Chem. Soc. Faraday Trans. I, 71 (1975) 298.
63. G.C. Allen, P.M. Tucker, A. Capon, and R. Parsons, J. Electroanal. Chem. 50 (1974) 335.
64. M. Peuckert and H. Ibach, Surf. Sci. 136 (1984) 319.
65. M. Peuckert, J. Electroanal. Chem. 185 (1985) 379.
66. J.D.E McIntyre and D.M. Kolb, Symp. Faraday Soc. 4 (1970) 99.
67. R. Kötz and H. Neff, Surf. Sci. 160 (1985) 517.
68. D. Sievert, D. Winkler, G. Scherer, A. Marek, and S. Stucki, in: Electrode Materials and Processes for Energy Conversion and Storage, S. Srinivasan, S. Wagner and H. Wroblowa (eds.),The Electrochemical Society Proceedings Series.
69. R. Kötz and S. Stucki, J. Appl. Electrochem. 17 (1987) 1190.
70. R. Kötz, S. Stucki, D. Scherson, and D.M. Kolb, J. Electroanal. Chem. 172 (1984) 211.
71. M. Wohlfahrt-Mehrens and J. Heitbaum, J. Electroanal. Chem. 237 (1987) 251.
72. J.F. Lopis and I.M. Tordesillas, in Encylopedia of Electrochemistry of the Elements, Vol. 6, A.J. Bard (ed.), p. 277, Marcel Dekker, New York, 1967.
73. M. Vukovic, H. Angerstein-Kozlowska, and B.E. Conway, J. Appl. Electrochem, 12 (1982) 193.
74. R. Kötz, unpublished results.
75. R.Y. Shaidullin, A.D. Semenova, G.D. Vovchenko, and Y.B. Vasil'ev, Elektrokhimiya, 17 (1981) 988.
76. H.Y. Hall and P.M.A. Sherwood, J. Chem. Soc., Faraday Trans. I, 80 (1984) 135.
77. J. Augustynski, M. Koudelka, J. Sanchez, and B.E. Conway, J. Electroanal. Chem. 160 (1984) 233.
78. M. Peuckert, Surf. Sci. 144 (1984) 451.
79. J. Mozota and B.E. Conway, Electrochimica Acta 28 (1983) 1.
80. R.S. Yeo, J. Orehotsky, W. Visscher, and S. Srinivasan, J. Electrochem. Soc. 128 (1981) 1900.
81. H. Miles, E.A. Klaus, B.P. Gunn, J.R. Locker, W.E. Serafin, and S. Srinivasan, Electrochimica Acta 23 (1978) 521.
82. R. Kötz and S. Stucki, J. Electrochem. Soc. 132 (1985) 103.
83. R. Kötz and S. Stucki, Electrochimica Acta 31 (1986) 1311.
84. C. Angelinetta, S. Trasatti, L.D. Atanasoska, Z.S. Minevski, and R.T. Atanasoski, Mater. Chem. Phys. 22 (1989) 231.
85. S. Gottesfeld and J.D.E. McIntyre, J. Electrochem. Soc. 126 (1979) 742.
86. W.C. Dautremont-Smith, L.M. Schiavone, S. Hackwood, G. Beni, and J.L. Shay, Solid State Ionics 2 (1981) 13.
87. L.D. Burke and D.P. Whelan, J. Electroanal. Chem. 162 (1984) 121.
88. B.E. Conway and J. Mozota, Electrochimica Acta 28 (1983) 9.
89. D.N. Buckley, L.D. Burke, and J.K. Mulcahy, J. Chem. Soc. Faraday Trans. I, 72 (1976) 1896
90. J.D.E. McIntyre, S. Basu, W.F. Peck, W.L. Brown, and W.M. Augustyniak, Phys. Rev. B25 (1982) 7242.
91. D. Michell, D.A.J. Rand, and R. Woods, J. Electroanal. Chem. 84 (1977) 117.
92. L.F. Mattheiss, Phys. Rev. B13 (1976) 2433.
93. A.K. Goel, G. Shorinko, and F.H. Pollak, Phys. Rev. B24 (1981) 7342.
94. S. Gottesfeld, J. Electrochem. Soc. 127 (1980) 1922.
95. C. Gutierrez, M. Sanchez, J.I. Pena, C. Martinez, and M.A. Martinez, J. Electrochem. Soc. 134 (1987) 2119.
96. M.L. Shek, P.M. Stefan, I. Lindau, and W.E. Spicer, Phys. Rev B27 (1983) 7277.
97. G.K. Wertheim, J. Electron Spec. Rel. Phen. 47 (1988) 271.
98. D.M. Kolb, in: Trends in Interfacial Electrochemistry, A.F. Silva (ed.), NATO ASI Series, p. 301, D. Reidel Publishing Company, 1986.
99. D.M. Kolb and R. Kötz, Surf. Sci 64 (1977) 689.
100. N. Furuya and S. Motoo. J. Electroanal. Chem. 72 (1976) 165, Idem, ibid. 98 (1979) 195.
101. A. Nidola and R. Shira, J. Electrochem, Soc. 133 (1986) 1653.

102.. P. Bindra, H. Gerischer, and D.M. Kolb, J. Electrochem. Soc. 124 (1977) 1012.
103. J.C.F. Boodts and S. Trasatti, J. Appl. Electrochem, 19 (1989) 255.
104. P.M.A. Sherwood, in: Contemporary Topics in Analytical and Clinical Chemistry, D.M. Hercules et al. (eds.), Vol. 4, Plenum (1982) p. 205.
105. N.S. McIntyre, in: Applied Electron Spectroscopy for Chemical Analysis, H. Windawi and F.F.L. Ho, (eds.), p. 89, John Wiley & Sons, New York, 1982.
106. P. Brüesch, K. Müller, A. Atrens, and H. Neff, Appl. Phys. A38 (1985) 1.
107. P. Brüesch, K. Müller, and H.R. Zeller, Surf. Sci. 169 (1986) L327.
108. S.C. Tjong and E. Yeager, J. Electrochem. Soc. 128 (1981) 2251.
109. O.J. Murphy, J. O'M. Bockris, T.E. Pou, L.L. Tongson, and MD. Monkowski, J. Electrochem. Soc. 130 (1983) 1792.
110. O.J. Murphy, J. O'M. Bockris, T.E. Pou, D.L. Cocke, and G. Sparrow, J.Electrochem. Soc. 129 (1982) 2149.
111. I. Olefjord, B. Brox, in: Passivity of Metals and Semiconductors, M. Froment, (ed.), Elsevier, Amsterdam 1983, p. 561.
112. S.C. Tjong, R.W. Hoffman, and E.B. Yeager, J. Electrochem. Soc. 129 (1982) 1662.
113. S. Sarangapani, J.R. Akridge and B. Schumm, (eds.), in: The Electrochemistry of Carbon, Proceedings Vol. 84–5, The Electrochemical Society Inc., 1984.
114. J.S. Mattson and H.B. Mark, Activated Carbon: Surface Chemistry and Adsorption from Solution, Marcel Dekker, 1971.
115. C. Kozlowski and P.M.A Sherwood, J. Chem. Soc. Faraday Trans. I, 80 (1984) 2099.
116. K.M. Sundberg, Lj. Anastasoska, R. Anastasoski, and W.H. Smyrl. J. Electroanal. Chem. 220 (1987) 161.
117. J.H. Wandrass, J.A. Gardella, N.L. Weinberg, M.E. Bolster, and L. Salvati, J. Electrochem. Soc. 134 (1987) 2734.
118. F. Hine, M. Yasuda, and M. Iwata, J. Electrochem. Soc. 121 (1974) 749.
119. H.E. Hintermann, A.C. Riddiford, R.D. Cowling, and J. Malyszko, Electrodepos, Surf. Treat. 1 (1972) 59.
120. J.P. Randin, in: Encyclopedia of Electrochemistry of the Elements, Vol VII-1, A.J. Bard (ed.), Marcel Dekker, Inc., 1976.
121. C.M. Alloca, W.S. Williams, and A.E. Kaloyeros, J. Electrochem. Soc. 134 (1987) 3170.
122. M. Cappadonia and R. Kötz, to be published.
123. N.R. Armstrong and R.K. Quinn, Surf. Sci. 67 (1977) 451.
124. C.N. Sayers and N.R. Armstrong, Surf. Sci. 77 (1978) 301.
125. J.L. Stickney, S.D. Rosasco, B.C. Schardt, T. Solomun, A.T. Hubbard, and B.A. Parkinson, Surf. Sci. 136 (1984) 15.
126. T. Solomun, R. McIntyre, W. Richtering, and H. Gerischer, Surf. Sci. 169 (1986) 414.
127. H. Goslowsky, H.M. Kühne, H. Neff, R. Kötz, and H.J. Lewerenz, Surf. Sci. 149 (1985) 191.
128. Y.Mirovsky, R. Tenne, D. Cahen, G. Sawatzky, and M. Polak, J. Electrochem. Soc. 132 (1985) 1070.
129. K.J. Bachmann, S. Menezes, R. Kötz, M. Fearheily, and H.J. Lewerenz, Surf. Sci. 138 (1984) 475.
130. H.J. Lewerenz and R. Kötz, J. Appl. Phys. 60 (1986) 1430.
131. D.M. Novak, B.V. Tilak, and B.E. Conway, in: Modern Aspects of Electrochemistry. J. O'M. Bockris, B.E. Conway and R.E. White (eds.), No. 14, p. 195. Plenum Press, 1982.
132. D.M. Kolb: in: Advances of Electrochemistry and Electrochemical Engineering, H. Gerischer and C.W. Tobias (eds.) Vol. 11, p. 125, John Wiley and Sons, 1978.
133. B.E. Conway, in: Progress in Surface Science, S.G. Davison (ed.), Vol. 16, p. 1. Pergamon Press, New York, 1984.
134. W.E. O'Grady, J. Electrochem. Soc. 127 (1980) 555.
135. K. Kuroda, B.D. Cahan, G. Nazri, E. Yeager, and T.E. Mitchell, J. Electrochem. Soc. 129 (1982) 2163.
136. E. Rach and J. Heitbaum, J. Electroanal. Chem. 205 (1986) 151.
137. J.B. Goodenough, A. Hamnett, B.J. Kennedy, R. Manoharan, and S.A. Weeks, J. Electroanal. Chem. 240 (1988) 133.
138. O.K.T. Wu, G.G. Peterson, W.J. LaRocca, and E.M. Butler Applications Surf. Sci. 11/12 (1982) 118.
139. E.M. Stuve, K. Bange, and J.K. Sass, in: Trends in Interfacial Electrochemistry, A.F. Silva (ed.), NATO ASI Series C, Vol. 179. p. 255, D. Reidel Publishing Company, 1986.

Progress in the Study of Methanol Oxidation by *In Situ*, *Ex Situ* and On-Line Methods

T. Iwasita-Vielstich

Institut für Physikalische Chemie der Universität Bonn, D-5300 Bonn 1, Federal Republic of Germany

Contents

Introduction

Among the substances candidate for direct energy conversion, methanol has held the attention of electrochemists for more than three decades. Voltammetric methods have contributed to building a general picture of the electrochemistry of methanol and related systems [1, 2]. In fact, the investigations have been handicapped by the lack of methods giving unequivocal information on the nature of reaction intermediates and products. Nevertheless, skillful use of electrochemical [3, 4] and analytical instrumental methods [5–7] resulted in very valuable contributions to the present knowledge on the electrochemistry of small organic molecules.

Coupling an electrochemical cell to an analytical device requires that hindering technical problems be overcome. In the last years there has been a considerable improvement in the combined use of electrochemical and analytical methods. So, for instance, it is now possible to analyze on-line electrode products during the simultaneous application of different potential or current programs. A great variety of techniques are based on the use of UHV for which the emersion of the electrode from the electrolytic solution is necessary. Other methods allow the *in situ* analysis of the electrode surface i.e the electrode reaction may take place almost undisturbed during surface examination. In the present contribution we shall confine ourselves to the application of some of those methods which have been shown to be very valuable for the study of organic electrode reactions.

– IR reflectance allows the *in situ* analysis of the electrode-electrolyte interface [8, 9]. The Fourier transform variant adds to this technique the advantage of very fast data collection [10].
– On line mass spectroscopy is a very sensitive method capable of giving on–line responses in less than 0.2 s [11, 12].
– The recently developed *ex situ* analysis of electrode ad-layers by thermal desorption mass spectroscopy has been demonstrated to be a powerful tool for the study of adsorbates [13, 14].

These new experimental approaches gave renewed motivation to the study of classical organic electrode reactions for direct electrochemical energy conversion. The present contribution intends to give a survey of the recent progress in the study of methanol oxidation attained by application of the above mentioned techniques.

1 The Experimental Techniques

1.1 Electrochemical On-Line Mass Spectroscopy

The possibility of measuring electrochemical reaction products by connecting the electrochemical cell to a mass spectrometer was first suggested by Bruckenstein [7]. The experimental arrangement consisted of attaching the working electrode to a porous membrane of small pore size and non-wetting properties, which was used as the interface between the cell and the MS inlet. The delay of response for the gaseous products entering the mass spectrometer was about 20 s. A considerable improvement was achieved by Wolter and Heitbaum [11] through the use of a turbo molecular pump system. This allows a rapid introduction of products into the ionization chamber (the time response is ca. 0.2 s) and a rapid elimination of the gases after their detection. In this way potentiodynamic current and mass intensity profiles may be correlated without major distortions for scan rates not higher than 50 mV/s. In order to emphasize the fact that the mass signal is instantaneously changing in the same way as the current does, the authors called this technique Differential Electrochemical Mass Spectroscopy (DEMS).

A further development using digital data storage allows the simultaneous measurement of different masses [12].

A quantitative correlation between the charges under the current and mass intensity signals can be carried out as suggested by Heitbaum and Wolter [11]. The magnitude of the mass intensity response depends not only on the electrochemical properties of the system under study but also on the permeability of the electrode to the volatile products in addition to mass spectrometer parameters. A calibration of the actual experimental setup is therefore necessary. The proportionality between mass intensity (MI) and faradaic current (I) can be formulated as follows:

$$MI = K(1/nF) A I \tag{1.1}$$

where K is a constant including all mass spectroscopic parameters, n is the number of electrons necessary to produce one molecule of product, F is the Faraday constant and A is the current efficiency.

Eq. (1.1) can be written in terms of the charges Q_I and Q_{MI} obtained by integration of current and mass intensity signals:

$$Q_{MI} = K^*(1/n) A Q_I \tag{1.2}$$

K^* includes here also the Faraday constant.

Eq. (1.2) can be used to establish the stoichiometric composition of adsorbed substances which are oxidized to some volatile products. This can be done by determining the number of electrons, n, required to form one molecule of an observed product [15]. For this purpose, K^* first needs to be known. For instance, in the case of CO_2 as a product, the calibration can be done by measuring the current and mass response of adsorbed CO for which $n = 2$.

1.1.1 Instrumentation

Schematic representation of the experimental setup is shown in Fig 1.1. The electrochemical system is coupled on-line to a Quadrupole Mass Spectrometer (Balzers QMS 311 or QMG 112). Volatile substances diffusing through the PTFE membrane enter into a first chamber where a pressure between 10^{-1} and 10^{-2} mbar is maintained by means of a turbomolecular pump. In this chamber most of the gases entering in the MS (mainly solvent molecules) are eliminated, a minor part enters in a second chamber where the analyzer is placed. A second turbo molecular pump evacuates this chamber promptly and the pressure can be controlled by changing the aperture between both chambers. Depending on the type of detector used (see below) pressures in the range 10^{-4}–10^{-5} mbar, (for Faraday Collector, FC), or 10^{-7}–10^{-9} mbar (for Secondary Electrton Multiplier, SEM) may be established.

The analyzer consists of an ion source, a mass filter (quadrupole system) and an ion collector. In an FC detector the ions strike a collector where they give up their charge. The current is converted to voltage and is then amplified. In the case of an SEM detector the ions impinging upon the collector are further accelerated to some keV, thus increasing the current gain [16]. Since this mode of detection requires much lower pressures, larger amounts of products must be eliminated in the first chamber: The advantage of the higher current gain is greatly diminished.

After amplification the data can be directly recorded or digitalized and stored. Several mass signals can be detected during a given experiment. Depending on the type of detector used and on the sensitivity of the measurement, the mass spectrometer

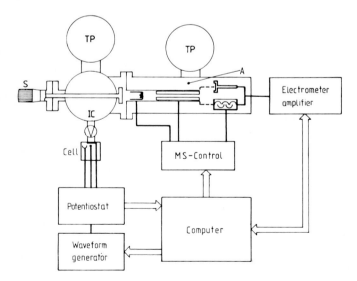

Fig. 1.1. Experimental setup for electrochemical on-line mass spectroscopic measurements with automatic data acquisition. TP = Turbo pump, IC = inlet chamber, A = analysis chamber, S = Screw mechanisms to control aperture between both chambers.

needs a given time to change from one mass to another. Typically, ca. 30 ms are required for the adjustment and registration of one mass intensity. If a potential scan of 10 mV/s is applied, then five different mass signals can be recorded in a potential range of 1.5 mV.

1.1.2 Cells, Electrodes

The cell shown in Fig. 1.2 was designed for on-line mass spectrometric measurements and electrolyte exchange under potential control. The solutions are filled through a glass tube into PTFE pot of 1.5 cm³, and the liquid (in excess) flows into a drain channel. Although solutions are deaerated before filling the cell, a capillary for additional inert gas bubbling is provided to stir the solutions during electrolyte exchange and to eliminate traces of air contamination diffusing through the joints.

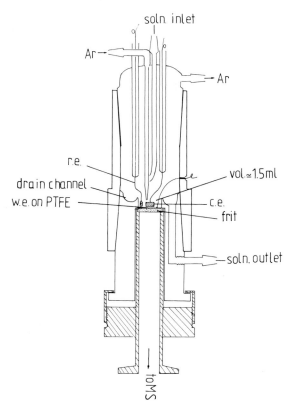

Fig. 1.2. Flow cell for adsorption measurements and on-line mass spectroscopy. The working electrode (w.e.) at the bottom of the cell is conneced directly to the MS; c.e. = counter electrode, r.e. = reference electrode.

Electrodes are placed well below the maximum liquid level so that solution exchange can be done under potential control. The working electrode is a porous noble metal layer prepared by applying a thin layer of a finely divided metal suspension on a PTFE membrane. Roughness factors in the range of 10 to 50 are the result, depending on the electrode thickness. The PTFE membrane with the metal layer facing the solution is fixed to a frit at the entrance to the mass spectrometer. Volatile solution components and gaseous reaction products can traverse the membrane and enter into the mass spectrometer. In order to minimize diffusion effects in the pores the electrodes should be made as thin as possible.

Heitbaum and co-workers are developing a porous rotating electrode for on-line mass spectroscopy [17]. In this way definite mass tranfer conditions in the solution can be established. Recent results on the hydrogen evolution reaction on platinum show H_2 mass intensity curves exhibiting a linear "limiting current" $-\omega^{1/2}$ behavior for low rotational speeds (up to 50 Hz) [17]. For studying fast reactions, however, diffusion in the pores cannot be neglected and besides this, higher rotational rates are desirable.

In principle, different reference electrodes may be used if the cell is provided with a separate compartment and a Luggin capillary. But if the flow cell technique is to be applied, it is more convenient to avoid the use of capillaries where the solution cannot be easily exchanged. Active bulk components could diffuse through the capillary and give rise to erroneous responses. A small palladium gauze charged with hydrogen directly immersed in the solution can be used as the reference electrode (PdH) [18].

A metal sheet pressed against the wall in the unique cell compartment is used as the counter electrode.

1.2 Electrochemical Thermal Desorption Mass Spectroscopy (ECTDMS)

This recently developed technique [13, 14] was applied to the analysis of adsorbates on electrodes after tranfer of the electrode from the electrochemical cell to the UHV analysis chamber of a quadrupole mass spectrometer.

1.2.1 Instrumentation

A schematic representation of the experimental arrangement is given in Fig. 1.3. The electrode is first transferred to a vacuum chamber where the liquid film covering its surface is eliminated by pumping down to 10^{-6} mbar with a turbo molecular pump system, and then to the UHV-chamber ($p = 10^{-8}$–10^{-9} mbar). Here the desorption is achieved by heating the electrode with the radiation of a white light beam, and the desorbing particles are focused to a grid ion source of a quadrupole mass spectrometer.

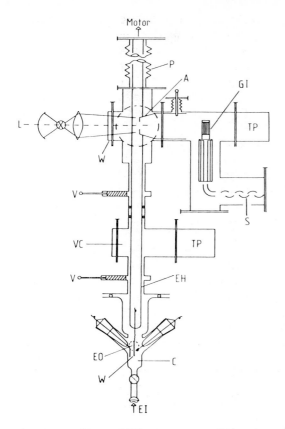

Fig. 1.3. Experimental setup for electrochemical thermal desorption mass spectroscopy (ECTDMS). C = electrochemical cell, W = working electrode, EI = electrolyte inlet, EO = electrolyte outlet, EH = electrode holder, V = valve, TP = turbo pump, VC = vacuum chamber, L = light source, W = window, P = protective jacket, A = aperture to analysis chamber, GI = grid ion source, S = SEM detector.

In principle, the analysis of molecules, ions and adsorbed intermediates is possible if they survive the emersion (no potential control) and UHV conditions (elimination of most of the solvent). The use of *ex situ* methods for the analysis of sub-monolayer quantities of oxygen-sensitive substances requires an extremely inert atmosphere when the electrode is emersed. In order to check whether a given adsorbate survives the experimental conditions, a control experiment must be carried out, as we describe here for adsorbed CO on Pt.

After adsorption of CO and solution exchange with pure base electrolyte, the oxidation of adsorbed CO during a triangular potential scan is observed (see Fig. 1.4a). In a second run after adsorption of CO the electrode is emersed and transferred to the UHV chamber in the same way as in the normal experimental procedure. The electrode is then transferred back to the cell and re-immersed in the base electrolyte. A potential scan is applied to oxidize the adsorbate. Fig. 1.4b shows

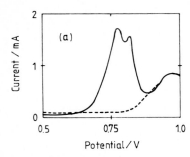 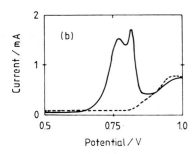

Fig. 1.4. Test for the survival of adsorbed CO on Pt after transfer in the UHV. *i–E* curve during the oxidation of CO_{ad} in 0.1 M H_2SO_4 (a) immediately after elimination of CO in the bulk; (b) after elimination of bulk CO and transfer in the UHV.

that in this experiment only minor changes occurred during the transfer procedure. The areas under the oxidation peaks are the same within the experimental error for the integration ($\pm$ 5%).

1.2.2 Cells, Electrodes

A one compartment flow cell of small volume (3–5 cm^3) may be appropriate for the ECTDMS technique. The main requirement is to purge exhaustively the whole system with an inert gas to avoid any contact between adsorbate and air. For this purpose glass or stainless steel tubes are recommended and joints must be tight and reduced to a minimum.

Electrodes of 2×2 cm^2 geometric area were used in the experiments. In the case of platinum, a pretreatment by fast triangular potential scans (200–300 V/s between 0.05 V and 1.5 V RHE) followed by heating up to 900 K in a 3×10^{-6} mbar O_2 atmosphere was carried out. Electrodes, pre-treated in this way, can be emersed with a thin liquid film, which can easily be evaporated in the vacuum chamber. The heat treatment drastically reduces contamination of the platinum electrode by carbon. The roughness factor is usually in the order of three.

1.3 *In Situ* IR Study of Electrode-Electrolyte Interfaces

In situ IR spectroscopy, a relatively new thin-layer technique, can identify, in principle, molecules with dipolar groups formed on electrodes during adsorption, oxidation or reduction of active species. The main experimental problem in measuring spectra of an electrode-electrolyte interface is associated with absorption of most of the IR radiation by the solvent (in reflectance spectroscopy). Even using a window-electrode distance of ca. 1 μm, the number of solvent layers is still 10^3. If the solvent is water the problem is particularly serious since water absorbs strongly in broad mid-IR regions. The solvent absorbance must be subtracted and the sensitivity to signals of surface species must be increased. This can be done by using a polarized IR beam.

The electromagnetic radiation may be separated into two components:

- linearly polarized light with the electric field vector parallel to the plane of reflection, "p-polarized" radiation and
- linearly polarized light with the electric field vector perpendicular to the plane of reflection, "s-polarized" radiation.

While s-polarized radiation approaches a phase change near 180° on reflection, the change in phase of the p-polarized light depends strongly on the angle of incidence [20]. Therefore, near the metal surface (in the order of the wavelength of IR) the s-polarized radiation is greatly diminished in intensity and the p-polarized is not [9]. This surface selection rule of metal surfaces results in an IR activity of adsorbed species only if $\delta\mu/\delta q \neq 0$ (μ = dipole moment, q = normal coordinate) for the vibrational mode perpendicular to the surface.

The problem of solvent absorption can be overcome by measuring the change in reflectivity of the electrode either by (a) modulating the state of light polarization between p-polarized and s-polarized radiation, or (b) using p-polarized radiation and taking spectra at two different electrode potentials.

In the first case, the difference in intensities $I_p - I_s$ is computed. Due to the surface selection rule, what results is a spectrum showing absorption bands of species on the surface.

In the second case the potential can be modulated between two values (a reference and a sample potential) while the spectral frequencies are slowly scanned, or else the spectral data can first be collected at a reference potential, after which this is stepped to the sampling value where a second spectrum is obtained. The change in reflectivity $\Delta R/R$ is then computed,

$$\Delta R/R = (R_2(v) - R_1(v))/R_1(v) \tag{1.3}$$

$R_1(v)$ and $R_2(v)$ are the reflectivities at the reference and sample potentials, respectively.

A change in potential can cause any of several effects, including migration of ions into or out of the thin layer, adsorption, desorption, and faradaic reactions consuming or producing species adsorbed on the surface or in solution. For these reasons a difference spectrum (see Eq. (1.3) can exhibit both negative bands due to species formed and positive bands due to species consumed at the sample potential.

The bond strengths of adsorbed species can be affected by a change of the electric field at the interface. In this case, shifts of the adsorbate vibrational frequencies are also observed [21]. According to Eq. (1.3) the frequency shift gives rise to bipolar bands (i.e., bands exhibiting both negative and positive parts).

1.3.1   The Different IR Technique Variations

A number of newly introduced abbreviations describes the special techniques used which depend also on the IR instrumentation (conventional dispersive optics or Fourier transform devices).

Electrochemically Modulated Infra-Red Spectroscopy (EMIRS) [23] consists of applying a square-wave potential modulation to the working electrode and analyzing the modulated part of the IR detector response using a dispersive instrument.

Linear Potential Sweep Infra-Red Spectroscopy (LPSIRS) is a variation of EMIRS techniques where the absorbance at a particular wavenumber is monitored during a potential sweep.

The use of Substractively Normalized Interfacial Fourier Transform Infra-Red Spectroscopy (SNIFTIRS) [24-26] allows very fast data collection. Polychromatic IR radiation reaches the electrode after passing through a Michelson interferometer. The data are recorded in the form of an "interferogram" of intensity of radiation versus the moving mirror position. One interferogram can be taken in fractions of a second and contains the encoded IR spectrum across the entire wavelength range. Signal averaging improves the signal to noise ratio. This averaging can be done in two ways:

– Few (4 to 8) [26] spectrometer scans are taken, averaged and stored while the electrode potential is held at a "reference" potential and then at a "working" potential. In this way only a short time is passed between the transition from the reference to the working condition. The sequence is repeated many times in order to improve the signal-to-noise ratio.

– A base interferogram is recorded at the reference potential applying a sufficient number of scans for averaging. Then the potential is stepped up to the working potential for the same number of scans [27, 30]. For this more simple procedure the abbreviation SPAIRS, Single Potential Alteration Infra-Red Spectroscopy has been suggested recently [30]. Instead of taking scans at a set working potential one can, of course, sample scans during a potential sweep as well and then integrate the signals obtained over a certain "working potential range" [30].

Polarization Modulated Infra-Red Reflection Absorption Spectroscopy (PMIRRAS or IRRAS or FT-IRRAS for Fourier transform variant) involves a modulation between s- and p-polarization of incident light at the set working potential [31]. A photo-elastic modulator is used to modulate the IR beam between the two polarization modes, e.g. at 80 kHz. This technique has the advantage of a very low signal to noise ratio but also a major drawback: the intensity of the reflected beam is different for s- and p-polarized light and the intensity difference $I_p - I_s$ contains information on both the surface layer and the attached solvent region [31]. This is problematic for highly absorbent solvent layers like aqueous solutions and reduces the advantages of PM techniques to application in suitable wave number regions.

The IR investigations described in this presentation are based on the SPAIRS variant of SNIFTIRS.

1.3.2 Instrumentation

Schematic representation of the experimental setup for an *in situ* IR study of the electrode-electrolyte interface is given in Fig. 1.5. From the radiation leaving the IR source only, the p-polarized light is used for the reflection-absorption experiment in

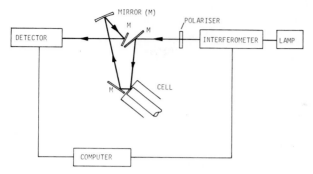

Fig. 1.5. Simplified block diagram of the experimental setup for SNIFTIRS.

the electrochemical cell. Taking into account the dependence of the electric field strength at the surface on the angle of incidence, values between 50° and 80° for this angle are chosen in the optical bench arrangement. IR windows of Si, CaF$_2$ or ZnSe material are used. The working electrode has to be carfeully polished and positioned parallel to the window in a distance of a few μm. Due to the small $\Delta R/R$ values of only 10^{-3} (potential modulation) to 10^{-2} (polarization modulation) a liquid nitrogen-cooled high sensitivity detector is used.

The experiments described here were performed with a Digilab FTS40 Fourier transform instrument, with a liquid nitrogen-cooled Mercury Cadmium Telluride, (MCT), detector. The instrument is provided with a computer for data acquisition, storage and mathematical treatment. P-polarized incident light was obtained by means of an Al wire-grid polarizer supported on a BaF$_2$ substrate.

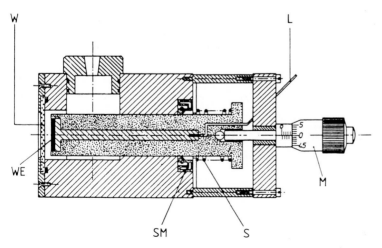

Fig. 1.6. Thin layer cell for *in situ* IR absorption–desorption measurements based in a model proposed by Seki et al. [32]. WE = working electrode, W = window, L = lead, M = micrometer screw, S = steel spring, SM = simmer gasket.

1.3.3   Cells, Electrodes

The working electrodes, 15 mm diameter discs of poly-crystalline platinum fixed in an araldite body were mechanically polished down to 0.05 μm, dipped in diluted HNO_3, followed by rinsing with an ultrasonic application.

The necessary thin-layer configuration was obtained by pressing the electrode against the optical window. Different cell constructions have been used. Fig. 1.6 shows a cell similar to one used by Seki et al. [32]. The micrometer screw attached to the working electrode allows a reproducible adjustment of electrode-window distances.

2   The Electrochemical Oxidation of Methanol

The most simple carbon containing fuels such as methanol, formaldehyde and formic acid present some thermodynamic properties which make them attractive as potential substances for direct electrochemical energy conversion. An inspection of the data in Table 2.1 shows that the standard open circuit potentials of fuel cells working with these substances are comparable to that of a hydrogen/oxygen fuel cell or even higher. For a 100% conversion to CO_2 they present also a high energy content per mole. Although the potential of these properties was recognized very early (1928 [33]), scientists didn't succeed in surmounting the barrier of kinetic problems of the electrochemical oxidation of methanol and related compounds, particularly in acidic media.

It is very well known that Pt is one of the best metal catalysts for hydrogen as well as for organic oxidations. Nevertheless, a comparison of the electrochemical behavior of hydrogen and any of these organic substances shows large differences. While hydrogen establishes its reversible thermodynamic potential with platinum in an aqueous acidic solution very quickly, the reversible potential of the other fuels could never be experimentally observed.

The behaviour of hydrogen and methanol at constant oxidation rates can be compared in Fig. 2.1. Using platinized platinum anodes a constant current density was applied in a 0.5 mol/L^{-1} H_2SO_4 solution containing (a) H_2 ($p = 1$ bar) and (b) 1 mol L^{-1} CH_3OH solution. While no more than 20 mV overpotential are

Table 2.1. Theoretical emf values at 25 °C [2].

Fuel	Reaction	$\Delta G°/$ kcal/mol	$\Delta E°/$ mV	Standard potential/nmV
Methanol	$CH_3OH + 3/2\ O_2 \rightarrow CO_2 + 2H_2O$	-166.77	1210	$+20$
Formaldehyde	$CH_2O + O_2 \rightarrow CO_2 + H_2O$	-124.7	1350	-120
Formic acid	$HCOOH + 1/2\ O_2 \rightarrow CO_2 + H_2O$	-68.2	1480	-250
Hydrogen	$H_2 + 1/2\ O_2 \rightarrow H_2O$	-56.69	1230	0

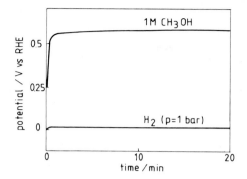

Fig. 2.1. Electrode potential during the oxidation of methanol and hydrogen on a platinized platinum electrode (real area $= 130 \, cm^2$) at constant current $i = 5$ mA. Base electrolyte: $= 0.5$ M H_2SO_4, room temperature.

required for the hydrogen reaction to be kept at a constant level of 5 mA, a potential shift of several hundred mV is necessary to keep methanol oxidation at the same rate. A technology using hydrogen in a fuel cell was relatively quickly developed and the acidic H_2/O_2 reaction on platinum catalysts runs successfully at MW levels in power plants [34]. On the other hand, the search for better catalysts for the methanol reaction is still a matter for investigation at the laboratory level.

The reason for this is that platinum is able to adsorb organic compounds dissociatively forming adsorbed hydrogen and organic residues and the oxidation mechanism of the latter involves extremely slow steps.

A cyclic voltammogram of a platinum electrode in $0.01 \, mol \, L^{-1}$ methanol solution is shown in Fig. 2.2 (full line). Its main features are very well known:

- the inhibition of hydrogen adsorption due to adsorbed organic residues,
- the relatively low oxidation currents in the double layer region during the potential scan in the positive direction and the reactivation peak during the negative scan,
- the oxidation peaks at ca. 0.85 V and 1.4 V.

Comparable patterns are followed by other organic substances such as formaldehyde and formic acid. All these substances are dissociatively adsorbed on platinum [4] and it was suggested that they build the same adsorption product [35].

Figure 2.2 shows also the oxidation pattern of adsorbed methanol in the absence of methanol in the solution (dashed curve). The experiment was performed using the flow cell procedure [36]. Methanol was adsorbed from a 10^{-2} M $CH_3OH/0.5$ M H_2SO_4 at 250 mV RHE for 10 min, then the solution was exchanged with pure base electrolyte under potential control, and a potential scan was applied.

Adsorbate oxidation at a reasonable rate on pure Pt electrodes occurs at potentials above ca. 500 mV RHE. The partial blockage of the surface is the reason for the low currents observed below this potential.

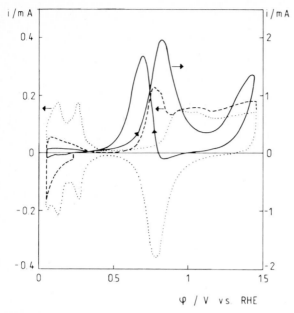

Fig. 2.2. Cyclic voltammogram of a polished Pt electrode in 10^{-2} M $CH_3OH/0.1$ M $HClO_4$ solution, (full line) and potentiodynamic oxidation of methanol adsorbate after solution exchange with base electrolyte (dashed line). Sweep rate: 60 mV/s, room temperature.

Besides this, the inhibiting effects are dependent on the nature of the Pt surface as has been demonstrated by investigations on single crystals [37, 38] and on polycrystalline material with preferred surface orientation [39].

The identification of adsorbed residues has received very much attention for more than 20 years. For methanol the following reactions have been proposed [40-43]:

$$CH_3OH \begin{cases} HCO_{ad} + 3\,H^+ + 3\,e^- & (2.1) \\ COH_{ad} + 3\,H^+ + 3\,e^- & (2.2) \\ CO_{ad} + 4\,H^+\,4\,e^- & (2.3) \end{cases}$$

For the final reaction to CO_2 an additional oxygen atom is needed. This O atom can be obtained only from a water molecule out of the electrolyte:

$$HCO_{ad} + H_2O \rightarrow CO_2 + 3\,H^+ + 3\,e^- \qquad (2.4)$$

$$COH_{ad} + H_2O \rightarrow CO_2 + 3\,H^+ + 3\,e^- \qquad (2.5)$$

$$CO_{ad} + H_2O \rightarrow CO_2 + 2\,H^+ + 2\,e^- \qquad (2.6)$$

Knowledge of the nature of adsorbed residues which act as a poison to the catalyst can orient the search for catalysts which may act either promoting the oxidation of the organic adsorbate or avoiding its formation by favoring alternative pathways.

2.1 The Adsorbed Intermediates

Different experimental approaches were applied in the past [6, 45] and in recent years [23, 46] to study the nature of the organic residue. But the results or their interpretation have been contradictory. Even at present, the application of modern analytical techniques and optimized electrochemical instruments have led to different results and all three particles given above, namely HCO, COH and CO, have been recently discussed as possible methanol intermediates [14, 15, 23, 46, 47]. We shall present here the results of recent investigations on the electrochemical oxidation of methanol by application of electrochemical thermal desorption mass spectroscopy (ECTDMS) on-line mass spectroscopy, and Fourier Transform IR-reflection-absorption spectroscopy (SNIFTIRS).

2.1.1 Examination of Adsorbed Layers by Thermal Desorption Mass Spectroscopy

2.1.1.1 Adsorption of CO and H_2 from the Gas Phase as a Test
for the Experimental Setup

Thermal desorption spectra of carbon monoxide on polycrystalline and on single crystal platinum are well known from experiments in the gas phase [48, 49]. The system is therefore appropriate to test the experimental setup.

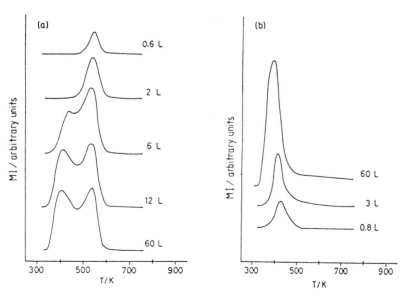

Fig. 2.3. Thermal desorption spectra after adsorption from the gas phase (a) adsorbed CO on Pt; (b) H_2 on Pt.

After cleaning the Pt electrode by heating it in an O_2-atmosphere (see Section 1.2) and in an UHV, variable quantities (between 0.6 L and 60 L) of carbon monoxide were adsorbed from the gas phase [50]. The thermodesorption spectra obtained are shown in Fig. 2.3a. Two different Pt–CO bonding states can be observed which are consecutively occupied. The places showing the stronger adsorption state (desorption peak at 530 K) are saturated first. At 6 L the second adsorption state can be recognized. This grows developing a well defined desorption peak at 420 K. These spectra are in good agreement with those reported in the literature [48, 49].

The results of a similar experiment with adsorbed hydrogen is shown in Fig. 2.3b. Only one desorption peak was observed in the temperature range studied [50]. The desorption peak temperature lies at 420 K for the experiment with 0.8 L and is shifted to lower temperatures as the H_2 concentration increases indicating second order desorption kinetics. Surface states with desorption temperatures at 165 K, 220 K, 280 K and 350 K were reported for the adsorption of H_2 and D_2 at 120 K [51]. Thermal desorption experiments after H_2 adsorption at 350 K show only one desorption state at ca. 450 K [52].

2.1.1.2   Adsorption from the Electrolyte Solution

A platinum electrode pretreated in the way as is described in Section 1.2 may show some minimal desorption of carbonaceous residues which may come from C-atoms diffusing from the bulk of Pt or from the rest of the gas in the UHV. A blank desorption experiment carried out by transferring a Pt electrode which was held at 450 mV in H_2SO_4 for 120 s is shown in Fig. 2.4(a).

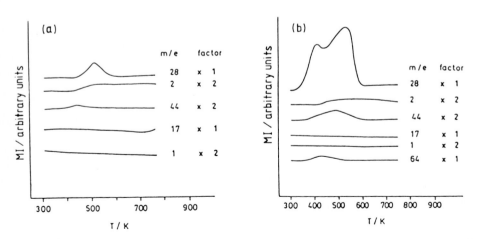

Fig. 2.4. (a) Thermal desorption blank experiment. The Pt electrode was held at 0.45 V vs. RHE in the base electrolyte (5×10^{-2} M H_2SO_4) during 120 s and then transferred to the UHV. (b) Thermal desorption spectra of adsorbed CO on Pt after adsorption from an aqueous solution. Temperature scan: 5 K/s.

A small peak for CO ($m/e = 28$) and a weak desorption signal for H_2 ($m/e = 17$ and 2) can be observed. No signals for adsorbed water ($m/e = 17$ and $m/e = 1$) are observed.

As we have shown in Section 1.2.1 carbon monoxide adsorbed on platinum can be transferred from the electrochemical cell to the UHV without detectable faradaic loss (see Fig. 1.4). Therefore this system can be taken as a model for the application of ECTDMS to the analysis of organic adsorbates.

Adsorbed carbon monoxide on platinum formed at 455 mV in H_2SO_4 presents a thermal desorption spectrum as shown in Fig. 2.4b. As in the case of CO adsorption from the gas phase, the desorption curve for $m/e = 28$ exhibits two peaks, one near 450 K for the weakly adsorbed CO and the other at 530 K for the strongly adsorbed CO species. The H_2 signal remains at the ground level. A slight increase in CO_2 concentration compared to the blank is observed, which could be due to a surface reaction with ions of the electrolyte. Small amounts of SO_2 ($m/e = 64$) are also observed.

2.1.1.3 Thermal Desorption of Methanol Adsorbate

In order to check the survival of methanol adsorbate to the transfer conditions, the following experiment was performed. After adsorption of methanol and solution exchange with base electrolyte, the Pt electrode was transferred to the UHV chamber over a period of ca. 10 min, then back to the cell where it was reimmersed into the pure supporting electrolyte. A voltammogram was run and compared with that of an usual flow cell experiment. The results, (see Fig. 2.5a,b), show that the transfer procedure is valid. The areas under the oxidation curve are the same. As in the case of adsorbed CO on Pt (see Fig. 1.4), the change in the double peak structure indicates that some surface re-distribution may occur.

The thermal desorption mass spectra of methanol adsorbate obtained from 5×10^{-3}M CH_3OH + 0.05M H_2SO_4 and in 0.5 M CH_3OH + 0.5 M H_2SO_4, are

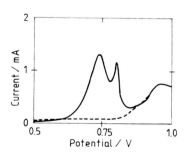
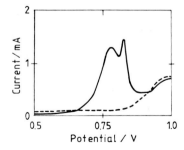

Fig. 2.5. Test for the survival of methanol adsorbate in the UHV: the potential scan was applied a) after adsorption in 5×10^{-3} M $CH_3OH/5 \times 10^{-2}$ M H_2SO_4 followed by electrolyte exchange with base solution; (b) after adsorption, transfer in the UHV chamber and reimmersion in base electrolyte; dotted line base voltammogram; sweep rate: 62.5 mV/s.

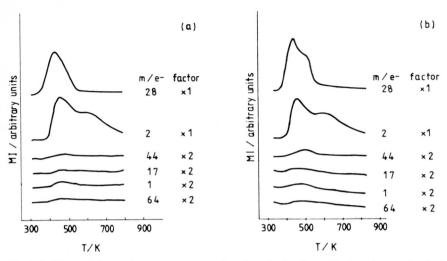

Fig. 2.6. Thermal desorption mass spectra of methanol adsorbate. Methanol was adsorbed at 0.4 V for 180 s from (a) 5×10^{-3} M CH_3OH + 0.5 M H_2SO_4 solution and b) 0.5 M CH_3OH + 0.5 M H_2SO_4 (b). The m/e signals corresponds to: CO (28), CO_2 (44), H_2O (17 and 1) and SO_2 (64); temperature scan: 5K/s.

shown in Fig. 2.6a,b. The mass signals for CO, ($m/e = 28$), show two distinct desorption states at $T_1 = 435$ K and $T_2 = 500$ K. For hydrogen, ($m/e = 2$), one desorption state at $T_1 = 455$ K is resolved in the temperature region where CO desorption occurs. Another peak is observed near 600 K. This shift of hydrogen desorption to higher temperatures and the appearance of other desorption states in this temperature interval is well known from studies of co-adsorption of H_2 and CO on polycrystalline platinum [48, 52] and Pt(111) [53]. This phenomenon was explained *via* a (COH) surface complex [48, 49]. A comparison with the blank experiment (Fig. 2.4a) indicates that H_2 is not produced by decomposition of co-adsorbed H_2O.

The signals for CO_2 ($m/e = 44$), for the fragments of desorbed H_2O ($m/e = 1$ and 17) and for SO_2 ($m/e = 64$) are very small. Other masses were not observed.

During the thermal desorption of adsorbed COH or HCO, CO and H_2 are produced. The CO desorption signal represents therefore the contributions of both (C, O) and (C, O, H) adsorbates. Taking into acccount that the area under the desorption curve is proportional to the integral of the partial pressure during the desorption and, consequently, to the surface concentration of the desorbing species, a calibration of the thermal desorption curve with known partial pressures of the corresponding gases was possible. Thus mole fractions were calculated and a quantitative interpretation of the data has been attempted [14].

Results of such evaluations are given in Fig. 2.7. Mole fractions are plotted as a function of the apparent degree of coverage θ' defined as $\theta' = Q_{CO}/Q_{COmax}$. Q_{COmax} and Q_{CO} are the areas under the desorption traces for CO at maximum coverage and at the actual coverage in the corresponding experiment, respectively.

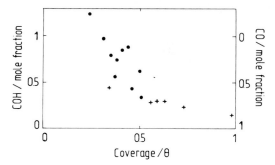

Fig. 2.7. Mole fraction of methanol adsorbate obtained from ECTDMS measurements as a function of coverage (see text); ● 5×10^{-3} M CH_3OH, + 0.5 M CH_3OH, 0.1 M H_2SO_4.

According to these results methanol adsorbate seems to consist of a mixture of (C, O) and (C, O, H) particles, the actual ratio depending on the concentration and total degree of coverage. This is in good agreement with coulometric determinations of the charge ratio for methanol adsorption, Q_{ad}, (see Eqs. 2.1 to 2.3) and adsorbate oxidation, Q_{ox} (see Eqs. 2.4 to 2.6) [14, 47]. These results will be discussed in Section 2.1.4.

The position of the H-atom in the (C, O, H) residue needs further investigation. A possible way is the use of isotope labeling to distinguish between hydrogen from the $-CH_3$ group from that of the $-OH$ group. Labeled chemicals have been successfully used to facilitate the interpretation of on-line mass spectra during the oxidation of formic acid [54], ethanol [55], urea [56] and glucose [57]. In the case of methanol adsorbate the investigation required the use of a flow cell to eliminate bulk contributions. The results are presented and discussed in the next section.

2.1.2 Investigation of Methanol Adsorbate by On-Line Mass Spectroscopy and Isotope Labeling

Charge measurements, as mentioned above, were also performed using the porous Pt electrodes required by the on-line MS technique. At low methanol concentrations $(10^{-2}$ M), the charge ratio Q_{ad}/Q_{ox}, near 1 indicates that (C, O, H) must be the predominant adsorbate composition [14, 47]. This result is in good agreement with that of Heitbaum and co-workers [15] who used Eq. 1.2 to determine the number of electrons, n, per CO_2 produced from methanol adsorbate. They found for n a value of 3, which would be in agreement with reactions 2.1 or 2.2 for methanol adsorption.

According to Eqs. (2.4 to 2.6) during adsorbate oxidation hydrogen ions are produced. These are partially formed from the splitting of water molecules but they may come also from the adsorbate. In principle a distinction can be made between both sources of hydrogen ions if, for instance, methanol labeled with deuterium is used and all other components of the solution contain light hydrogen. In considering the results of such experiments it must be taken into account that hydrogen atoms associated to the methyl group do not undergo isotopic exchange with the solution while those of the −OH group do.

In a previous experiment the conditions were optimized to improve the sensitivity for HD detection [47]. At first a potential step program was applied to the electrode in the solution containing 10^{-2} M CD_3OH in the base electrolyte (10^{-4} M $HClO_4$ + 0.1 M $NaClO_4$). During a first pulse to a potential of 0.97 V vs. a Pd-H electrode D^+ ions were formed because of the bulk oxidation of CD_3OH. After 0.5 s, a second pulse to −0.44 V was applied and the mass intensity signal for HD was observed. The result is shown in Fig. 2.8. For comparison, a blank experiment was performed by switching the potential from open circuit to −0.44 V (i.e. without methanol oxidation). This experiment demonstrates that the detection of HD (if formed), is possible within tenths of a second.

A flow cell procedure was then applied. The experiment consisted of (a) adsorption of methanol (in a solution containing deuterated methanol and light hydrogen base electrolyte), (b) solution exchange with base electrolyte and (c) application of two potential steps, one of short duration to oxidize the adsorbed residue and then a second one in the negative direction to reduce the ions H^+ and/or D^+ formed. During this time the mass intensity signals for HD, ($m/e = 3$) and for CO_2 ($m/e = 44$) were

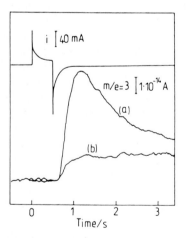
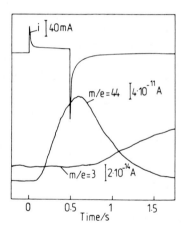

Fig. 2.8. (a)On-line mass spectroscopic detection of HD ($m/e = 3$) produced by application of a potential step to 0.97 V vs. PdH to a Pt electrode in 10^{-2} M $CD_3OH/10^{-4}$ M/$HClO_4$/0.1 M $NaClO_4/H_2O$ during 0.5 s, followed by a potential step to −0.44 V vs. Pd-H. For comparison: HD signal without methanol oxidation when switching the potential from open circuit to H_2 evolution at −0.44 V (b). Upper part: recording of current.

Fig. 2.9. Current transient and mass signal responses during oxidation of methanol adsorbate in pure base electrolyte (flow cell procedure). Methanol was adsorbed from a 10^{-2} M CD_3OH + 10^{-4} M $HClO_4$ + 0.1 M $NaClO_4$. $E_{ad} = 356$ mV; $t_{ad} = 400$ s. Potential step to 975 mV vs. Pd-H for 0.5s to produce CO_2 ($m/e = 44$) and hydrogen ions, followed by a step to −574 mV vs. Pd-H to detect HD ($m/e = 3$).

monitored. Since one is working with monolayer quantities of adsorbed substance, the mass for CO_2 ($m/e = 44$) was also taken to prove that the potential and the length of the oxidation step were enough to produce detectable quantities of products. Fig. 2.9 shows both signals ($m/e = 44$ and $m/e = 3$). A comparison with a blank experiment indicates that the HD mass signal observed was produced from the natural deuterium content in the system.

These results lead to the conclusion that the H-containing adsorbate can only be COH, i.e. hydrogen must be bound to the oxygen atom and can, therefore, exchange with bulk components.

2.1.3   FTIR Measurements of Adsorbed Methanol

Most results of *in situ* IR studies on Pt in acidic methanol solutions so far have been obtained using a relatively fast (8.5–13.6) Hz) modulation of electrode potential. As already pointed out by Bockris [27], collection of spectral data alternatively at two potentials is not appropriate for processes which are not reversible to follow the change of potential. In this study the SPAIRS version of SNIFTIRS was performed by stepping the potential from a reference potential in the anodic direction, allowing sufficient time at each potential to reach stationary conditions.

The working electrode was activated by repetitive cycling between the onset of hydrogen and oxygen evolution in the 0.1 M $HClO_4$ base electrolyte. Activation in the presence of methanol was avoided because the CO_2 produced can form reduced CO_2 at potentials in the hydrogen region. For the same reason not more than one series of spectra at potentials higher than 0.4 V was taken in a given solution. Spectra taken in the base electrolyte at potentials between 0 V and 1.5 V vs. RHE were used as a criterion for the absence of impurities with optical IR activity.

For adsorption, the potential was held at 0.35 V in the base electrolyte. The methanol containing solution (from 0.01 M to 5 M) was allowed to flow into the cell. After 15 min the electrode was pushed against the window (CaF_2). The measurements started after a sufficient purging of the gas atmosphere in the IR box. Spectra were taken at potentials between 0 V and 1.5 V RHE with a delay of 1 min after setting each potential.

Fig. 2.10 shows spectra taken in a 10^{-2} M CH_3OH solution. The bipolar band[1] near 2060 cm^{-1} corresponds to the C–O stretching vibration of linearly adsorbed CO (l–CO) [23]. Another band is observed at ca. 1870 cm^{-1}. According to data for the gas/solid phase [58] this band was assigned to the C–O stretching vibration of bridge bonded CO. Recently Beden et al. also observed this band by applying a 13.6 Hz modulation of the electrode potential [59]. In analogy to the interpretation of data for CO adsorption at the gas/solid phase [60], Pons et al. [61] suggested that a band at 1850 cm^{-1} corresponds to CO triple coordination. On Pd the band for bridge bonded CO is at ca. 1980 cm^{-1}. The lower CO frequency (1850 cm^{-1}) observed on Pt could be

---

[1] The formation of a bipolar band is due to a shift of the absorption frequency with the potential and the procedure used to evaluate the relative reflectance. (see Eq. 1.3).

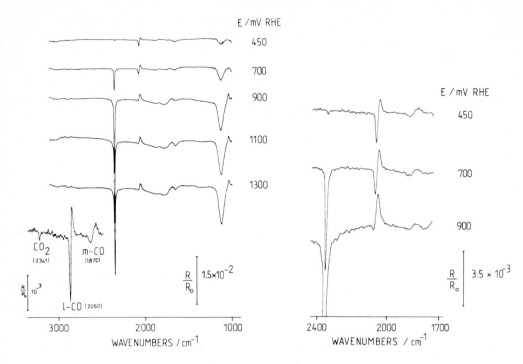

Fig. 2.10. Potential step SNIFTIRS spectra from a polished polycrystalline Pt electrode, immersed in 10^{-2} M $CH_3OH/0.1$ M $HClO_4$ electrolyte. All spectra (90 scans each, 8 cm^{-1} resolution) were normalized to the base spectrum collected at 0 mV vs. RHE. Insert: part of the curve at 450 mV in expanded scale.

Fig. 2.11. Potential step SNIFTIRS spectra from a polycrystalline crystalline Pt electrode, in 10^{-2} M $CH_3OH/0.1$ M $HClO_4$. 2400–1700 cm^{-1} spectral region showing the changes in the bands for linear and multi bonded CO on Pt.

attributed to a stronger CO adsorption due to a triple-coordinated species [62]). As no definitive proofs are available for the one or the other form we prefer to call this CO species multi-bonded carbon monoxide (m-CO).

Both bands on Pt have a bipolar character up to a potential of 700 mV (Fig. 2.11). In these spectra the formation of CO_2 as final product can also be observed (band at 2341 cm^{-1}). The intensity of the linear adsorbed CO band is slightly higher at 700 mV than at 450 mV while that of the multi-bonded CO slightly decreases. Nevertheless in both cases, the bipolar character is still observed at 700 mV. The amount of bulk methanol in the thin layer is already consumed at 900 mV and only the positive going bands[2] for CO at 2045 cm^{-1} and 1823 cm^{-1} are then observed. This is not the case in

---

[2] Substances formed at a given potential give rise to a diminution of the reflected intensity as compared with that at the reference potential. As a consequence, the relative reflection band presents a minimum ("negative going band"). On the contrary, the reflected intensity for substances consumed at a given potential present a maximum ("positive going band").

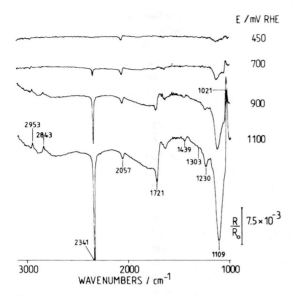

Fig. 2.12. Potential step SNIFTIRS spectra from a polished polycrystalline Pt electrode, in 3 M $CH_3OH/0.1$ M $HClO_4$. Reference spectrum taken at 50 mV. Other details as in Fig. 2.10.

more concentrated solutions where the negative going part of the band for linear adsorbed CO is observed at high anodic potentials (Fig. 2.12). New experiments are required to establish whether the negative part of this band indicates that CO_{ad} is produced at high potentials also, or CO_{ad} formed at low potentials remains un-oxidized at the center of the electrode due to the high ohmic drop in the thin layer. Other bands in the spectra of Figs 2.10 and 2.12 will be discussed in connection with the identification of bulk products.

The identification of COH is difficult. The intensity of the C-O stretching vibration of this particle can be weak and the expected frequency [63] is in a region of relatively high noise (below 1380 cm^{-1}). An additional difficulty is the fact that the absorption bands of ClO_4^-, SO_4^{2-} and HSO_4^- are also in this frequency region. Recent SNIFTIRS experiments employing Pt deposited on a glassy carbon substrate and 0.5 M $CH_3OH/0.1$ M H_2SO_4 solution showed a new feature near 1200 cm^{-1} [28]. Measurements in HF electrolyte could probably help to clarify this question.

2.1.4   Conclusions on the Nature of the Adsorbed Residue of Methanol

The results of the different techniques given above can be summarized as follows.

The ratio Q_{ad}/Q_{ox} varies between 1 and 2 depending on methanol concentration and degree of coverage. In principle this can be interpreted in terms of a mixture of particles of composition (C, O) and (C, O, H). The nature of the Pt surface seems to influence also the charge ratio. In principle, the following configurations could be expected:

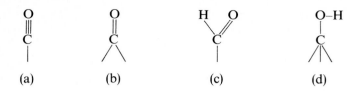

(a)   (b)   (c)   (d)

1. (C, O) species. All four methods indicate the presence of CO as an adsorbed residue of methanol. According to charge measurements, on-line mass spectroscopy and ECTDMS, CO predominates at high methanol concentrations and high degrees of coverage. *In situ* SNIFTIRS show that from the two CO forms given above the linear bonded form (a), prevails.
2. (C, H, O) species. This composition is found at low methanol concentrations and low degrees of coverage. The presence of hydrogen in the adsorbate was demonstrated by ECTDMS. A quantitative analysis of the thermodesorption spectra give results in good agreement with those of charge measurments. On-line mass spectroscopy experiments with D-labeled methanol can be explained on the basis of C–O–H (d) particle rather than a H–C=O one (c). The difficulties in demonstrating a C–O–H adsorbate by SNIFTIRS have already been discussed above.

A COH species requires three adjacent adsorption sites. Under conditions of large degrees of coverage, the site requirement should be an important factor in determining the kind of adsorbate formed.

For the formation of COH Bagotzky et al. have suggested [64] a stepwise mechanism:

$$CH_3OH \rightarrow -CH_2OH + H^+ + e^- \tag{2.7}$$

$$-CH_2OH \rightarrow \;>\!CHOH + H^+ + e^- \tag{2.8}$$

$$>\!CHOH \rightarrow \;\geqslant\!COH + H^+ + e^- \tag{2.9}$$

The less stable intermediates may react as well to form CO, for example via:

$$>\!CHOH \xrightarrow{\text{two steps}} -CO + 2H^+ + 2\,e^- \tag{2.10}$$

Which reaction channel is followed depends on the availability of neighboring free places. This could explain the effect of methanol concentration on adsorbate composition. It has been observed that the initial rate of adsorption is strongly enhanced by increasing methanol concentration [14]. From the adsorption steps given above, the first one, Eq. (2.7), is directly affected by the bulk concentration. At high methanol concentrations the Pt surface becomes very quickly covered with species like CH_2OH or CHOH. Further reaction to a more stable state such as COH is inhibited because of the lack of free adjacent sites. Under these conditions CO should be formed with a greater probability.

The degree of coverage, however, seems to influence the adsorbate composition at low methanol concentrations also. In particular, on smooth platinum, the dependence of surface composition on θ is observed at concentrations as low as 5×10^{-3} M (Fig. 2.7). In this case it could be possible that COH can be formed as long as adjacent sites are available. At high coverage (by all species involved in the adsorption process), the formation of COH should be geometerically prevented.

In the case of porous electrodes, diffusion in the pores can retard the rate of adsorption, in particular at low methanol concentrations. This could be the reason why at a concentration of 10^{-2} M mainly COH is formed independently of θ [14].

2.2 The Bulk Products of Methanol Oxidation

Using on-line mass spectroscopy [65] carbon dioxide and formic acid were demonstrated as soluble products of methanol oxidation. The former gives the most intense MS signal according to the fact that it is the main product. There are two main problems to detect formic acid as such. In the presence of carbon dioxide most of the m/e signals of HCOOH overlap with signals of the major product. Besides this, in the presence of methanol, formic acid reacts to form the methyl ester:

$$HCOOH + CH_3OH \rightleftharpoons HCOOCH_3 + H_2O \qquad (2.11)$$

Since this is an equilibrium reaction and the concentrations of all substances involved are changing during the potential scan, it is difficult to predict what percentage of reacting methanol follows this path. Nevertheless the situation presents some advantages at least. We can use the reaction product of Eq. 2.11 to demonstrate the existence of a parallel pathway involving formic acid.

In Fig. 2.13 the mass intensities for carbon dioxide ($m/e = 44$) and methyl formate ($m/e = 60$) during a potential scan are given. While the signal for CO_2 follows the current pattern in the whole potential range that for $HCOOCH_3$ does not. This indicates the existence of parallel pathways. Methyl formate was also detected as an electrolysis product in long duration experiments [66].

IR investigations of methanol oxidation have been mainly devoted to demonstrate the nature of strongly adsorbed residues produced during the adsorption of methanol [21, 59, 67, 68]. Bulk oxidation products were not investigated by *in situ* IR spectroscopy.

Because of the close distance between electrode and window the concentration of methanol in the thin electrolyte layer diminishes at positive potentials and can only slowly be supplied by diffusion. In order to have measurable quantities of formic acid (or methyl formate) one has to work with methanol concentrations in the order of 1 M or more.

In Fig. 2.12 characteristic IR reflectance spectra of concentrated (3 M) methanol on platinum at potentials between 0.45 V and 1.3 V are shown. The single beam spectrum at 0 mV was taken as background. The following characteristic bands allow the identification of bulk products:

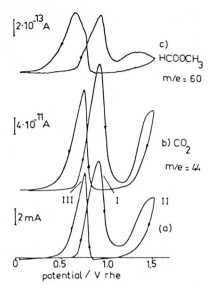

Fig. 2.13. Current and mass intensity signals, ($m/e = 44$, CO_2 and $m/e = 60$, methylformate), during a potential scan of a porous Pt electrode in 0.1 M CH_3OH / 1 M $HClO_4$. Sweep rate: 20 m/s; surface roughness: ca. 50.

1. The asymmetric strectching band of CO_2, which has a strong intensity at 2341 cm^{-1}. This band can be observed in diluted solutions when the potential is 0.45 V or higher (see insert in Fig. 2.10). In concentrated methanol solutions the oxidation to CO_2 starts at even higher potentials (Fig. 2.12). This is probably due to a stronger poisoning effect.

Other bands in this spectrum may be associated with the presence of formic acid and methyl formate.

2. The band at 1721 cm^{-1} may be due to the C=O stretching vibration of a carboxylic acid or ester. This band is observed in pure formic acid and in pure methyl formate at 1730 cm^{-1} [70]. The slight shift to lower frequency can be due to interaction with water.
3. The C–O–C stretching vibration in carboxylic esters is characterized by an absorption band between 1100 cm^{-1} and 1280 cm^{-1} [70]. In the case of pure methyl formate one observes a broad band of strong intensity between 1150 and 1220 cm^{-1}. Considering the shift in aqueous solution the band at 1230 cm^{-1} in Fig. 2.12 could be attributed to the C–O–C stretching vibration of methylformate.
4. Carboxylic acids present two bands at 1430 cm^{-1} and 1300 cm^{-1} which are related to the coupled C–O stretching and O–H deformation vibrations [70]. In the spectrum of Fig. 2.12 the bands at 1439 cm^{-1} and 1303 cm^{-1} could be related to the presence of formic acid.

Positive going bands indicate the consumption of species, which are present in higher concentration at the potential of the background spectrum:

5. Methanol bands at 1021 cm^{-1}, 2843 cm^{-1} and 2953 cm^{-1} appear in all spectra taken at potentials where the bulk oxidation is important (i.e above 900 mV).

All bands discussed here were also observed with 1 M CH$_3$OH solutions. In order to check that the above discussed bands are due to bulk products accumulated in the thin layer, one series of experiments was performed stepping the potential from 1500 mV down to 450 mV after taking the background spectrum at 0 mV. In this case the bands are observed at all potentials showing that stable bulk products are accumulated in the interface.

At low concentrations (Fig. 2.10) most of methanol between window and electrode is consumed early during the potential step sequence, therefore probably undetectable ester quantities can be formed. Nevertheless the 1721 cm^{-1} band of HCOOH is actually weakly apparent as a step at potentials above 900 mV.

The SNIFTIRS results presented here confirm the presence of formic acid and methyl formate as by-products of methanol oxidation. Other by-products such as formaldehyde could not be detected under our experimental conditions. In fact, formaldehyde hydrolyses (99.99%) in aqueous solutions to a gemdiol: H$_2$C(OH)$_2$, and the typical aldehyde bands are, therefore, not expected.

Formaldehyde can react with methanol (to a very small extent and in solutions with a low water content [71]) to form methylal:

$$HCHO + 2\ CH_3OH \rightleftharpoons H_2C(OCH_3)_2 + H_2O \qquad (2.12)$$

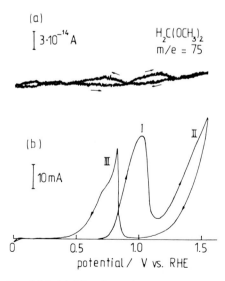

Fig. 2.14. (a) Mass intensity signal ($m/e = 75$, methylal) and (b) current, during a potential scan of a porous Pt electrode in a 44% (v/v) CH$_3$OH/1 M H$_2$SO$_4$ solution. Scan rate = 20 mV/s.

T. Iwasita-Vielstich

An MS experiment was performed using a solution containing 44 vol % CH_3OH. Small quantities of methylal were detected, (Fig. 2.14). Other masses of formaldehyde are overlapped with those of methanol and CO_2.

3 On-Line Mass Spectroscopy as a Tool for the Study of Exchange Reactions between Adsorbates and Bulk Solution Components

Exchange reactions between bulk and adsorbed substances can be studied by on-line mass spectroscopy and isotope labeling. In this section the results on the interaction of methanol and carbon monoxide in solution with adsorbed methanol and carbon monoxide on platinum are reported [72]. A flow cell for on-line MS measurements (Fig. 1.2) was used. ^{13}C-labeled methanol was absorbed until the Pt surface became saturated. After solution exchange with base electrolyte a potential scan was applied. Parallel to the current-potential curve the mass intensity-potential for $^{13}CO_2$ was monitored. Both curves are given in Fig. 3.1a,b. A second scan was always taken to check the absence of bulk substances.

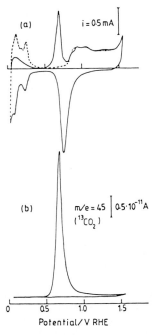

Fig. 3.1. Current (a), and mass intensity for $^{13}CO_2$ production (b) during the potentiodynamic oxidation of methanol adsorbate (flow cell procedure, $E_{ad} = 0.2$ V RHE, see text). Scan rate: 10 mV/s.

3.1 Interaction of Methanol Adsorbate with Bulk Methanol

The stability of methanol adsorbate is very well known. The following experiment checks the degree of irreversibility of the adsorption reaction on platinum. After adsorption of $^{13}CH_3OH$ the solution was exchanged with $^{12}CH_3OH$. After 10 min of interaction with the ^{13}C-adsorbate, bulk methanol was eliminated by substituting the electrolyte with 0.5 M H_2SO_4. A subsequent potential scan showed no change in mass signal ($^{13}CO_2$) when compared with the experiment with ^{13}C-methanol only. The $^{12}CO_2$-signal observed corresponded to small quantities (< 1%) of $^{12}CH_3OH$ already present in ^{13}C-methanol solution. No displacement of methanol adsorbate by bulk methanol was observed.

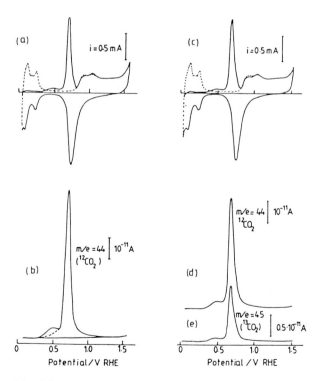

Fig. 3.2. Current and mass intensity signal during the oxidation of CO adsorbate; adsorption potential 0.2 V RHE, scan rate : 10 mV/s. (a) current during $^{12}CO_{ad}$ oxidation. (b) Mass intensity for $^{12}CO_2$ during scan (a). Full line: after displacement of dissolved CO by bubbling argon during 5 min; broken line: after 80 min bubbling argon. (c) Current response after $^{12}CO_{ad}$ of interaction with ^{13}CO. (d) $^{12}CO_2$-mass intensity signal during scan (c) (due to rest of initially adsorbed CO). (e) mass intensity response for $^{13}CO_2$ (due to exchanged CO).

3.2   Adsorbed Carbon Monoxide and Dissolved CO

A series of adsorption experiments with ^{12}CO was performed. After adsorption at a potential of 200 mV RHE bulk carbon monoxide was eliminated from the solution by bubbling argon gas through it for 5 min. A potential scan (current and CO_2 mass signal) of adsorbed CO is shown in Fig. 3.2a,b. Bubbling argon for a longer time causes the desorption of a small fraction of adsorbed CO. The dashed line in Fig. 3.2b shows the diminution of the CO_2 signal after 80 min of argon bubbling (the same result was obtained after 2.5 h). The CO species oxidized under the small peak at ca. 500 mV were described in the literature as weakly adsorbed CO [73, 74]. The rest of CO seems to be more strongly adsorbed.

These results seemed to indicate that CO adsorption on platinum is an "irreversible" process, i.e. there is no equilibrium of the form $Pt\text{–}CO_{ad} \rightleftharpoons CO_{sol} + Pt$. The purpose of the following experiment was to check this hypothesis.

^{12}CO was adsorbed during 5 min. at 0.2 V (saturation coverage) and then eliminated from the bulk by bubbling Ar for 5 min. Then the solution was exchanged

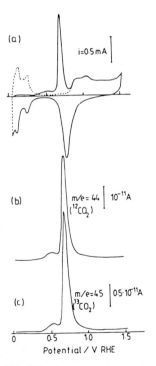

Fig. 3.3. Current and mass intensity signal showing the effect of the interaction of bulk ^{12}CO with ^{13}C-methanol adsorbate (flow cell procedure). (a) Current due to the oxidation of methanol and CO adsorbates. (b) Mass intensity for $^{12}CO_2$ due to CO_{ad}, (c) mass intensity for $^{13}CO_2$ (due to rest of adsorbed methanol.

with a ^{13}CO containing electrolyte, after 10 min. of interaction ^{13}CO was displaced from the solution with argon. A voltammogram was run (Fig. 3.2c) and mass signals 44 and 45 were simultaneously monitored (Fig. 3.2d,e).

A comparison of the areas under the curves for $m/e = 44$ in Fig. 3.2d and b indicates an exchange of ca. 20% between adsorbed and bulk CO after 10 min of interaction. When considering these results quantitatively one has to consider that the concentration of ^{13}CO solution was below saturation and this can limit the rate of exchange. Possible reasons for these results will be discussed later in this chapter.

3.3 Adsorbed Methanol Interacting with Dissolved CO

Methanol was adsorbed from a 10^{-2} M $^{13}CH_3OH$/0.5 M H_2SO_4 solution at 0.2 V RHE during 10 min and then eliminated from bulk by a 0.5 M H_2SO_4 flow under potential control. Then the solution was exchanged with pure supporting electrolyte, and ^{12}CO was bubbled through for 5 min. After elimination of bulk CO with Ar (5 min), a voltammogram was run (Fig. 3.3a). Carbon dioxide formation was followed by monitoring $m/e = 44$ ($^{12}CO_2$) and $m/e = 45$ ($^{13}CO_2$), Fig. 3.3b,c. A comparison with the result in Fig. 3.1b shows that interaction between methanol adsorbate and bulk CO results in a diminution (15–30%) of $m/e = 45$ (methanol adsorbate), parallel to a partial displacement of the latter to the places where weakly chemisorbed CO is observed (pre-peak at 0.5 V).

3.4 Discussion of the Observed Exchange Effects

3.4.1 The CO_{bulk}/CO_{ad} Exchange

Experiments in the gas phase show that CO desorption can be accelerated in the presence of CO in the bulk. This was demonstrated in measurements of isotopic exchange of CO on Pt [75], Pd [76, 77], Rh [76, 78] and Ni [79].

In principle, it could be expected that the energy released during the adsorption of a molecule is transferred to a vibrational mode of a neighboring Pt–CO bond, thus reducing the activation energy for desorption. Zhdanov [80] has found that for an effective use of the energy released, (a) the time for energy transfer in the adsorption layer must be shorter than the time required for relaxation into inactive degrees of freedom, and (b) the microscopic rate of desorption from a given excited state must be faster than the rate of relaxation into the ground state. Usually condition (b) is not fulfilled [80], i.e. the probability for such a coupled adsorption–desorption process is very low.

Exepriments in the gas phase have supplied us with considerable knowledge on the state of adsorbed carbon monoxide on platinum.

- It was established that an equilibrium is maintained between CO adsorbed on steps and on terraces [81], a fact that is in agreement with the high mobility ascribed to chemisorbed CO on Pt.
- It has been proposed that the precursor state [81, 82] for the adsorption–desorption reaction consists of weakly physisorbed CO. This can be CO sitting on an occupied site (CO_{ad}–CO) or on an sterically unfavorable Pt site. According to Ertl [81], the desorption process occurs through a "trapping" mechanism on such sites: if the surface is saturated by chemisorbed CO the "desorption channel involves either a CO_{ad}–CO potential well or a Pt–CO attractive well which is sterically weakened by the presence of pre-absorbed CO".
- For adsorbed CO on platinum a dependence of the desorption energy with coverage was observed [83, 84]. According to data of McCabe and Schmidt for the gas phase [83] the desorption energy, E_d, decreases by a factor of 2 going from zero to saturation coverage (Fig. 3.4). It can be clearly seen that the decrease of E_d with θ is particularly pronounced at high coverages. This has been explained in terms of repulsive interactions between adsorbed molecules [85].

Unfortunately, respective data for CO adsorption on Pt from an electrolyte solution are not available.

Although the desorption energy at the metal-electrolyte interface should be different from that of the metal-gas phase, there is no reason to expect a different behavior for the change of E_d with θ.

If we assume a similar behavior in the electrochemical cell a considerable decrease of adsorption energy should be expected as saturation of the surface is reached in the presence of bulk CO. Under these conditions weakly physisorbed CO should act as a

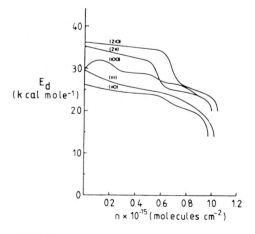

Fig. 3.4. Heat of desorption for CO on single crystal platinum as a function of coverage (taken from [83]).

precursor for the adsorption–desorption mechanism. The high mobility of chemisorbed CO on Pt should also operate in the condensed phase. The moving particles may be trapped in the potential well of physisorbed CO and then desorb giving rise to the observed isotopic exchange.

After saturation of the Pt surface it is possible for physisorbed CO to be washed up during electrolyte exchange without producing any detectable increase in the current for hydrogen adsorption. However, the surface coverage can diminish slightly below the saturation value. Experiments of H_2 evolution on electrodes covered with adsorbed CO [73] support this idea. When a Pt electrode is saturated with adsorbed CO and this is then eliminated from the solution, the overpotential for H_2 evolution is large. But it is even larger if, additionally CO remains in the bulk of the solution [74]. It can be concluded that in the absence of bulk CO physisorbed CO desorbs and a large increase of E_d can be produced (Fig. 3.4) thus inhibiting the desorption of chemisorbed CO.

Weakly chemisorbed CO producing the pre-peak at 0.5 V during oxidation of CO_{ad} on Pt (Fig. 3.2a) is not the species we are referring to as the precursor state for adsorption–desorption. After 10 min of bubbling argon, weakly chemisorbed CO is still present on the electrode surface. It is also not desorbed after solution exchange with base electrolyte. On the other hand, in the presence of bulk CO, after 10 min of interaction with a dilute CO solution, almost 20% of CO_{ad} can be exchanged. We may conclude that the desorption energy of weakly chemisorbed CO is much higher than that of physisorbed CO.

3.4.2   Interactions Involving Methanol Adsorbate

It has been shown in Section 2.1.4 that methanol adsorbate formed from dilute solutions on a porous Pt surface, consists of CO_{ad} and COH_{ad} in a ratio CO:COH of ca. 20–30% [14]. The results of isotopic exchange with bulk CO seem to indicate that only the fraction present as CO_{ad} can be desorbed and replaced by bulk CO. Probably the same arguments as in the case of pure CO_{ad} can apply. COH_{ad} seems to be more strongly bound to the Pt surface and cannot be desorbed.

Obviously, desorption of COH_{ad} at a given constant potential requires the break of the triple bond of COH_{ad} to Pt and the ionization of hydrogen. As was shown above in thermal desorption experiments H_2 and CO are formed. The desorption process of COH_{ad} could require a higher activation energy than the desorption of CO_{ad}.

4   The Electrocatalysis of Methanol Oxidation

Binary combinations of platinum and less noble metals (Pt/Sn, Pt/Re, Pt/Ru, Pt/Pd) have electrocatalytic effects upon the rate of oxidation of methanol and parent compounds [86–94]. Current enhancements up to two orders of magnitude or even more have been observed during the oxidation of formic acid on platinum in the presence of lead [93, 94]. For methanol oxidation, tin adsorbed on Pt showed the best catalytic effects. Current enhancements by a factor of 50–100 were reported [92].

The mechanism of the cocatalytic effect is still a matter of investigation. For most of the systems of interest in electrocatalysis, data for characterization of the surface by means of spectroscopic UHV methods are still missing. Also measurements of changes in the electronic properties of the metal in the presence of adatoms in addition to more intensive application of *in situ* and on-line methods are desirable for a systematic search of new catalytic materials.

The use of conventional electrochemical methods to study the effect of metal adatoms on the electrochemical oxidation of an organic adsorbate may be in some cases of limited value. Very often, in the potential region of interest the current due to the oxidation of an organic residue is masked by faradaic or capacitive responses of the cocatalyst itself. The use of on-line mass spectroscopy overcomes this problem by allowing the observation of the mass signal-potential response for the CO_2 produced during the oxidation of the adsorbed organic residue.

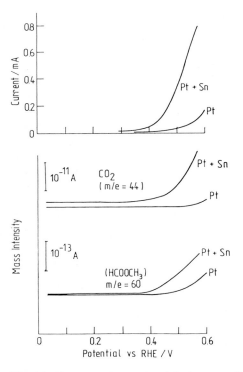

Fig. 4.1. Current and mass signal during an on-line mass spectroscopic experiment showing the effect of adsorbed tin on platinum upon methanol electrooxidation. 1 M CH_3OH/0.5 M H_2SO_4; sweep rate: 10 mV/s, 24 °C.

Another advantage of the on-line MS method is the possibility to observe separately the catalyst effects on parallel reactions. This is illustrated in Fig. 4.1 for methanol electroxidation on Pt with 60% of tin. Current and mass signals for CO_2 and for $HCOOCH_3$ (methylformate) during the oxidation of 1 M CH_3OH on a porous platinum electrode were recorded.

As we have pointed out in Section 1.1 the mass signal response for a given reaction is proportional to the respective current (see Eq. 1.1). The result in Fig. 4.1 shows that at 0.58 V an increase in current by a factor of 7.5 is obtained. At the same time, the mass signal for CO_2 increases by a factor of 13 and that for $HCOOCH_3$ by a factor of 3. This result indicates that tin affects the current efficiency for both parallel reaction pathways to a different extent.

The catalytic properties of a Pt/Sn combination were observed on different kinds of electrode materials: alloys [90], electro co-deposits of Pt and Sn [89, 90], under-potential deposited tin [42] or a mixture of tin oxide and platinum deposited on glass [95]. All different materials present a marked influence on methanol electrooxidation.

4.1   The Oxidation State of Adsorbed Sn on Pt

Using the flow cell technique current, transients during the adsorption of tin at potentials between 0.1 and 0.7 V were measured. Fig. 4.2a shows a typical i–t transient when Sn(IV) ions are added to the solution containing only the supporting electrolyte. Negative currents during the adsorption of Sn(IV) indicate a variable degree of Sn(IV) reduction in the potential range studied. The situation is somewhat more complicated for Sn(II). Below 0.25 V the adsorption current is negative (Fig. 4.2b, dashed line). At more positive potentials both negative and positive currents are observed (Fig. 4.2b, full line). A corresponding behavior was observed by Szabo [96], who measured potential shifts towards negative and then positive values during the adsorption of tin (II) which were attributed to a disproportionation reaction.

A voltammogram after exchange of the solution with base electrolyte is shown in Fig. 4.3. The degree of coverage by tin, θ_{Sn}, calculated from the charge relations in the H-region is shown in Fig. 4.4a.

In order to assign an oxidation number to the adsorbed species, knowledge of the number of Pt sites occupied by one tin adatom, S, is necessary. This was calculated by different authors. Motoo [97] and Sobkowski [98] reported a value of 2, while Szabo found a value of 2.2 [96]. The method used by Sobkowski has the advantage, for our purposes, of making no assumption on the oxidation state of the adsorbed tin layer. The number of electrons associated with the formation of adsorbed tin, n, can be calculated as follows:

$$n = \frac{Q_{ad,\,Sn} \cdot S}{(Q_H - Q_H^\circ)/\theta_H} \tag{4.1}$$

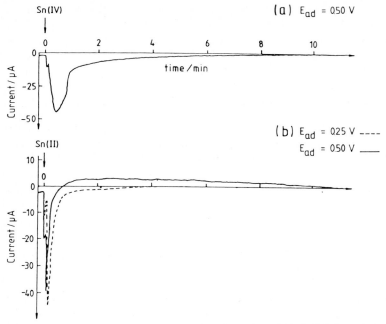

Fig. 4.2. Current-time transient during the adsorption of (a) Sn(IV) from 4×10^{-4} M Sn(SO$_4$)$_2$ in 0.5 M H$_2$SO$_4$ and (b) Sn(II) from 4×10^{-4} SnSO$_4$ M in 0.5 M H$_2$SO$_4$ on porous platinum at constant potentials E_{ad} as indicated.

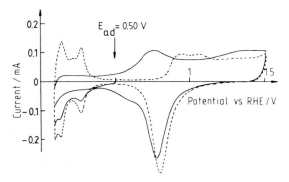

Fig. 4.3. Cyclic voltammogram for adsorbed tin on platinum, v = 10 mV/s. Experimental procedure: 11 min. adsorption from a 4×10^{-4} M Sn(SO$_4$)$_2$ solution in 0.5 M H$_2$SO$_4$ at 0.5 V followed by electrolyte replacement with pure supporting electrolyte.

$Q_{ad,Sn}$ is the charge under the adsorption transient. In order to account for the total surface area, the charge ΔQ_H must be divided by θ_H (i.e. the coverage degree by H atoms in the absence of co-adsorbed tin, at the lower potential limit used for the integration). A value of $\theta_H = 0.77$ at 0.08 V was taken from [99].

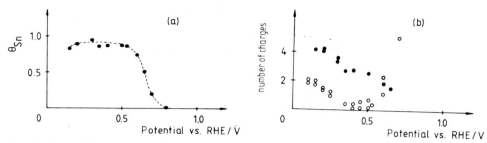

Fig. 4.4. Potential dependence of (a) the coverage degree of platinum by tin and (b) charge transferred during tin adsorption on platinum (according to Eq. 4.1). Tin was adsorbed from $Sn(SO_4)_2$ (●) and from $SnSO_4$ (○).

Assuming for the number of sites occupied by one tin particle, $S = 2$, n values were calculated for Sn(IV) and Sn(II) adsorption on platinum. These are plotted as a function of potential in Fig. 4.4b.

For Sn(IV) adsorption a value of around 4 is reached at potentials below 0.25 V indicating the reduction to Sn(0). As the adsorption potential is made more positive the number of charges for tin adsorption decreases, indicating that increasing amounts of Sn(II) species are formed. Above 0.6 V, $n < 2$ indicates that tin is partially adsorbed as a mixture of Sn(II) and Sn(IV).

During Sn(II) adsorption at potentials above 0.25 V reduction and oxidation transients were observed (Fig. 4.2b). The number of electrons per adsorbed tin species was calculated taking the algebraic sum of positive and negative charges. This results in a net reduction process below 0.4 V and a net oxidation process above this potential. Starting with a value of 2 at the lower potentials, [reduction to Sn(0)], n decreases as the adsorption potential is made more positive. A minimum close to $n = 0$) is reached near 0.4 V, indicating an apparent oxidation state of 2. From this potential a predominance of positive charge makes n values increase again as the potential is further increased.

The number of electrons required to oxidize one adsorbed tin ion was given in the literature as calculated from the charge under the voltammogram. A value of 2 was reported [98]. Since tin is desorbed as Sn(IV) ions, this result led to the interpretation that the adsorbed species consist of Sn(II). According to the present results, however, tin can be on the electrode surface in different redox states depending on the potential. This means that the charge for the oxidation of adsorbed tin during a potential scan should vary with the cathodic limit of the voltammogrm. This is indeed the case, as shown by Mayer and co-workers [100], who also present evidence to show that below 0.2 V RHE a platinum-tin alloy is formed. These results agree with the present observation that tin is reduced to Sn(0) below 0.25 V. We have made no attempt to integrate the oxidation charge for tin in the voltammogram. Overlap with Pt oxide formation leads to a considerable uncertainty in the evaluations (Fig. 4.3).

4.2   The Interaction of Sn with Adsorbed Methanol

After adsorption of methanol at 0.4 V and electrolyte exchange with base solution to eliminate bulk methanol (flow cell), a step to potentials between 0.25 V and 0.6 V was applied, then Sn(IV) or Sn(II) was added and the CO_2 mass intensity signal was monitored.

The MS response after *Sn(IV) addition* is given in Fig. 4.5 for two oxidation potentials E_{ox}. The CO_2 signal grows and passes through a maximum after some minutes. A more pronounced response is observed at higher potentials. The lowest potential at which this effect can be observed is ca. 0.425 V. Blank experiments (without addition of tin, dashed curve in Fig. 4.5) show a CO_2 production at potentials above 0.4 V, but this was always lower than in the presence of tin.

Fig. 4.6a,b shows the effect of *Sn(II) addition* on current and CO_2-mass signal respectively. The enhancement of CO_2 production is clearly more pronounced than that produced by Sn(IV) addition. This effect can be observed at potentials as low as 0.25 V (lower than in the case of Sn(IV)).

The current-time response of the system during Sn(II) addition presents the same features as the mass intensity-time curve. For comparison the i–t curve for a blank experiment (only adsorbed methanol being present, no tin addition) is also shown in Fig. 4.6a. The observed response is not simply the sum of the individual responses of Sn(II) (Fig. 4.2b) and adsorbed methanol (see dashed curve in Fig. 4.6a), to the applied potential step.

A voltammogram recorded after 15 min of interaction with tin is shown in Fig. 4.7a, together with the mass signal response for CO_2. The current signal shows

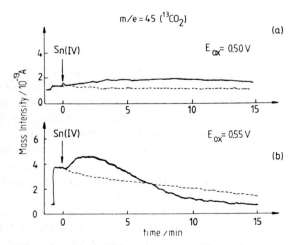

Fig. 4.5. Mass spectroscopic detection of carbon dioxide during methanol adsorbate oxidation and Sn(IV) injection. Porous Pt electrode, real area: 12.3 cm^2. Procedure: after methanol adsorption at 0.4 V from 10^{-2} M $^{13}CH_3OH/0.5$ M H_2SO_4, the electrolyte was exchanged with 0.5 M H_2SO_4, then potential step to E_{ox} was applied and Sn(IV) was added. Dashed line: no tin added.

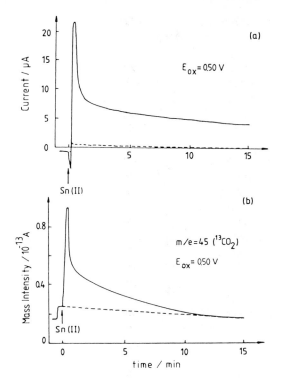

Fig. 4.6. (a) Current-time and (b) corresponding mass spectroscopic signal for CO_2 during methanol adsorbate oxidation after Sn(II) injection. Porous Pt electrode, real area: 35.0 cm^2. Procedure as in Fig. 4.5. Dashed line: no tin added.

the inhibition of H adsorption and the overlapped oxidation of adsorbed tin and the rest of the methanol intermediate which was not oxidized during the preceding step. From the current response, the effect of adsorbed tin on the current-potential response for methanol adsorbate oxidation cannot be seen. However, the on-line MS response allows one to observe the oxidation of the organic remainder alone. This is illustrated in Fig. 4.7b for a case where Sn(IV) had been added at 0.475 V, (full line). For comparison the results of experiments in the absence of tin are also included (dashed line). The oxidation of methanol adsorbate starts at 0.35 V in the presence of adsorbed tin, i.e. 0.15 V more negative than without tin. An identical potential displacemnt was observed after interaction with Sn(II). Once tin is absorbed on the Pt surface the oxidation state depends on the electrode potential. No difference should be expected in the kind of tin ions in solution from which adsorption took place, as indeed is the case.

We can now discuss the different explanations of the cocatalytic tin effect on methanol oxidation given in the literature.

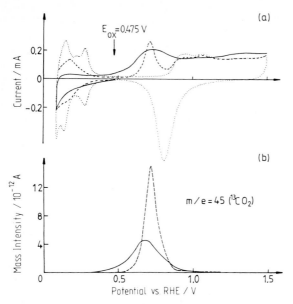

Fig. 4.7. Current (a) and mass intensity (b) voltammograms for methanol adsorbate oxidation without tin (dashed line) and with tin (full line). Base electrolyte: dotted line. The voltammograms were recorded after applying a potential step to $E_{ox} = 0.475$ V during 15 min.

a) Based on the values of the standard potential of the redox couple $Sn(OH)_4/Sn(OH)_2$ ($E^0 = 0.075$ V), Cathro [89] suggested that a direct redox reaction of $Sn(OH)_4$ with the strongly adsorbed residue from methanol takes place. The reduced oxide is then electrochemically reoxidized.

This theory does not fit with the results given above showing a higher rate of CO_2 production for Sn(II) than for Sn(IV) addition.

b) Janssen et al. [90] reported that "tin atoms are present in a zero-valent state at the Pt surface at potentials where methanol oxidation takes place". Changes in the adsorption properties of platinum due to a strong interaction with adsorbed tin are responsible for stronger adsorption of H_2O presumably in the form of OH. The results presented above show that in the potential region of interest adsorbed tin on platinum is more likely in the oxidation state II.

c) According to Lamy and co-workers [88], the presence of adsorbed Sn impedes the formation of organic rests like COH due to the occupation of neighboring Pt sites (COH is supposed to occupy three Pt atoms).

This effect, called the "third body effect" by Conway and co-workers [101], is however controversial [102]. The main argument against this theory is the fact that there is a specificity of catalytic behavior for each kind of metal adatom. Even adatoms producing similar geometrical blocking effects, present different catalytic properties. So, for instance, tin and lead [97] occupy two Pt atoms, but tin produces higher effects than lead on the current for bulk methanol oxidation at low potentials [88].

A recent study of the interactions of tin and lead with the strongly adsorbed intermediate of formic acid [103] showed that although tin is more effective than lead in catalyzing the oxidation of adsorbates, lead is a better catalyst than tin for the bulk oxidation of formic acid.

d) According to XPS data for a Pt-tin oxide electrode [95] ionic Pt^{2+} and Pt^{4+} species should be present in the catalyst. This result was interpreted in terms of a stabilization of platinum oxides in the presence of SnO_2. Platinum orbitals are occupied by OH or O bonds and are not available to adsorb organic residues.

This point of view is however disputable. The adsorption step is necessary for the oxidation process. Metals which are not able to adsorb an organic substance are also not able to oxidize it.

e) The importance of the presence of oxygen associated to the co-adsorbed metal was recognized early by Koch [86] and later emphasized by Motoo [87, 97]. It was suggested that Sn, Ge and Pb adsorbed on Pt are able to enhance the catalytic activity of Pt upon organic oxidation, by forming oxides from which adsorbed intermediates should take the oxygen atom to form CO_2 at relatively low potentials.

In view of the fact that the oxidation of methanol adsorbate requires an additional oxygen atom (see Eqs. 2.4 to 2.6) this theory seems quite plausible to explain the data given above.

4.3   Conclusions on the Co-Catalysis of Methanol Oxidation by Tin

As we have seen in Section 2.1.4, depending upon the concentration, methanol adsorbate seems to consist of variable amounts of COH and CO species. Oxidation to CO_2 requires the splitting of H_2O molecules which could deliver an oxygen atom to form CO_2. The stability of H_2O makes the oxidation process difficult. Its weak adsorption on platinum does not contribute to improve the situation.

The electronic configuration of tin after loosing two p electrons to form Sn^{2+} is $5s^2$. The σ-donor tendency of tin is relatively small but can be enhanced in the presence of a powerful acceptor, especially if the empty $5d$ orbitals of tin can act as π-acceptors [104]. This could be the situation for adsorbed Sn(II). Platinum may act as an acceptor of the tin σ electrons while $5d$ orbitals of the latter accept π-electrons from H_2O to build an adsorbed $Sn(OH)^+$ complex:

$$(Pt)Sn_{ad}^{2+} + H_2O \rightleftharpoons (Pt)Sn(OH)_{ad}^{+} + H^+ \qquad (4.2)$$

The complex $Sn(OH)^+$ given here is purely formal since a further reaction could result in $Sn(OH)_2$ or in the dehydrated form SnO. What shall be emphasized is that the theories explaining the catalytic effect of tin through oxygen-containing species can be justified by taking into account the chemical properties of Sn(II).

In this way Sn(II) would bring oxygen-containing species to the Pt interface which are stronger adsorbed than H_2O and can therefore act as oxygen donors more easily during methanol adsorbate oxidation:

$$(Pt)Sn(OH)^+_{ad} + (Pt)COH \rightarrow CO_2 + (Pt)Sn^{2+}_{ad} + 2H^+ + 3e^- \qquad (4.3)$$

or

$$(Pt)Sn(OH)^+_{ad} + (Pt)CO \rightarrow CO_2 + (Pt)Sn^{2+}_{ad} + H^+ + 2e^- \qquad (4.4)$$

The fact that electrodes prepared with SnO_2 [95] also show catalytic properties upon methanol oxidation does not invalidate the result that Sn(II) species are actually responsible for the observed effects. It could be possible for SnO_2 to be reduced to SnO (or $SnOH^+$) at the potentials where the catalytic effect is observed.

5   References

1. "Electrochemical Processes in Fuel cells", M.W. Breiter, Springer Verlag, Berlin, 1969.
2. "Fuel Cells", W. Vielstich, Wiley, New York, 1970.
3. V.S. Bagotzky and Yu.B. Vassiliev, Electrochim. Acta 27 (1966) 1439.
4. W. Güther and W. Vielstich, Electrochim. Acta 27 (1982) 811.
5. R.E. Schmith, H.B. Urbach, J. H. Harrison, and N.L. Hatfield, J. Phys. Chem. 71 (1967) 1250.
6. M.W. Breiter, Electrochim. Acta 12 (1967) 1213.
7. L. Grambow and S. Bruckenstein, Electrochim. Acta 22 (1977) 377.
8. A. Bewick, Trends in Interf. Electrochem., A. F. Silva (ed.), 331–358 (1986).
9. P.A. Christensen and A. Hamnet in "Comprehensive Chemical Kinetics", C.H. Bamford and R.G. Compton (eds), in press.
10. S. Pons, T. Davidson, and A. Bewick, J. Electroanal. Chem. 160 (1984) 63.
11. O. Wolter and J. Heitbaum, Ber. Bunsenges. Phys. Chem. 88, (1984) 2.
12. B. Bittins-Cattaneo, E. Cattaneo, W. Vielstich, and P. Königshoven, "Electroanal. Chem.", A. J. Bard (ed.), Marcel Dekker, in press.
13. S. Wilhelm, W. Vielstich, H.W. Buschmann, and T. Iwasita, J. Electroanal. Chem. 229 (1987) 377.
14. S. Wilhelm, T. Iwasita, and W. Vielstich, J. Electroanal. Chem. 238 (1987) 383.
15. J. Willsau, O. Wolter, and J. Heitbaum, J. Electroanal. Chem. 185 (1985) 163.
16. "Partial pressure measurement in vacuum technology", Balzers A. G. (ed.) Liechtenstein.
17. D. Tegmeyer and J. Heitbaum, 87. Meeting Bunsenges., 1988, Passau, Abstr. 171.
18. "Reference Electrodes: Theorie and Practice", D.J. Ives and G.J. Janz, Academic Press, New York, 1961.
19. S. Wilhelm, H.W. Buschmann and W. Vielstich, in: Dechema Monographien, Vol. 112 J. Russow (ed.), VCH, 1988; pp. 113–124.
20. R. Greenler, J. Chem. Phys. 44 (1966) 310.
21. A. Bewick and S. Pons, Adv. in IR and Raman Spec. 12 (1985) 1.
22. H. Nakajima. H. Kita, K. Kunimatsu, and A. Aramata, J. Electroanal. Chem. 201 (1986) 175.
23. B. Beden, C. Lamy, A. Bewick, and K. Kunimatsu, J. Electroanal. Chem. 121 (1981) 343.
24. S. Pons, T. Davidson, and A. Bewick, J. Am. Chem. Soc. 105 (1983) 1802.
25. T. Davidson, S. Pons, A. Bewick, and P.P. Schmit, J. Electroanal. Chem. 125 (1981) 237.
26. S. Pons, J. Electroanal. Chem., J. Electroanal. Chem. 150 (1983) 495.
27. M.A. Habib and J. O'M. Bockris, J. Electroanal. Chem. 180 (1984) 287.
28. P.A. Christensen, W. Vielstich, S.A. Weeks, and A. Hamnet, J. Electroanal. Chem. 242 (1988) 327.

29. P.A. Christensen, A. Hamnet, and S.A. Weeks, J. Electroanal. Chem., 250 (1988) 127.
30. D. Corrigan and M. Weaver, J. Electroanal. Chem. 241 (1988) 143.
31. K. Kunimatsu, W.G. Golden, H. Seki, and M.R. Philpott, Langmuir 1 (1985) 245.
32. H. Seki, K. Kunimatsu, and W.G. Golden, Appl. Spectr. 39 (1985) 437.
33. E. Müller, Z. Elektrochem. 28 (1928) 101.
34. D.S. Cameron, Platinum Metals Rev. 29 (1985) 107.
35. A. Capon and R. Parsons, J. Electroanal. Chem. 44 (1973) 1.
36. P. Stonehart and G. Kohlmayr, Electrochim. Acta 17 (1972) 369.
37. R.R. Adzic, A. V. Tripkovic, and W.E. O'Grady, Nature 296 (1982) 137.
38. C. Lamy, J.M. Leger, J. Clavilier, and R. Parsons, J. Electroanal. Chem. 150 (1983) 71.
39. E.P.M. Leiva, E. Santos, R.M. Cerviño, M.C. Giordano, and A.J. Arvia, Electrochim. Acta 30 (1985) 1111.
40. O.A. Petry, B.I. Podlovchenko, A.N. Frumkin, and Hira Lal, J. Electroanal. Chem. 10 (1965) 253.
41. T.Biegler and D.F.A. Koch, J. Electrochem. Soc. 114 (1967) 904.
42. M. Watanabe, Y. Furuuchi, and S. Motoo, J. Electroanal. Chem. 191 (1985) 367.
43. V.S. Bagotzky and Yu. B. Vassiliev, Electrochim. Acta 12 (1967) 1323.
44. B.I. Podlovchenko and E.P. Gorgonova, Dokl. Akad. Nauk SSSR 156 (1964) 673.
45. V.E. Kazarinov, G.Ya. Tysiachnaya, and V.N. Andreev, J. Electroanal. Chem. 65 (1975) 391.
46. J. Willsau and J. Heitbaum. J. Electroanal. Chem. 185 (1986) 181.
47. T. Iwasita, W. Vielstich, and E. Santos, J. Electroanal. Chem. 229 (1987) 367.
48. J.H. Craig, Surf. Sci. 111 (1981) L695.
49. D.M. Collins, W.E. Spicer, Surf. Sci. 69 (1977) 85.
50. S. Wilhelm, Doctoral Thesis, University of Bonn, 1988.
51. Y.K. Peng and P.T. Dawson, Can. J. Chem. 53 (1975) 298.
52. J.H. Craig, Appl. Surf. Sci. 10 (1982) 315.
53. V.H. Baldwin and J.B. Hudson, J. Vacuum Techol. 8 (1971) 49.
54. O. Wolter, J. Willsau, and J. Heitbaum, J. Electrochem. Soc. 132 (1985) 1635.
55. S. Wilhelm, B. Bittins-Cattaneo, E. Cattaneo, and W. Vielstich, Ber. Bunsenges. Phys. Chem., in press.
56. A. Bolzan and T. Iwasita, Electrochim. Acta 33 (1988) 109.
57. A. Bolzan, T. Iwasita, and W. Vielstich, J. Electrochem. Soc. 000.
58. R.A. Shigeishi and D.A. King, Surf. Sci. 58 (1976) 379.
59. B. Beden, S. Juanto, J.M. Leger, and C. Lamy, J. Electroanal. Chem. 238 (1987) 323.
60. N. Sheppard and T.T. Nguyen in Adv. in Infrared and Raman Spectroscopy, R.H.H. Clark and R.E. Hester (eds.), Heyden, London, 1978, vol. 5, p. 67.
61. J.K. Foley, C. Korzeniewski, J.L. Daschbach, and S. Pons, in Adv. in Electroanalytical Chem., A.J. Bard (ed.), vol. 14, p. 310, Marcel Dekker, N.Y. 1986.
62. G. Blyholder, J. Phys. Chem. 79 (1975) 756.
63. J.M. Bowman, J. S.. Bittman, and L . Harding. J. Chem. Phys. 85 (1986) 911.
64. V.S. Bagotzky, Yu.B. Vassiliev, and O.A. Khazova, J. Electroanal. Chem. 81 (1977) 229.
65. T. Iwasita and W. Vielstich, J. Electroanal. Chem. 201 (1986) 403.
66. K. Ota, Y. Nakagawa, and M. Takahashi, J. Electroanal. Chem. 179 (1984) 179.
67. K. Kunimatsu, J. Electroanal. Chem. 145 (1983) 219.
68. B. Beden, F. Hahn, S. Juanto, C. Lamy, and J.M. Leger, J. Electroanal. Chem. 225 (1987) 215.
69. S. Juanto, B. Beden, F. Hahn, J.M. Leger, and C. Lamy, J. Electroanal. Chem. 237 (1987) 119.
70. Sadtler Standard Grating Spectra. Sadtler Research Laboraties Inc., Philadelphia, 1975.
71. M.Delepine, Bull. Soc. Chim. France 25 (1901) 346.
72. T. Iwasita and U. Vogel, Electrochim. Acta 33 (1988) 557.
73. M.W. Breiter, J. Phys. Chem. 72 (1968) 1305.
74. E.P.M. Leiva, E. Santos, and T. Iwasita, J. Electroanal. Chem. 215 (1986) 357.
75. Ref. 17 in V.P. Zhdanov, Catalysis Rev., 24 (1982) 373.
76. T. Matsushima, J. Catalysis 64 (1980) 38.
77. T. Yamada, T. Onishi, and K. Tamaru, Surface Sci. 133 (1983) 533.
78. J.T. Yates Jr., T.M. Duncan, and R.W. Vaughan, J. Chem. Phys 71 (1979) 3908.
79. J.T. Yates Jr. and D.W. Goodman, J. Phys. Chem. 73 (1980) 5371.

80. V.P. Zhdanov, Dokl. Akad. Nauk. SSSR 254 (1980) 392.
81. C.T. Campbell, G. Ertl, H. Kuipers, and J. Segner, Surface Sci. 107 (1981) 207.
82. D.A. King, Surface Sci. 64 (1977) 4314.
83. R.W. McCabe and L.D. Schmidt, Surface Sci. 66 (1977) 101.
84. G. Ertl, M. Newmann, and K.M. Streit, Surface Sci. 64 (1977) 393.
85. C. Tracy and P.W. Palmberg, J. Chem. Phys. 51 (1969) 4852.
86. D.F.A. Koch, Australian Pàt. 46123, 1964.
87. S. Motoo and M. Watanabe, J. Electroanal. Chem. 69 (1976) 429.
88. B. Beden, F. Kadirgan, C. Lamy, and J. M. Leger, J. Electroanal. Chem. 127 (1981) 75.
89. K.J. Cathro, J. Electrochem. Soc. 116 (1969) 1608.
90. M.M.P. Janssen and J. Moolhuysen, J. Catalysis 46 (1976) 289.
91. M.M.P. Janssen and J. Moolhuysen, Electrochim. Acta 21 (1976) 861.
92. M.M.P. Janssen and J. Moolhuysen, Electrochim. Acta 21 (1976) 869
93. E. Schwarzer and W. Vielstich, 3rd International Fuel Cells Symposium, Bruxelles 1969, Presses Academiques Europennes, Bruxelles, 1969, p. 220.
94. R.R. Adzic, D.N. Simic, A.R. Despic, and D.M. Drazic, J. Electroanal. Chem. 65 (1975) 587.
95. A. Katayama, J. Phys. Chem. 84 (1980) 376.
96. S. Szabo, J. Electroanal. Chem. 172 (1984) 359.
97. N. Furuya and S. Motoo, J. Electroanal. Chem. 98(1979) 195.
98. J. Sobkowski, K. Franaszczuk, and A. Piasecki, J. Electroanal. Chem. 196 (1985) 145.
99. T. Biegler, D.A.J. Rand, and R. Woods, J. Electroanal. Chem. 29 (1971) 269.
100. M.M. Stefenel, T. Chierchie, and C. Mayer, Z. Phys. Chem. Neue Folge 135 (1983) 251.
101. H. Angerstein-Kozlowska, D. Mac Dougal, and B.E. Conway, J. Electrochem. Soc. 120 (1973) 756.
102. A. Castro Luna, T. Iwasita, and W. Vielstich, J. Electroanal. Chem. 196 (1985) 301.
103. D. Zylka, T. Iwasita, and B. Bittins-Cattaneo, in preparation.
104. D. Donaldson, In: Progress in Inorganic Chemistry, F.A. Cotton (ed.), Vol. 8, p. 296. Interscience Pub. N.Y., 1967.

Theory and Experimental Aspects of the Rotating Hemispherical Electrode

Der-Tau Chin

Department of Chemical Engineering, Clarkson University,
Potsdam, New York 13676, USA

Contents

1 Introduction

The rotating hemispherical electrode (RHSE) was originally proposed by the author in 1971 as an analytical tool for studying high-rate corrosion and dissolution reactions [13]. Since then, much work has been published in the literature. The RHSE has a uniform primary current distribution, and its surface geometry is not easily deformed by metal deposition and dissolution reactions. These features have made the RHSE a complementary tool to the rotating disk electrode (RDE).

The common RDE is made of a circular metal disk embedded in a cylindrical or bell-shaped insulator. The surface of the metal disk exposed to the electrolyte is flush with the insulation surface, and the entire metal-insulation plane is rotated about the symmetry axis in the electrolyte. Although this arrangement permits the RDE to have a uniform limiting current distribution, its primary and secondary current distributions are nonuniform because of an edge effect by the insulation plane [44]. Serious experimental errors may be introduced by the RDE in the determination of electrode kinetic parameters below the limiting current density. A second drawback of the RDE is the requirement that the metal surface must be flush with the insulation surface on its rotating plane. This requirement impedes the application of RDE and ring-disk electrode to high-rate electrodissolution reactions, such as electrochemical machining, electropolishing, and high-rate corrosion processes. The metal surface being dissolved will recede from the rotating plane and forms a step between the metal disk and insulation. This surface irregularity generates flow turbulence and invalidates the quantitative interpretation of the RDE data.

The above problems may be alleviated by the use of a rotating hemispherical electrode (RHSE). In this geometry, the flat circular disk on the RDE is replaced with a metal hemisphere as shown in Fig. 1(a). The theory, experimental setup and methods of application to electrochemical studies are similar to those of the RDE. The advantages of the RHSE are:

1. The primary current distribution is uniform on the hemisphere. Numerical calculations using the potential theory have shown that the current distribution is essentially uniform on the RHSE if the current density is less than 68% of the average limiting current density [47].
2. During the dissolution reactions, the RHSE merely reduces in size. Thus, the change in the flow characteristics will be insignificant within a reasonably short duration of an experiment.
3. The RHSE protrudes from the surface of an inert support rod. The electrode can be easily replaced with a new one, and only one support rod is needed. This feature is advantageous if the hemisphere is used in conjunction with a concentric ring electrode located on the support surface to form a rotating ring-hemisphere electrode (RRHSE) for studying reaction intermediates of a corrosion reaction [20]. The fabrication of a replaceable central disk electrode in the rotating ring-disk setup is difficult because of the requirement of a flat surface on its rotating plane.
4. For electrochemical reactions involving gaseous reactants and products, the RHSE has an advantage over a downward-faced RDE in sweeping away gas bubbles from its surface because of combined action of centrifugal and buoyant forces.

This article presents a brief account of theory and practical aspects of rotating hemispherical electrodes. The fluid flow around the RHSE, mass transfer correlations, potential profile, and electrochemical application to the investigations of diffusivity, reaction rate constants, intermediate reaction products, passivity, and AC techniques are reviewed in the following sections.

Fig. 1. Flow near a rotating hemisphere electrode . (a) Dye movement at $Re = 1300$. (b, c) Spiral flow patterns etched on a copper hemisphere.

2 Theory

2.1 Fluid Flow

The behavior of a rotating sphere or hemisphere in an otherwise undisturbed fluid is like a centrifugal fan. It causes an inflow of the fluid along the axis of rotation toward the spherical surface as shown in Fig. 1(a). Near the surface, the fluid flows in a spiral-like motion towards the equator as shown in Fig. 1(b) and (c). On a rotating sphere, two identical flow streams develop on the opposite hemispheres. The two streams interact with each other at the equator, where they form a thin swirling jet toward the bulk fluid. The Reynolds number for the rotating sphere or hemisphere is defined as:

$$Re = a^2\Omega/v \tag{1}$$

where a is the radius of the sphere, Ω is its angular velocity, and v is the kinematic viscosity of the fluid.

The problem of a slowly rotating sphere was first considered by Stokes [58], Lamb [41], and Bickley [6]. They estimated the location where the transition from inflow along the pole to outflow near the equator takes place at low Reylonds numbers. Howarth [33] introduced the concept of laminar boundary layer flow on a rotating sphere for large Reynolds numbers, and obtained an approximate solution using the Karman-Pohlhausen integral method. He showed that the motion near the pole of rotation is similar to that on an infinitely large rotating disk. Since then, the analysis

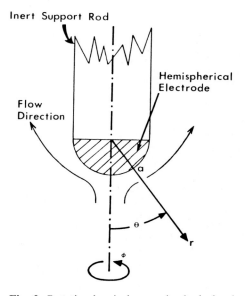

Fig. 2. Rotating hemisphere and spherical polar coordinates.

of laminar boundary layer on a rotating sphere has been further investigated by many others including Nigam [46], Fox [30], Banks [2, 4], Manohar [42], Singh [56], and Dennis et al. [28]. Experimental investigations of flow boundary layers on a rotating sphere have been carried out by Kobashi [38], Bowden and Lord [7], Kreith et al. [40], Sawatzki [53], and Kohama and Kobayashi [39]. Using a visual technique and hot wire anemometry, these investigations revealed that in laminar flow and at sufficiently high Reynolds numbers, the outflowing swirling jet is confined to a thin region of less than 2° latitude from the equator. The transition from laminar to turbulent flow occur approximately at a Reynolds number of 15,000–40,000.

To describe the velocity profile in laminar flow, let us consider a hemisphere of radius a, which is mounted on a cylindrical support as shown in Fig. 2 and is rotating in an otherwise undisturbed fluid about its symmetric axis. The fluid domain around the hemisphere may be specified by a set of spherical polar coordinates, r, θ, ϕ, where r is the radial distance from the center of the hemisphere, θ is the meridional angle measured from the axis of rotation, and ϕ is the azimuthal angle. The velocity components along the r, θ, and ϕ directions, are designated by V_r, V_θ, and V_ϕ. It is assumed that the fluid is incompressible with constant properties and the Reynolds number is sufficiently high to permit the application of boundary layer approximation [54]. Under these conditions, the laminar boundary layer equations describing the steady-state axisymmetric fluid motion near the spherical surface may be written as:

Equation of continuity

$$\frac{\partial V_r}{\partial r} + \frac{1}{a}\frac{\partial V_\theta}{\partial \theta} + \frac{\cot\theta}{a} V_\theta = 0 \tag{2}$$

θ-momentum

$$V_r \frac{\partial V_\theta}{\partial r} + \frac{V_\theta}{a}\frac{\partial V_\theta}{\partial \theta} - \frac{V_\phi^2 \cot\theta}{a} = v\frac{\partial^2 V_\theta}{\partial r^2} \tag{3}$$

ϕ-momentum

$$V_r \frac{\partial V_\phi}{\partial r} + \frac{V_\theta}{a}\frac{\partial V_\phi}{\partial \theta} + \frac{V_\theta V_\phi}{a}\cot\theta = v\frac{\partial^2 V_\phi}{\partial r^2} \tag{4}$$

with the *boundary conditions*

$$\left. \begin{array}{lll} \text{at } r & = a & V_r = V_\theta = 0 \\ & & V_\phi = a\Omega\sin\theta \\ \text{at } r & \to \infty & V_\theta = V_\phi = 0 \\ \text{at } \theta & = 0 & V_r, V_\theta, V_\phi \text{ finite} \end{array} \right\} \tag{5}$$

Howarth [33] showed that the above partial differential equations may be transformed into a set of ordinary differential equations by introducing the following dimensionless variables and velocity components:

Dimensionless radius

$$\eta = (\Omega/v)^{1/2}(r - a) \tag{6}$$

Dimensionless meridional velocity

$$F(\eta, \theta) = (V_\theta/a\Omega) = [\theta F_1(\eta) + \theta^3 F_3(\eta)$$
$$+ \theta^5 F_5(\eta) + \theta^7 F_7(\eta) + \cdots] \tag{7}$$

Dimensionless azimuthal velocity

$$G(\eta, \theta) = (V_\phi/a\Omega) = [\theta G_1(\eta) + \theta^3 G_3(\eta)$$
$$+ \theta^5 G_5(\eta) + \theta^7 G_7(\eta) + \cdots] \tag{8}$$

Dimensionless radial velocity

$$H(\eta, \theta) = (V_r/\sqrt{v\Omega}) = [H_1(\eta) + \theta^2 H_3(\eta)$$
$$+ \theta^4 H_5(\eta) + \theta^6 H_7(\eta) + \cdots] \tag{9}$$

Substituting Eqs. (6)–(9) into Eqs. (2)–(5), one has:

$$\left.\begin{array}{l} F_1^2 + H_1 F_1' - G_1^2 = F_1'', \\ 2F_1 G_1 + H_1 G_1' = G_1'', \\ 2F_1 + H_1' = 0. \end{array}\right\} \tag{10}$$

$$\left.\begin{array}{l} 4F_1 F_3 + H_1 F_3' + H_3 F_1' - 2G_1 G_3 + G_1^2/3 = F_3'', \\ 4F_1 G_3 + 2F_3 G_1 + H_1 G_3' + H_3 G_1' - F_1 G_1/3 = G_3'', \\ 4F_3 + H_3' - F_1/3 = 0 \end{array}\right\} \tag{11}$$

$$\left.\begin{array}{l} 6F_1 F_5 + 3F_3^2 + H_1 F_5' + H_3 F_3' + H_5 F_1' - 2G_1 G_5 - G_3^2 \\ \qquad + 2G_1 G_3/3 + G_1/45 = F_5'', \\ 6F_1 G_5 + 4F_3 G_3 + 2F_5 G_1 + H_1 G_5' + H_3 G_3' + H_5 G_1' - F_1 G_3/3 - F_3 G_1/3 \\ \qquad - F_1 G_1/45 = G_5'', \\ \qquad 6F_5 + H_5' - F_1/45 - F_3/3 = 0 \end{array}\right\} \tag{12}$$

$$8F_1 F_7 + 8F_3 F_5 + H_1 F_7' + H_3 F_5' + H_5 F_3' + H_7 F_1' - 2G_1 G_7 - 2G_3 G_5$$
$$\qquad + G_3^2/3 + 2G_1 G_5/3 + 2G_1 G_3/45 + 2G_1^2/945 = F_7'',$$
$$8F_1 G_7 + 6F_3 G_5 + 4F_5 G_3 + 2F_7 G_1 + H_1 G_7' + H_3 G_5' + H_5 G_3' + H_7 G_1' - F_1 G_5/3$$
$$\qquad - F_3 G_3/3 - F_5 G_1/3 - F_1 G_3/45 - F_3 G_1/45 - 2G_1 F_1/945 = G_7''. \tag{13}$$
$$\qquad 8F_7 + H_7' - 2F_1/945 - F_3/45 - F_5/3 = 0$$

with the boundary conditions

$$
\begin{aligned}
\text{at } \eta = 0 \quad & F_n = H_n = 0 \\
& G_n = (-1)^{(n-1)/2}/n! \\
\text{at } \eta \to \infty \quad & F_n = G_n = 0
\end{aligned}
\Biggr\}
\tag{14}
$$

where the prime, $'$, refers to the differentiation with respect to η. The set of ordinary differential Eqs. (10)–(14) has been integrated numerically by Banks [2]. His results for the radial velocity component, $H_n(\infty)$, and the meridional and azimuthal velocity gradients, $F_n'(0)$ and $G_n'(0)$, are listed in Table 1 for $n = 1, 3, 5$, and 7. With the exception of the radial velocity component, V_r, at the equator, the series expansions of Eqs. (7)–(9) are everywhere convergent. The first four terms in the series agreed with the experimental velocity profile for θ up to 1.31 (or 75°) [53]. Note that the results for the first term in the series expansions, $F_1'(0)$, $G_1'(0)$, and $H_1(\infty)$, are identical to those on a rotating disk [51]. Thus, the velocity profile at the pole of a rotating sphere is the same as that on a rotating disk.

Table 1. Numerical results for the first four terms in the power series expansions of the velocity profile near a rotating sphere [2].

n	$F_n'(0)$	$G_n'(0)$	$H_n(\infty)$
1	0.51023	−0.61592	−0.88445
3	−0.22129	0.24765	0.16070
5	0.02071	−0.02569	0.00084
7	−0.00189	0.00181	0.00084

For large values of θ, Manohar [42], and Banks [4] solved the boundary layer Eqs. (2)–(5) numerically with a finite difference method. Manohar's results for the meridional and azimuthal velocity gradients on the spherical surface have been curve-fitted by Newman [45] and Chin [18] to follow the following equations in the regime of $0 < \theta \leqslant \pi/2$:

$$
F'(0) = \frac{v^{1/2}}{a\Omega^{3/2}} \frac{\partial V_\theta}{\partial r}\bigg|_{r=a}
$$
$$
= 0.51023 - 0.18088\,\theta^3 - 0.04041\sin^3\theta
\tag{15}
$$

$$
G'(0) = \frac{v^{1/2}}{a\Omega^{3/2}} \frac{\partial V_\phi}{\partial r}\bigg|_{r=a}
$$
$$
= 0.61592\,\theta - 0.18946\,\theta^3 - 0.05819\sin^3\theta
\tag{16}
$$

The torque required to maintain the hemisphere at a constant speed of rotation may be calculated by

$$
T_q = 2\pi a^3 \int_0^{\pi/2} \left(-\mu\frac{\partial V_\phi}{\partial r} \right)_{r=a} \sin^2\theta\, d\theta
\tag{17}
$$

In fluid mechanics, it is customary to express the torque requirement in terms of a turning moment coefficient C_M defined as

$$C_M = 2T_q/(\tfrac{1}{2}\rho a^5 \Omega^2)$$

$$= -(8\pi/Re^{1/2}) \int_0^{\pi/2} G'(0)\sin^2\theta\,d\theta \tag{18}$$

Substituting Eq. (16) into Eq. (18), and carrying out the integration, one obtains

$$C_M = 6.53/Re^{1/2}, \quad \text{(for } Re < 4\times 10^4) \tag{19}$$

Banks [2] used the results of the four term series expansion for $G'(0)$ listed in Table 1, and obtained $C_M = 6.54/Re^{1/2}$, in good agreement with the numerical results of Manohar [42].

The transition from laminar to turbulent flow on a rotating sphere occurs approximately at $Re = 1.5 \sim 4.0 \times 10^4$. Experimental work by Kohama and Kobayashi [39] revealed that at a suitable rotational speed, the laminar, transitional, and turbulent flow conditions can simultaneously exist on the spherical surface. The regime near the pole of rotation is laminar whereas that near the equator is turbulent. Between the laminar and turbulent flow regimes is a transition regime, where spiral vortices stationary relative to the surface have been observed. The direction of these spiral vortices is about $4 \sim 14°$ from the negative direction of the azimuthal angle, ϕ. The phenomenon is similar to the flow transition on a rotating disk [19].

For turbulent flow on a rotating sphere or hemisphere, Sawatzki [53] and Chin [22] have analyzed the governing equations using the Karman-Pohlhausen momentum integral method. The turbulent boundary layer was assumed to originate at the pole of rotation, and the meridional and azimuthal velocity profiles were approximated with the one-seventh power law. Their results can be summarized by the following equations:

Dimensionless radius

$$\eta = (r-a)/\delta \tag{20}$$

Meridional velocity

$$V_\theta = \alpha a\Omega \sin\theta . \eta^{1/7}(1-\eta) \tag{21}$$

Azimuthal velocity

$$V_\phi = a\Omega \sin\theta(1 - \eta^{1/7}) \tag{22}$$

Thickness of turbulent momentum boundary layer

$$\delta = \frac{a}{Re^{1/5}}\Delta(\theta) \tag{23}$$

Wall sear stress component, $\tau_{r\theta}$

$$\tau_{r\theta} = -0.0225\rho\alpha(1+\alpha^2)^{3/8}(a\Omega \sin\theta)^{7/4} v^{1/4}\delta^{-1/4} \tag{24}$$

Wall shear stress component, $\tau_{r\phi}$

$$\tau_{r\phi} = 0.0225\rho(1 + \alpha^2)^{3/8}(a\Omega\sin\theta)^{7/4}\nu^{1/4}\delta^{-1/4} \qquad (25)$$

Turning moment coefficient

$$C_M = 0.348/Re^{1/5} \qquad (26)$$

where α and Δ are dimensionless functions of θ. The quantity, Δ, may be considered as the dimensionless thickness of the momentum boundary layer. Near the spherical surface, the direction of local fluid velocity relative to the surface is at an angle of $\tan^{-1}(1/\alpha)$ from the θ-direction and is at an angle of $\cot^{-1}(1 - \alpha)$ from the negative of the ϕ-coordinate. The values of α and Δ as calculated by Chin [22] are plotted in Fig. 3 for a range of θ from 0 to $\pi/2$.

Fig. 4 shows a comparison between the experimental measurements of Sawatzki [53] and theoretical predictions for the turning moment coefficient, C_M, over a wide range of Reynolds numbers from 1 to 10^7. Eq. (26) for turbulent flow agrees with the data to within 4% for $Re > 10^5$. Eq. (19) for laminar flow agrees with the measurements of Sawatzki (53) and Bowden and Lord [7] to within $4 \sim 10\%$ in the regime of

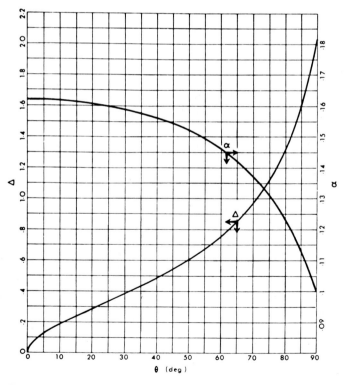

Fig. 3. Numerical values of Δ and α for the solution of turbulent flow boundary layer on a rotating hemisphere. The value of meridional angle, θ, is given in degrees. From [22].

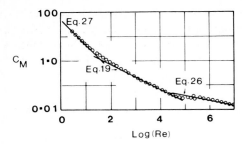

Fig. 4. Comparison between the theory and experimental data [53] for the turning moment coefficient.

$200 < Re < 4 \times 10^4$. For $Re < 200$, Eq. (19) starts to deviate from the experimental data due to the limitation of the boundary layer approximation ($a \gg \delta$) used in the analysis. Lamb's analytical results [41] for a slowly rotating sphere,

$$C_M = 50.3/Re \tag{27}$$

agrees with the data in the flow regime of $1 < Re < 20$.

2.2   Mass Transfer

Analysis of heat and mass transfer to a rotating sphere in laminar flow has been made by Baxter and Davies [5], Singh [56], Banks [3], Chin [13, 16], Newman [45], and Wein [60]. Banks obtained a numerical solution of the thermal boundary layer at the Prandtl numbers of 0.7–1.0. Baxter and Davies' work was confined to large Prandtl numbers; they obtained an implicit solution based on Howarth's [33] approximate momentum solutions. Wein carried out an analysis of mass transfer to a slowly rotating sphere for $Re \to 0$. Chin used Banks' [2] power series expressions for the velocity distributions and solved the convective diffusion equation with an asymptotic expansion method for a range of Schmidt numbers from 0.7 to infinity. Newman calculated the rate of mass transfer using Manohar's [42] numerical results of wall shear stress on the sphere surface. The experimental measurement of heat transfer for a rotating sphere was carried out by Kreith et al. [40]. The rate of mass transfer to a rotating hemisphere was measured by Chin [14, 22] using an electrochemical limiting current technique, and by Kim and Jorne [36] using mass transfer limited corrosion reactions.

2.2.1   Analytical Solution for Mass Transfer in Laminar Flow

To consider the convective mass transfer problem of a rotating hemisphere electrode, we assume that sufficient inert salts are present in the electrolyte that the migrational flux of the diffusing ion in the electric field is negligible. The electrode has a radius a, and is rotating in an electrolyte of constant properties at a sufficiently high rotational speed to permit the application of the boundary layer approximation. The differential equation describing the axisymmetric concentration distribution of a diffusing species in laminar flow may be expressed as:

$$V_r \frac{\partial C}{\partial r} + \frac{V_\theta}{a} \frac{\partial C}{\partial \theta} = D \frac{\partial^2 C}{\partial r^2} \tag{28}$$

with the boundary conditions

$$
\left.
\begin{array}{ll}
\text{at } r = a & C = C_s \\
\text{at } r \to \infty & C = C_\infty \\
\text{at } \theta = 0 & C \text{ finite}
\end{array}
\right\} \tag{29}
$$

where C_s and C_∞ refer respectively to the surface and bulk concentrations of the diffusing species; and D is its diffusivity in the electrolyte. From Eqs. (7)–(9) and Banks' analytical results listed in Table 1, one may express the meridional velocity, V_θ, and radial velocity, V_r, in the following forms for small distance from the spherical surface.

$$V_\theta = a\Omega[\theta(0.51023\,\eta - \tfrac{1}{2}\eta^2 + 0.20531\,\eta^3 + \cdots)$$
$$+ \theta^3(-0.22129\eta + \tfrac{1}{3}\eta^2 - 0.18520\eta^3 + \cdots) + \cdots] \tag{30}$$

$$V_r = (\nu\Omega)^{1/2}[(-0.51023\eta + \tfrac{1}{3}\eta^2 - 0.10265\,\eta^4 + \cdots)$$
$$+ \theta^2(0.52762\,\eta - \tfrac{1}{2}\eta^3 + 0.20231\,\eta^4 + \cdots) + \cdots] \tag{31}$$

with $\eta = (\Omega/\nu)^{1/2}(r - a)$. Eq. (28) may be reduced to a set of ordinary differential equations by expanding a dimensionless concentration, $\Phi = (C - C_\infty)/(C_s - C_\infty)$, into a power series of θ:

$$\Phi = \frac{C - C_\infty}{C_s - C_\infty} = \Phi_1(\eta) + \theta^2 \Phi_3(\eta) + \cdots \tag{32}$$

Substituting Eqs. (30)–(32) into Eq. (28), one obtains

Zeroth-order equation

$$(-0.51023\,\eta^2 + \tfrac{1}{3}\eta^3 - 0.10265\,\eta^4 + \cdots)\frac{d\Phi_1}{d\eta} = \frac{1}{Sc}\frac{d^2\Phi_1}{d\eta^2} \tag{33}$$

First-order equation

$$(0.52762\,\eta^2 - \tfrac{1}{2}\eta^3 + 0.20231\,\eta^4 + \cdots)\frac{d\Phi_1}{d\eta}$$

$$+ (-0.51023\,\eta^2 + \tfrac{1}{3}\eta^3 - 0.10265\,\eta^4 + \cdots)\frac{d\Phi_3}{d\eta}$$

$$+ 2(0.51023\,\eta - \tfrac{1}{2}\eta^2 + 0.20531\,\eta^3 + \cdots)\Phi_3 = \frac{1}{Sc}\frac{d^2\Phi_3}{d\eta^2} \tag{34}$$

with the *boundary conditions*

$$\text{at } \eta = 0, \quad \Phi_1 = 1, \Phi_3 = 0 \left.\right\}$$
$$\text{at } \eta \to \infty, \quad \Phi_1 = \Phi_3 = 0 \left.\right\} \tag{35}$$

Eqs. (33)–(35) have been solved analytically by Chin [13]. For large Schmidt numbers, Sc, the concentration distribution in the viscinity of the rotating sphere is given by:

$$\Phi = \frac{C - C_\infty}{C_s - C_\infty} = 1 - 0.62045 \, Sc^{1/3} \int_0^\eta \exp(-0.17008 \, Sc\eta^3) \, d\eta$$
$$- 0.12833 \, Sc^{1/3} \, \theta^2 \, \eta \exp(-0.17008 \, Sc \, \eta^3) + 0(\theta^4) \tag{36}$$

The local mass transfer rate is related to the concentration gradient by

$$j_m = -D \left(\frac{\partial C}{\partial r} \right)_{r=a} = k_m (C_s - C_\infty) \tag{37}$$

where k_m is the local mass transfer coefficient.

For an electrochemical reaction of the type

$$\text{Ox} + ne^- \to \text{Red} \tag{38}$$

taking place on a rotating hemisphere, the limiting current density on the spherical surface is related to the mass transfer coefficient of the diffusing species, Ox, by

$$i_{\lim} = nF|j_m| = nF k_m C_\infty \tag{39}$$

Using Eqs. (36)–(39), one obtains the following expressions for the local Sherwood number and local limiting current density on the speherical surface

$$Sh_{\text{loc}} = \left(\frac{k_m a}{D} \right) = Re^{1/2} Sc^{1/3} (0.62045 - 0.12833 \, \theta^2), \quad \text{(for } Sc \gg 1) \tag{40}$$

$$i_{\lim} = nFC_\infty D^{2/3} v^{-1/6} \Omega^{1/2} (0.62045 - 0.12833 \, \theta^2) \tag{41}$$

The average Sherwood number, Sh_{av}, and average limiting current density, $i_{\lim, \text{av}}$, can be obtained by integrating Eqs. (40)–(41) over the electrode surface. The results for a hemisphere, whose entire surface is subject to mass transfer are:

$$Sh_{\text{av}} = \left(\frac{k_{m, \text{av}} a}{D} \right) = 0.474 \, Re^{1/2} \, Sc^{1/3}, \quad \text{(for } Sc \gg 1) \tag{42}$$

$$i_{\lim, \text{av}} = 0.474 nFC_\infty D^{2/3} v^{-1/6} \Omega^{1/2} \tag{43}$$

where $k_{m, \text{av}}$ is the average mass transfer coefficient on the rotating hemisphere.

Eqs. (40)–(41) are obtained from the analytical solution using the first two terms in the θ-series expansion of the concentration profile. As a result, they are accurate only for small values of meridional angle, θ. To correct for large values of θ, Newman [45] used Lighthill's transformation and Eq. (15) for the meridional velocity gradient to calculate the local mass transfer rate as $Sc \to \infty$. His numerical result is plotted in Fig. 5 in the form of $Sh_{\text{loc}}/Re^{1/2} Sc^{1/3}$ vs. θ as the thin solid line. The dashed line is

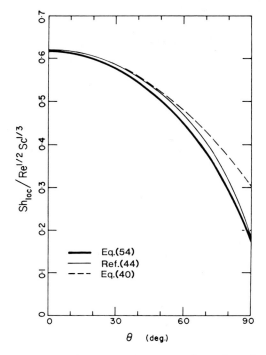

Fig. 5. Local mass transfer rate on the surface of a rotating hemisphere in laminar flow. Here the meridional angle, θ, is given in degrees.

calculated from the result of θ-expansion analysis, Eq. (40). Also shown in the figure is the local laminar mass transfer rate using Chilton–Colburn's analogy discussed in the next section. Eq. (40) starts to deviate from the numerical results when $\theta > 1.05$ (or 60°), and considerably overestimates the rate of mass transfer near the equator. The average Sherwood number and average limiting current of a rotating hemisphere as corrected by Newman are:

$$Sh_{av} = 0.451\,Re^{1/2}\,Sc^{1/3}, \quad (\text{for } Sc \gg 1) \tag{44}$$

$$i_{lim,\,av} = 0.451\,nFC_{\infty}\,D^{2/3}v^{-1/6}\Omega^{1/2} \tag{45}$$

Eqs. (40)–(45) describe the behavior of a mass transport process occurring at very large Schmidt numbers. For finite values of Sc, Chin [16] extended the θ-series expansion analysis and obtained the following asymptotic series for Sh_{av} on a rotating hemisphere:

$$Sh_{av} = 0.47396\,Re^{1/2}\,Sc^{1/3}(1 - 0.28549\,Sc^{-1/3} - 0.5097\,Sc^{-2/3}) \tag{46}$$

This correlation has been shown to agree with Bank's (3) numerical results for $Sc = 0.7$ and 1.0. At high Schmidt numbers, the second and third terms are negligible, and Eq. (46) asymptotically approaches to the values predicted by Eq. (42).

2.2.2   Mass Transfer in Turbulent Flow

For turbulent flow, we shall use the Chilton–Colburn analogy [12] to derive an expression for mass transfer to the spherical surface. This analogy is based on an investigation of heat and mass transfer to a flat plate situated in a uniform flow stream. At high Schmidt numbers, the local mass transfer rate is related to the local wall shear stress by

$$Sh_x = Re_x Sc^{1/3}(f_x/2) \tag{47}$$

where Sh_x, Re_x, and f_x are the local Sherwood number, local Reynolds number and local friction coefficient, respectively, based on the surface distance, x, from the leading edge of the flow boundary layer. Although this correlation is semi-empirical in nature, it has been used successfully to estimate the rate of heat and mass transfer for a number of flow geometries, including channel flow [34], rotating disk [29], and impinging jet [26].

For the flow induced by the rotating hemisphere, the leading edge of the flow boundary layer occurs at the pole of rotation. The surface distance, x, from the leading edge is equal to $a\theta$. One may also take $a\Omega\sin\theta$ as the characteristic velocity for every local point on the spherical surface. Thus, the quantities, Sh_x, Re_x, and f_x can be expressed as

$$Sh_x = k_m a\theta/D \tag{48}$$

$$Re_x = a^2\Omega\theta\sin\theta/v \tag{49}$$

$$f_x = \tau_{r\phi}/(\tfrac{1}{2}\rho a^2\Omega^2\sin^2\theta) \tag{50}$$

For turbulent flow, the local wall shear stress, $\tau_{r\phi}$, is given by Eq. (25). Substituting Eqs. (48)–(50) into Eq. (47) and making use of Eq. (25), one arrives at an expression for the Sherwood number based upon the radius of the rotating hemisphere:

$$Sh_{\mathrm{loc}}(\text{turbulent}) = 0.0225\, Re^{4/5}\, Sc^{1/3}\,\frac{(1+\alpha^2)^{3/8}\sin^{3/4}\theta}{\Delta^{1/4}}, \quad (\text{for } Sc \gg 1) \tag{51}$$

Here, the dimensionless thickness of the momentum boundary layer Δ, and the dimensionless quantity, α, relating to the direction of local flow are functions of θ; they are shown graphically in Fig. 3. The average Sherwood number may be obtained by integrating Eq. (51) over the surface; the result for a hemisphere may be given as [22]:

$$Sh_{\mathrm{av}}(\text{turbulent}) = 0.0198\, Re^{4/5}\, Sc^{1/3}, \quad (\text{for } Sv \gg 1) \tag{52}$$

The Chilton–Colburn analogy can be also used to estimate the local mass transfer rate in laminar flow where the wall shear stress is related to the azimuthal velocity gradient by

$$\tau_{r\phi} = -\mu\left(\frac{\partial V_\phi}{\partial r}\right)_{r=a} = -\frac{a\Omega^{3/2}}{v^{1/2}}\,G'(0) \tag{53}$$

Substituting Eqs. (48)–(50), and (52) into Eq. (47) and making use of Eq. (16), one has

$$Sh_{loc}(\text{laminar}) = Re^{1/2} Sc^{1/3} \frac{0.61592\theta - 0.18946\,\theta^3 - 0.05819\sin^3\theta}{\sin\theta},$$

(for $Sc \gg 1$)   (54)

Eq. (54) is plotted in Fig. 5 as the thick solid curve. It is seen that results of Chilton–Colburn analogy for laminar flow is in good agreement with the two previous theories based on more rigorous mathematical treatment. The maximum deviation of Eq. (54) from Newman's curve, which occurs at $\theta = 1.4$ (or 80°), is less than 9%. Eq. (54) gives $0.433 Re^{1/2} Sc^{1/3}$ dependence for the average Sherwood number on a rotating hemisphere, in good agreement with the theoretical results of Eqs. (42) and (44).

2.2.3 Experimental Verification of Mass Transfer Theory

The experimental measurement of mass transfer to a rotating hemisphere has been made by Chin [14, 22] using the limiting current technique over the Reynolds number range of $10^2 \sim 10^5$ and by Kim and Jorne [36] over the Reynolds number range of $Re = 200 \sim 6000$. Fig. 6 shows a comparison between the experimental results and theoretical predictions in both laminar and turbulent flow regimes. The experimental data for a range of Sc from 920 to 6300 are plotted in the form of a log–log of $Sh_{av}/Sc^{1/3}$ vs. Re. Two kinds of rotating hemisphere geometries are shown in the figure. The data with a support rod of an equal radius are represented by the open symbols, and those obtained with a support rod of a larger radius are represented by the filled symbols. For comparison, the heat transfer measurement on a rotating sphere in turbulent flow [40] are also given in the figure as the thin dashed line. The theoretical predictions of Eqs. (42) and (44) for laminar flow, and of Eq. (52) for turbulent flow are given in the figure as the thick solid lines.

In laminar flow, the size of support rod does not affect the mass transfer rate on the hemisphere, and both Eqs. (42) and (44) agree well with the experimental data. Although the $0.474 Re^{1/2}$ line seems to give a better fit of the data in the figure, the $0.451 Re^{1/2}$ relation agrees with Kim and Jorne's data in the Reynolds number range of $200 \sim 6000$. Recently, Razafiarimanana et al. [49] reported a $0.46 Re^{1/2}$ dependence for $Sh_{av}/Sc^{1/3}$ on a rotating sphere in laminar flow.

In turbulent flow, the edge effect due to the shape of the support rod is quite significant as shown in Fig. 6. The data obtained with a support rod of equal radius agree with the theoretical prediction of Eq. (52). The point of transition with this geometry occurs at $Re = 40000$. However, the use of a larger radius support rod arbitrarily introduces an outflowing radial stream at the equator. The radial stream reduces the stability of the boundary layer, and the transition from laminar to turbulent flow occurs earlier at $Re = 15000$. Thus, the turbulent mass transfer data with the larger radius support rod deviate considerably from the theoretical prediction of Eq. (52); a least square fit of the data results in a $0.092 Re^{0.67}$ dependence for

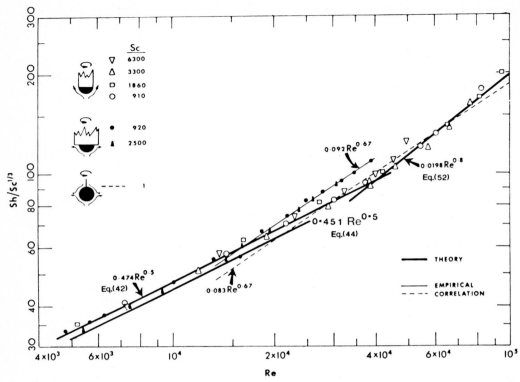

Fig. 6. Comparison between mass transfer correlations and results of limiting current measurements [14, 22].

$Sh_{av}/Sc^{1/3}$ as indicated by the thin solid line. This 0.67 power of Re agrees with the result of a turbulent heat transfer measurement on a rotating sphere [40]. Since the flow induced by a rotating sphere is also characterized by an outflowing radial jet at the equator caused by the collosion of two opposing flow boundary layers on the sphere, the 0.67 power dependence on Re is clearly related to the radial flow stream away from the equator.

2.3 Potential Profile and Current Distribution

In electrochemistry, spherical and hemispherical electrodes have been commonly used in the laboratory investigations. The spherical geometry has the advantage that in the absence of mass transfer effect, its primary and secondary current distributions are uniform. However, the limiting current distribution on a rotating sphere is not uniform. The limiting current density is highest at the pole, and decreases with increasing θ toward the equator as discussed in the preceeding section. The situation is contrary to that of a rotating disk electrode where the limiting current distribution is uniform and primary current distribution is nonuniform.

Let us consider a sphere or a hemisphere electrode mounted on an inert support rod of a larger radius as shown in Fig. 1(a). The electrode is rotating about its axis at a sufficiently high rotational speed to permit the application of boundary layer theory. The flow is laminar, and the solution contains excess supporting electrolyte such that the migrational effect of the diffusing ion is negligible. The reference electrode is located at a large distance from the sphere and its electric potential is taken to be zero. The potential profile of the electrolyte in the neighborhood of the spherical electrode may be obtained by solving the axisymmetric Laplace equation in spherical co-ordinates:

$$\frac{\partial}{\partial r}\left(r^2 \frac{\partial \Phi}{\partial r}\right) + \frac{1}{\sin^2 \theta} \frac{\partial^2 \Phi}{\partial \theta^2} = 0 \tag{55}$$

with the boundary conditions

$$\left.\begin{array}{ll} \text{at } r = a & \partial \Phi / \partial r = -i(\theta)/\kappa \\ \text{at } r \to \infty & \Phi = 0 \\ \text{at } \theta = 0 & \partial \Phi / \partial \theta = 0 \\ \text{at } \theta = \pi/2 & \partial \Phi / \partial \theta = 0 \end{array}\right\} \tag{56}$$

where κ is the electric conductivity of solution, and $i(\theta)$ is the local current density on the electrode surface. Eqs. (55)–(56) may be solved by the method of separation of variables; the result for the electric potential of electrolyte on the electrode surface is

$$\Phi_s = \eta_{\text{ohm}} = \frac{a}{\kappa} \sum_{m=0}^{\infty} \frac{4m+1}{2m+1} P_{2m}(\cos \theta) \int_0^1 i(\theta) P_{2m}(\cos \theta) \, d(\cos \theta) \tag{57}$$

where $P_{2m}(\cos \theta)$ is the Legendre polynomial of order $2m$. Since the electric potential of the electrolyte on the reference plane is taken to be zero, the value of Φ_s is numerical equal to the ohmic potential drop between the sphere and the reference electrode.

The other potential losses required to drive an electrode reaction are the activation overpotential, η_a, and concentration overpotential, η_{conc}. The problem of current distribution is then governed Eq. (57) as well as by the following equations:

Voltage balance

$$\eta_{\text{total}} = \eta_a(\theta) + \eta_{\text{conc}}(\theta) + \Phi_s(\theta) \tag{58}$$

Electrode kinetics

$$i(\theta) = i_o \left[\frac{C_s(\theta)}{C_\infty}\right]^\gamma \left\{\exp\left[\frac{\alpha_a nF}{RT} \eta_a(\theta)\right] - \exp\left[-\frac{\alpha_c nF}{RT} \eta_a(\theta)\right]\right\} \tag{59}$$

Here i_o is the exchange current density of the electrode reaction based on the bulk concentration; α_a and α_c are the anodic and cathodic charge transfer coefficients, respectively; and γ is a dimensionless kinetic parameter.

Concentration overpotential

$$\eta_{\text{conc}}(\theta) = \frac{RT}{nF} \ln \frac{C_s(\theta)}{C_\infty} \tag{60}$$

Convective diffusion

$$V_r \frac{\partial C}{\partial r} + \frac{V_\theta}{a} \frac{\partial C}{\partial \theta} = D \frac{\partial^2 C}{\partial r^2} \tag{28}$$

with the boundary conditions

$$\left.\begin{array}{lll}
\text{at } r = a & \partial c/\partial r = -i(\theta)/nFD \\
\text{at } r & C = C_\infty \\
\text{at } \theta = 0 & \partial c/\partial \theta = 0
\end{array}\right\} \tag{61}$$

Eqs. (28) and (57)–(61) must be simultaneously solved for six unknowns, $i(\theta)$, $C(r, \theta)$, $C_s(\theta)$, $\eta_a(\theta)$, $\eta_{\text{conc}}(\theta)$, and $\Phi_s(\theta)$. In numerical calculations it is useful to introduce the following dimensionless parameters:

$$N \left(\begin{array}{c} \text{dimensionless limiting} \\ \text{current density} \end{array}\right) = \frac{n^2 F^2 DC_\infty}{9RT\kappa} Re^{1/2} Sc^{1/3} \tag{62}$$

$$J \left(\begin{array}{c} \text{dimensionless exchange} \\ \text{current density} \end{array}\right) = nFai_o/RT\kappa \tag{63}$$

The dimensionless limiting current density N represents the ratio of ohmic potential drop to the concentration overpotential at the electrode. A large value of N implies that the ohmic resistance tends to be the controlling factor for the current distribution. For small values of N, the concentration overpotential is large and the mass transfer tends to be the rate-limiting step of the overall process. The dimensionless exchange current density J represents the ratio of the ohmic potential drop to the activation overpotential. When both N and J approach infinity, one obtains the geometrically dependent primary current distribution.

Nisancisglu and Newman [47] have calculated the current distribution on a rotating sphere for a first order cathodic reaction having $\alpha_a = \alpha_c = \gamma = 1/2$. Their results for the condition of constant surface flux at very high rotational speed ($N \to \infty$) are shown in Fig. 7. It is seen that a uniform current density can be maintained on the surface of a rotating sphere so long as the current density is smaller than 68% of the average limiting current density ($i/i_{\text{lim, av}} < 0.68$). Above this value, the surface concentration of a reactant will become zero at a certain meridional angle, beyond which the current drops and follows the limiting current curve as shown in the figure. The maximum potential variation on the spherical surface occurs at the limiting current densities. Nisancioglu and Newman [47] have estimated the maximum surface potential difference between the pole and equator to be:

$$\Delta\Phi_s = 0.546 \, ai_{\text{lim, av}}/\kappa \tag{64}$$

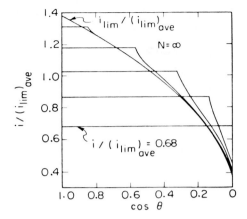

Fig. 7. Current distribution on a rotating hemispherical electrode at high rotational speed. From [47].

This formula is useful in the design calculation to determine the electrode size for a maximum allowable surface potential variation during the experiments.

3 Experimental Considerations

The experimental setup of a rotating hemispherical electrode (RHSE) is similar to that of a rotating disk electrode [50]. The basic system consists of a removable hemispherical electrode, and a variable speed rotator equipped with a provision, such as the slip-ring contact, to make electric connection to the hemispherical electrode during the experiments.

   The application of RHSE is primarily in the laminar boundary layer flow regime of $Re < 15000$, where the edge effect is negligible and the mass transfer theory has been confirmed by experimental investigations. An important consideration in the design of a practical RHSE system is to conform to the theoretical requirement that the boundary layer thickness be thin in comparison to the radius of the RHSE ($\delta \ll a$). This condition can be met by choosing either a RHSE of a sufficiently large radius, or by maintaining a high speed of rotation. From the results of the turning moment measurements shown in Fig. 4, one may take $Re = 200$ as the lower limit where the boundary layer approximation is valid. Thus the useful flow regime for electrochemical application is:

$$200 < Re < 15000 \tag{65}$$

$$\text{or} \quad 14(v/\Omega)^{1/2} < a < 122(v/\Omega)^{1/2} \tag{66}$$

The minimum electrode size for a given rotational speed may be estimated using the lower limit of Eq. (66). For the typical aqueous electrolyte, $v \sim 0.01 \, \text{cm}^2/\text{s}$, the minimum radius of RHSE is then approximately equal to 0.4 cm at 100 rpm, 0.15 cm at 1000 rpm, and 0.04 cm at 10000 rpm.

Fig. 8(a) shows the design of a rotating hemisphere electrode used in the author's laboratory [14]. It is composed of a hemisphere electrode, an arcylic support rod, and a tappered brass holder to be attached to a high speed rotator [Fig. 8(b)]. The electrode is machined into the form of a metal screw with a hemispherical head, and is threaded into the inert acrylic support rod of a larger radius. The design has the advantages that

1. the hemisphere electrode is replaceable; and
2. the primary current distribution is uniform because the insulation plane is perpendicular to the electrode surface at the equator.

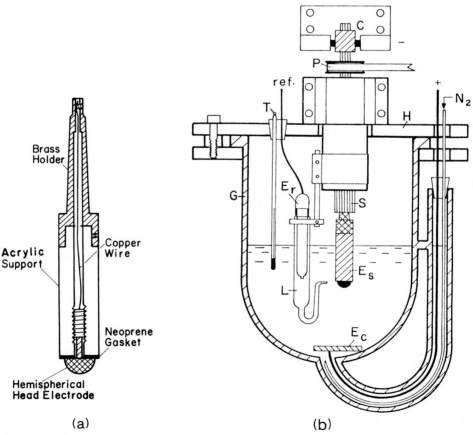

(a)   (b)

Fig. 8. Construction of a rotating hemisphere electrode (a); and cell setup (b). From [14].

The electric connection to the electrode is made by connecting a copper wire from the brass holder to the threaded portion of the hemisphere electrode. The brass holder is machined to fit snuggly into the steel shaft of a rotator. The rotation of the electrode is provided by a timing pulley connected to a variable speed DC motor. A graphite slip-ring contact located on the top of the shaft is used to provided electric contact to the RHSE during the experiments.

The surface preparation of a RHSE is more difficult than the planar rotating disk electrode. The procedure generally involves: (a) hand-rubbing with a polishing paper; and (b) electropolishing in a suitable electrolyte. These are then followed by regular pretreatments pertaining to individual electrochemical systems. Care must be taken not to change the sphericity of the electrode surface. The sphericity can be checked by measuring radii at the pole and at the equator; a difference of 10% will cause approximately 20% error in the surface area calculation.

Fig. 9 shows the other shapes of rotating spherical electrode which have been reported in the literature. Shape (a) is a rotating spherical bead or micro-sphere electrode for electrokinetic investigations [59]. Shape (b) is a hemisphere mounted on a support rod of equal radius [9, 11, 22]. Shape (c) is the hemisphere electrode mounted on a support rod of larger radius as discussed in the preceeding paragraphs. Shape (d) is the rotating ring-hemisphere electrode for studying intermediates of an electrode reaction [20]. Shape (e) is a rotating dropping mercury electrode proposed

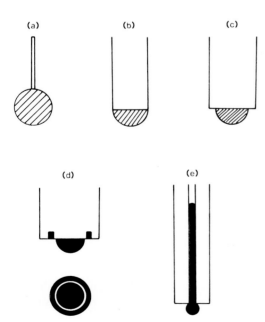

Fig. 9. Types of rotating spherical electrodes reported in the literature. (a) Rotating micro-sphere electrode, (b, c) rotating hemisphere electodes; (d) rotating ring-hemisphere electrodes; (e) rotating dropping mercury electrode.

for polarographic investigations [43]; its limiting current behavior has been found to follow the mass transfer theory described in Section 2. Only Shapes (a), (c) and (d) possess a uniform primary current distribution as discussed in the theory section.

The position of a reference electrode for the RHSE is not as crucial as for the rotating disk electrode because of the uniform potential distribution near the surface. To minimize the flow disturbances which might be introduced by a reference capillary, it is advisable to place the reference tip near the equator rather than near the pole of rotation. For a reference electrode located at a large distance from the RHSE, the ohmic potential drop may be estimated from Eq. (57) as (47):

$$\eta_{ohm} = ia/\kappa \qquad (67)$$

where i is the average current density on the RHSE, a is the radius of RHSE, and κ is the solution conductivity. This equation may be used to calculate the ohmic potential compensations for the measurement of polarization curves with a RHSE.

4 Applications

The basic theory of mass transfer to a RHSE is similar to that of a RDE. In laminar flow, the limiting current densities on both electrodes are proportional to the square-root of rotational speed; they differ only in the numerical values of a proportional constant in the mass transfer equations. Thus, the methods of application of a RHSE for electrochemical studies are identical to those of the RDE. The basic procedure involves a potential sweep measurement to determine a series of current density vs. electrode potential curves at various rotational speeds. The portion of the curves in the limiting current regime where the current is independent of the potential, may be used to determine the diffusivity or concentration of a diffusing ion in the electrolyte. The current-potential curves below the limiting current potentials are used for evaluating kinetic information of the electrode reaction.

4.1 Diffusivity Measurements

The diffusivity of a diffusing ion may be obtained by plotting the mass transfer rate or limiting current density against the square-root of angular velocity. According to Eqs. (44)–(45), this would result in a straight line passing through the point of origin. The slope of the straight line is:

$$\left\{ \begin{array}{l} \text{slope of} \\ i_{lim} \text{ vs. } \Omega^{1/2} \end{array} \right\} = 0.45 n F D^{2/3} v^{-1/6} C_\infty \qquad (68)$$

This expression contains four quantities n, D, v, and C_∞. Since n is normally known for a given electrode reaction, and v can be experimentally determined with a viscometer, the slope permits one to determine the concentration of the diffusing ion, C_∞, if its diffusivity, D is known. Conversely, one may use the slope to determine the diffusivity, D, if the bulk concentration, C_∞ can be measured by the other analytical methods.

Kim and Jorne [35] have used a zinc rotating hemisphere to determine the diffusivity of dissolved chlorine in aqueous solutions of $ZnCl_2$, $ZnSO_4$ and $HClO_4$. The corrosion of zinc in chlorine-contaminated aqueous solutions is controlled by the diffusion of dissolved chlorine to the zinc surface. A plot of the zinc corrosion rate against the square-root of rotational speed resulted in a straight line passing through the origin. In 37% $ZnCl_2$, the diffusivity of Cl_2 was found to be 6.33×10^{-6} cm^2/s in close agreement with the literature value of 6.64×10^{-6} cm^2/s [35].

4.2 Determination of Reaction Order and Reaction Rate Constants

Similar to the rotating disk, the RHSE has the ability to determine the reaction order and reaction rate constants of an electrode reaction. Consider an electrochemical reaction of the type

$$Ox + ne^- \rightarrow Red \tag{38}$$

taking place on the surface of a RHSE below its limiting current potentials. The current for a given rotational speed may be expressed by a rate law of the type:

$$i = nFk[Ox]_s^m \tag{69}$$

where k is the reaction rate constant at a given temperature and electrode potential, m is the reaction order, and $[Ox]_s$ is the surface concentration of the reactant, Ox. As the angular velocity of the RHSE is increased, the rate of transport of Ox species to the electrode surface increases, and eventually the rate of mass transfer becomes insignificant in determining the overall rate of the electrode reaction. Under these circumstances, the surface concentration of Ox becomes equal to the bulk concentration, and the current density would tend to a maximum value, i_{max}, independent of further increase in rotational speed

$$i_{max}(@\Omega \rightarrow \infty) = nFk[Ox]_\infty^m \tag{70}$$

It may be shown that the ratio of i/i_{max} for a given rotational speed is related to the limiting current density, i_{lim}, at that rotational speed by

$$\frac{i}{i_{max}} = \left(1 - \frac{i}{i_{lim}}\right)^m \tag{71}$$

Thus a plot of $\log i$ against $\log(1 - i/i_{lim})$ would result in a straight line having a slope and an intercept

$$\left\{\begin{array}{l} \text{slope of } \log i \\ \text{vs. } \log(1 - i/i_{lim}) \end{array}\right\} = m \tag{72}$$

$$\left\{\begin{array}{l} \text{Intercept of } \log i \\ \text{vs. } \log(1 - i/i_{lim}) \end{array}\right\} = \log i_{max} = \log(nFk[Ox]_\infty) \tag{73}$$

The method permits the simultaneous determination of reaction order, m, and reaction rate constant, k, from the slope and the intercept of the straight line. The procedure can be repeated for various potential values below the limiting current plateau to yield k as a function of electrode potential. The exchange current density and the Tafel slope of the electrode reaction can be then evaluated from the k vs. potential curves.

For a first order irreversible reaction, $m = 1$, Eq. (71) may be rearranged to:

$$\frac{1}{i} = A + \frac{B}{\Omega^{1/2}} \quad (\text{for } m = 1) \tag{74}$$

with

$$A = \frac{1}{i_{max}} = \frac{1}{nFk[Ox]_\infty} \tag{75}$$

$$B = \frac{1}{0.45nFD_{Ox}^{2/3} \, v^{-1/6}[Ox]_\infty} \tag{76}$$

Thus a plot of $1/i$ vs. $1/\Omega^{1/2}$ would give a straight line, and the reaction rate constant, k, and diffusivity, D, can be respectively determined from the intercept, A, and the slope, B, of the straight line. Equations similar to those of the RDE may be also derived for other type electrochemical reaction using the procedures illustrated in Ref. [48, 50].

Kim and Jorne [37] have used a rotating zinc hemisphere to study the kinetics of zinc dissolution and deposition reactions in concentrated zinc chloride solutions. The electrodeposition reaction of cadmium on mercury was used by Mortko and Cover [43] in their investigation of a rotating dropping mercury electrode; their data behaved according to Eqs. (74)–(76).

4.3   Rotating Ring-Hemisphere Electrode

The hemispherical electrode may be coupled with a ring [20] to form a rotating ring-hemisphere electrode (RRHSE) as shown as Fig. 9(d). The ability of this combination to detect intermediate reaction products is demonstrated in Fig. 10, where a series of cathodic sweep curves for the reduction of Cu^{2+} in acidic cupric chloride solution are given for a RRHSE composed of a gold hemisphere electrode of 0.765 cm in diameter, and a platinum ring electrode of 0.798 cm ID × 1.003 cm OD. Two waves are obtained for the hemisphere current during the scans [Fig. 10(a)]. The first wave corresponds to the reduction of Cu^{2+} to Cu^+. The second wave, which begins at a hemisphere potential of − 0.2 V vs. SCE, corresponds to the reduction of Cu^+ to metallic copper on the hemispherical electrode. Fig. 10(b) shows a plot of ring current against the hemisphere potential. During the experiment, the ring was maintained at a limiting current potential of 0.4 V vs. SCE to oxidize any Cu^+ carried by the electrolyte flow to the ring surface. The ring current is seen to increase with the hemisphere potential as the first reaction wave occurs on the hemisphere electrode, and then levels off when the reaction on the hemisphere reaches a limiting current plateau. The level of the ring current increases correspondingly with the level of the hemisphere current as the speed of rotation increases. When the hemisphere reaches

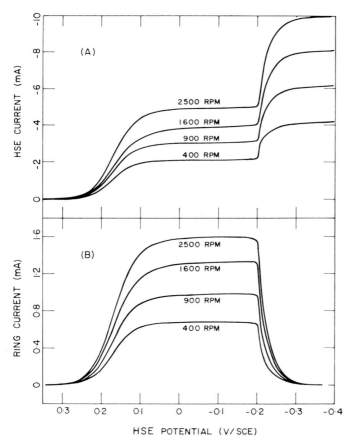

Fig. 10. Cathodic scan curves for a platinum-ring-gold-hemisphere in $CuCl_2$–HCl solution. The ring potential was maintained at 0.4 V vs. SCE to detect Cu^+ ion. From [20].

the potential where the second wave reaction starts to occur, the ring current drops immediately, and becomes zero at a potential corresponding to the limiting current for the reduction of Cu^+ to Cu on the hemisphere electrode. The collection efficiency of cuprous ion, calculated by dividing the magnitude of the ring current plateau by the magnitude of the first hemisphere current plateau is 33%, independent of the speed of rotation.

The rotating ring-hemisphere electrode has been used by Chin [21] to study the dissolution of iron in neutral sulfate solutions, and by Zou and Chin [61, 62] to identify the corrosion products of iron in concentrated sodium hydroxide solutions.

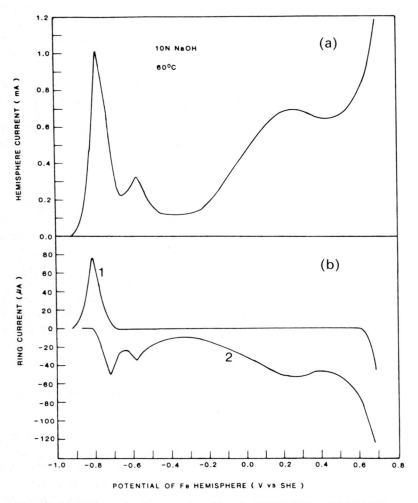

Fig. 11. RRHSE data for the corrosion of iron in 10 N NaOH at 60 °C. (a) Polarization curve of iron hemisphere. (b) Ring current vs. hemisphere potential. Curve 1 is for the detection of $HFeO_2^-$ ion, and curve 2 is for detecting FeO_2^- ion. From [62].

Fig. 11 shows a set of RRHSE data [62] obtained with a platinum ring and an iron hemisphere. Fig. 11(a) shows the anodic polarization curve of the iron hemisphere in 10 N NaOH at 60 °C. The ring current vs. hemisphere potential curves are given in Fig. 11(b). Two different ring-potential settings were used in this experiment. Curve 1 was obtained when the ring potential was set at 0.26 V vs. SHE; at this potential any soluble Fe(II) species arriving at the ring would be oxidized to a Fe(III) species. Curve 2 was the ring current when the ring potential was controlled at -0.87 V vs. SHE to reduce soluble Fe(III) species at the iron hemisphere during the potential scan. The results indicate that in the active potential regime of $-0.91 \sim -0.76$ V vs. SHE, the soluble corrosion product is $HFeO_2^-$ species. In the passive potential regime of $-0.67 \sim 0.5$ V vs. SHE, no Fe(II) species was detected, and the main corrosion product is FeO_2^- species. In the transpassive potentials of greater than 0.5 V vs. SHE, both soluble oxygen and a Fe(VI) species, FeO_4^{2-}, were detected by the ring electrode. The detailed mechanism concerning the corrosion of iron in concentrated alkaline solutions is given in Ref. [62].

The RRHSE possesses a collection efficiency value comparable to that of a rotating ring-disk electrode. Chin [20] used a fixed size platinum ring electrode and varied the ring-hemisphere distance by using different sizes of gold hemispheres. He measured the collection efficiency of the RRHSE by carrying out the reduction of $Fe(CN)^{3-}$ to $Fe(CN)^{4-}$ at the central hemisphere, and the oxidation of $Fe(CN)^{4-}$ to $Fe(CN)^{3-}$ at the ring electrode. For comparison he also calculated the collection efficiency from the ring-disk theory [1] using the projected area of the hemisphere. Except within the regime where the inner radius of the ring was less than 1.08 radius of hemisphere, the measured data were within $2 \sim 3\%$ of the ring-disk theoretical values.

Easy replacement of the central hemispherical electrode is an important advantage of the RRHSE. It offers an alternative choice where the use of the ring-disk electrode would fail to give a meaningful result (such as in electrochemical machining), and where frequent replacements of the disk electrode would be needed. However, in view of the lack of a mathematical theory, each RRHSE should be calibrated beforehand with a known electrochemical reaction. The ring-disk theory can be used as an approximation for the estimate of the collection efficiency if the inner radius of the ring electrode is 1.08 times greater than the radius of the hemispherical electrode.

4.4 Alternating Current Electrode Processes

The rotating sphere and hemisphere electrodes are particularly suitable for investigating periodic electrochemical processes. When an alternating current (AC) in the forms of sinusoidal, rectangular and triangular waves, is applied to an electrode, it causes periodic concentration changes of reacting species in the electrolyte. At sufficiently high AC frequencies, the periodic concentration profile occurs only near the electrode surface and within a thin regime of the Nernst diffusion layer where the contribution of convective mass flux is negligible as compared to the diffusional mass flux [32]. This implies that a uniform periodic concentration profile can be obtained on a rotating sphere owing to the symmetric diffusion of reacting species to the electrode surface. One would therefore have an interesting situation where the primary, secondary and tertiary current distributions are all uniform on the electrode.

The theoretical analysis of periodic concentration profile on a rotating sphere has been made by Chin [24] who solved the nonsteady state convective mass transfer equation for the application of sinusoidal current on the electrode. Cheng and Chin [8, 10, 11] later proposed a stationary film model for the analysis of mass transfer when direct current (DC) is superimposed upon sinusoidal, triangular and rectangular AC. In this model, the concentration of a diffusing ion is separated into two independent components: a time-invariant DC component, and a periodic AC component. The DC concentration distribution is obtained by solving the steady state mass transport equation as discussed in the theory section. The periodic AC concentration distribution is analyzed by the solution to the one-dimensional transient diffusion equation based on the concept of the Nernst diffusion layer. The approximation greatly simplifies the mathematical complexities of the problem. A comparison of the numerical results with Chin's vigorous solutions [24] indicates that the film model is a good approximation in the regime of a dimensionless AC frequency $K = (\omega/\Omega)Sc^{1/3}$ greater than 2 and less than 0.01. For sinusoidal AC of the type

$$\tilde{i} = i_p \exp(j\omega t) \tag{77}$$

flowing across the electrode/electrolyte interface, the film model gives the following surface concentration on a RHSE in laminar flow:

$$\frac{C_s}{C_\infty} = 1 + \frac{i_p}{nFC_\infty D^{2/3} \nu^{-1/6} \Omega^{1/2}}$$

$$\cdot \left(\frac{1}{2K^{1/2}} \right) \frac{\sqrt{\sinh^2(2\beta_1) + \sin^2(2\beta_1)}}{\cos^2(\beta_1) + \sinh^2(\beta_1)} \exp[j(\omega t + \Delta_s)] \tag{78}$$

where $j = \sqrt{-1}$; i_p is the peak AC current density; ω is its frequency in rad/s; t is the time. The dimensionless frequency K, and quantity β_1, are given as

$$K = \frac{\omega}{\Omega} Sc^{1/3} \tag{79}$$

$$\beta_1 = (K/2)^{1/2}(1.6117 + 0.2435\,\theta^2) \tag{80}$$

The phase shift, Δ_s, between the applied sinusoidal AC and the periodical surface concentration is:

$$\Delta_s = \tan^{-1}\left[\frac{\sin(2\beta_1)}{\sinh(2\beta_1)} \right] - \pi/4 \tag{81}$$

The concentration overpotential is:

$$\tilde{\eta}_{conc} = \frac{RT}{nF} \ln \frac{C_s}{C_\infty} \tag{82}$$

The Warburg impedance is related to the concentration overpotential and applied AC by

$$Z_w = \tilde{\eta}_{conc}/\tilde{i} \qquad (83)$$

For small AC current densities, Eq. (83) may be linearized to:

$$Z_w = \frac{RT}{n^2 F^2 C_\infty D^{2/3} v^{-1/6} \Omega^{1/2}}$$

$$\cdot \left(\frac{1}{2K^{1/2}}\right) \frac{\sqrt{\sinh^2(2\beta_1) + \sin^2(2\beta_1)}}{\cos^2(\beta_1) + \sinh^2(\beta_1)} \exp(j\Delta_s) \qquad (84)$$

Although Eqs. (78)–(80) indicate that C_s is a function of the meridional angle θ, numerical calculations [8, 24] reveal that C_s is essentially uniform when K becomes greater than 10. The periodic concentration overpotential, phase shift, Δ_s, and limiting AC current density corresponding to $\eta_{conc} \to \infty$, have been experimentally measured by Cheng and Chin [9–11] using the reduction of ferricyanide ion on a gold plated hemisphere electrode in both laminar and turbulent flows. The measurements agreed with the predictions of the film model for a range of K from 1 to 240.

The rotating hemisphere electrode has been used to investigate the effect of AC on the electrodissolution and deposition reactions of zinc in zinc chloride [25] and copper in acid copper sulfate solutions [55]. AC was found to increase the rate of nucleation and produce more uniform deposit on the zinc electrode. The corrosion of an iron rotating hemisphere in dilute sulfuric acid was investigated by Haili [31] using the AC impedance measurement.

4.5   Passivation Phenomena

Not all applications of the RHSE need quantitative interpretation with the convective diffusion theory. The RHSE may be used in place of a stationary electrode because of improved experimental reproducibility. For the electrode reactions involving gaseous reactants and products, the spherical profile is useful in sweeping away gas bubbles from the electrode surface. The uniform potential and current distribution on a RHSE is suitable for investigating passivation phenomena, particularly those preceded by a metal dissolution reaction at high current densities. Shown in Fig. 12(a) is an example of this where the steady-state anodic polarization curves are given for a mild steel hemisphere in neutral sodium sulfate solutions at three rotational speeds [15]. The curves at zero and 400 rpm exhibit three distinct potential states, i.e., the active dissolution, the passive and the transpassive regions. The current density does not seem to vary with the rotational speed; however, the critical potential for the transition from the active state to the passive state is shifted toward the noble direction as the speed of rotation increases. Consequently, the span of passive potentials diminishes with increaing rotational speeds. At 1600 rpm, no passive region

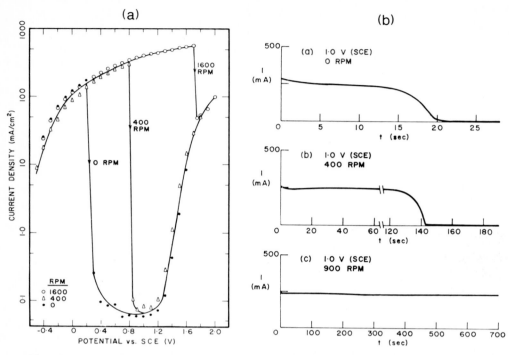

Fig. 12. Steady-state anodic polarization curves (a), and potentiostatic transient curves (b), of a mild steel hemisphere in neutral Na_2SO_4 solution. From [15].

is observed, and the current drops directly from the active dissolution state to the transpassive state at 1.7 V vs. SCE. The phenomenon is supplemented with a potentiostatic transient measurement, in which the potential of the hemisphere was stepped up from the rest potential to a value in the passive potential region, and the resulting current was recorded as a function of time on a strip-chart recorder. The results are shown in Fig. 12(b) for three rotational speeds. The curves are characterized by an initial constant current plateau followed by a rapid drop to a very small steady-state value. The magnitude of the current plateau is equal to the magnitude of the active dissolution current on the steady-state anodic polarization curves. Evidently, the plateau current represents the iron dissolution rate prior to the passivation, and the current drop is caused by decreasing dissolution rate due to the spreading of an anodic film over the electrode surface. At a constant electrode potential, the time required for passivation to occur becomes longer as the speed of rotation increases. These results indicate that the onset of a passive film for mild steel in Na_2SO_4 solution is caused by a dissolution-precipitation mechanism. The mechanism suggests that the initial stage of the passivity involves the dissolution of iron at a high rate. Since the solubility of $FeSO_4$ is small, the electrolyte near the electrode surface soon become super-saturated with the ferrous salt which then precipitates out on the electrode surface to block the further dissolution of the iron electrode. The passivity of many transitional metals has been found to follow the dissolution-precipitation mechanism, including iron in sulfate, perchlorate and chloride solutions [15]. Russell and Newman [52] used a RHSE to study the current oscillation caused by the periodic breakdown and re-precipitation of a salt on iron in sulfuric acid. Other examples of passivation studies using a rotating hemisphere electrode may be found in Ref. [17, 23, 31].

5   Concluding Remarks

In conclusion, we have discussed the theory, experimental practice, and electro-chemical applications of the rotating spherical electrodes. The advantage of a RHSE lies in its ability to maintain a stable electrode configuration and to provide uniform potential and current distributions. Easy replacement of the central hemisphere in the RRHSE has proved it to be a versatile tool in studying corrosion and electro-dissolution reactions. On the other hand, the surface preparation of a RHSE is inherently more difficult than a planar electrode. Although the ring-hemisphere electrode possesses a collection efficiency comparable in value to that of a rotating ring-disk electrode, no mathematical theory has been thus far developed. To quantita-tively evaluate the rate constants of intermediate reaction steps with a ring-hemi-sphere it is necessary to calibrate the collection efficiency with a known redox reaction.

The RHSE has the same limitation as the rotating disk that it cannot be used to study very fast electrochemical reactions. Since the evaluation of kinetic data with a RHSE requires a potential sweep to gradually change the reaction rate from the state of charge-transfer control to the state of mass transport control, the reaction rate constant thus determined can never exceed the rate of mass transfer to the electrode surface. An upper limit can be estimated by using Eq. (44). If one uses a typical Schmidt number of $Sc \sim 1000$, a diffusivity $D \sim 10^{-5}$ cm/s, a nominal hemisphere radius $a \sim 0.3$ cm, and a practically achievable rotational speed of 10000 rpm ($Re \sim 10^4$), the mass transfer coefficient in laminar flow may be estimated to be:

$$k_{m,\,av} \sim 0.015 \text{ cm/s} \tag{85}$$

This value represents the upper limit of a first order reaction rate constant, k, which may be determined by the RHSE. This limit is approximately one order of magnitude smaller that of a rotating electrode. One way to extend the upper limit is to combine the RHSE with an AC electrochemical technique, such as the AC impedance and faradaic rectification metods. Since the AC current distribution is uniform on a RHSE, accurate kinetic data may be obtained for the fast electrochemical reactions with a RHSE.

6　References

1. W.J. Albery, S. Bruckenstein, and D.T. Napp, Trans. Faraday Soc. 62, 1932 (1966).
2. W.H.H. Banks, Quart. J. Mech. Appl. Math. 18, 443 (1965).
3. W.H.H. Banks, ZAMP 16, 780 (1965).
4. W.H.H. Banks, Acta Mech. 24, 273 (1976).
5. C.B. Baxter and D.R. Davies, Quart, J. Mech. Appl. Math. 13, 247 (1960).
6. W.G. Bickley, Phil. Mag. 25, 746 (1938).
7. F.P. Bowden and R.G. Lord, Proc. Roy. Soc. A271, 143 (1963).
8. C.Y. Cheng and D-T. Chin. AIChE J. 30, 757 (1984).
9. C.Y. Cheng and D-T. Chin, AIChE J. 30, 765 (1984).
10. C.Y. Cheng and D-T. Chin, AIChE J. 31, 1372 (1985).
11. C.Y. Cheng and D-T. Chin, Chem. Eng. Commun. 36, 17 (1985); 38, 181 (1985).
12. T.H. Chilton and A.P. Colburn, Ind. Eng. Chem. 26, 1183 (1934).
13. D-T. Chin, J. Electrochem. Soc. 118, 1434 (1971).
14. D-T. Chin, J. Electrochem. Soc. 118, 1764 (1971).
15. D-T. Chin, J. Electrochem. Soc. 119, 1043 (1972).
16. D-T. Chin, J. Electrochem. Soc. 119, 1049 (1972).
17. D-T. Chin, J. Electrochem. Soc. 119, 1181 (1972).
18. D-T. Chin, J. Electrochem. Soc. 119, 1699 (1972).
19. D-T. Chin and M. Litt, J. Fluid Mech. 54, 613 (1972).
20. D-T. Chin, J. Electrochem. Soc. 120, 631 (1973).
21. D-T. Chin, J. Electrochem. Soc. 121, 1593 (1974).
22. D-T. Chin, AIChE J. 20, 245 (1974).
23. D-T. Chin, AIChE J. 23, 434 (1977).
24. D-T. Chin, J. Electrochem. Soc. 127, 2162 (1980).
25. D-T. Chin and S. Venkatesh, J. Electrochem. Soc. 128, 1439 (1981).
26. D-T. Chin and K.L. Hsueh, Electrochim. Acta 31, 561 (1986).
27. W.G. Cochran, Proc. Cambridge Phil. Soc 30, 365 (1934).
28. S.C.R. Dennis, S.N. Singh, and D.B. Ingham, J. Fluid Mech. 101, 257 (1980).
29. L.A. Dorfman, Soviet Phys. Dokladay 3, 248 (1958).
30. J. Fox, NASA TN D-2491 (1964).
31. C.C. Haili, Ph.D. Thesis, University of California, Berkeley (1987).
32. J.M. Hale, J. Electroanal. Chem. 6, 187 (1963).
33. L. Howarth, Phil. Mag. Ser. 7, 42, 1308 (1951).
34. D.W. Hubbard and E. N. Lightfoot, I & EC Fundamentals 5, 370 (1966).
35. J.T. Kim and J. Jorne, J. Electrochem. 125, 89 (1978).
36. J.T. Kim and J. Jorne, J. Electrochem. 126, 1937 (1979).
37. J.T. Kim and J. Jorne, J. Electrochem. 127, 8, (1980).
38. Y. Kobashi, J. Sci. Hiroshima Univ. A20, 149 (1957).
39. Y. Kohama and R. Kobayashi, J. Fluid Mech. 137, 153 (1983).
40. F. Kreith, K.G. Roberts, J.A. Sullivan, and S.N. Sinha, Int. J. Heat Mass Transfer 6, 881 (1963).
41. H. Limb, "Hydrodynamics," pp. 558–589, Cambridge University Press, 1932.
42. R. Manohar, ZAMP 18, 320 (1967).
43. H.J. Mortko and R.E. Cover, Anal. Chem. 51, 1144 (1979).
44. J. Newman, J. Electrochem. Soc. 113, 501 (1966).
45. J. Newman, J. Electrochem. Soc. 119, 69 (1972).
46. S.D. Nigam, ZAMP 15, 151 (1954).
47. K. Nisancioglu and J. Newman, J. Electrochem. 121, 241 (1974).
48. F. Opekar and P. Beran, J. Electroanal. Chem. 69, 1 (1976).
49. M.T. Razafiarimanana, M. Daguenet, G. LePalec, and F. Coeuret, Electrochim. Acta 32, 1103 (1987).
50. A.C. Riddford, in "Advances in Electrochemistry and Electrochemical Engineering, Vol. 4," edited by P. Delahay, pp. 47–116, Interscience, New York (1966).

51. M.H. Rogers and G.N. Lance, J. Fluid Mech. 7, 617 (1960).
52. P. Russell and J. Newman, J. Electrochem. Soc. 133, 2093 (1986).
53. O. Sawatzki, Acta Mech. 9, 159 (1970).
54. H. Schlichting, "Boundary Layer Theory," pp. 107–127, McGraw Hill, New York (1968).
55. R. Sethi and D-T. Chin, J. Electroanal. Chem. 160, 79 (1984).
56. S.N. Singh, Appl. Sci. Res. A9, 780 (1960).
57. S.N. Singh, Phys. Fluid 13, 2452 (1970).
58. G.G. Stokes, Trans. Phil. Soc. 8, 287 (1945).
59. O. Wein, N.A. Pokryvailo, and Z.P. Shulman, Elektrohimiya 18, 1613 (1982).
60. O. Wein, Coll. Czech. Commun. 48, 1571 (1983).
61. J.Y. Zou and D-T. Chin, Electrochim. Acta 32, 1751 (1987).
62. J.Y. Zou and D-T. Chin, Electrochim. Acta 33, 477 (1988).

Electrochemical Separation of Gases

Jack Winnick

Georgia Institute of Technology, School of Chemical Engineering,
225 North Avenue, Atlanta, Georgia 30332, USA

Contents

1 Introduction

Gas separation through membranes achieved commercialization after the introduction of the Prism process by Monsanto a decade ago. Originated for hydrogen recovery, high area membrane equipment is now used for other gases, notably CO_2 [1]. Hydrogen, carbon dioxide, and other components are now being removed from mixtures on an industrial scale [2, 3].

"Membrane separations" to most scientists and engineers equates to a separation brought about by the application of a pressure difference across the membrane with the higher pressure on the mixture side. The thermodynamic basis for the separation is the inequality in the chemical potential, μ_i, across the membrane for each component:

$$\mu_i - \mu_i' = \Delta\mu_i = RT \ln (a_i/a_i') \tag{1}$$

where the prime (') phase is on the low pressure side. The activity of each component is affected by the applied pressure as:

$$\frac{\mathrm{d}\ln a_i}{\mathrm{d}P} = \bar{V}_i/RT \tag{2}$$

where $\bar{V}_i$ is the partial molar volume. All components experience a driving force for transfer, kinetic factors determining the rate of transport for each. The composition and thickness of the membrane then establish the permeability for each and thus the selectivity. Since no membrane shows zero permeability for any component, absolute selectivity is not possible.

A rather different picture emerges with membranes that use an electric field. With charged or chargeable species it is the electrochemical potential, $\bar{\mu}_i$ which determines the driving force:

$$\Delta\bar{\mu}_i = \Delta\mu_i + z_i F \Delta\phi \tag{3}$$

where $\Delta\phi$ is the electric potential difference across a membrane.

Consider a gaseous species i transferring across a standard membrane; to simplify, we assume $a_i = p_i = y_i P$ and $\bar{V}_i = RT/P$. To transfer i from, say, a 5% mixture to a pure stream at one atm requires a minimum of 20 atm. On the other hand, a voltage difference of only 40 mV, if z is 2, provides the same effect. The contrast becomes more impressive when lower concentrations are considered, say 500 ppm, as might be encountered in contaminant removal. The pressure ratio needed is now 2,000, but the voltage requirement is just 130 mV. Thus, there is a clear advantage, at least at first glance, to the application of an electrochemical drive for membrane separation of gases. It is also clear, however, that such application is not universal; the gas must be capable of charge transfer. To date, the technique has proven technically successful with six gases: H_2, O_2, CO_2, Cl_2, H_2S and SO_2. The process developed for each is necessarily specific, especially in terms of materials: however, there is a striking similarity in the equipment design and overall concept.

2 Hydrogen

As the most electroactive gas, H_2 was the first object of electrochemical separation. In an unusual circumstance, separation was not proposed from a mixture of chemically dissimilar species such as N_2, CO_2, etc., but rather, from deuterium, D_2. It had been noted as early as the late 1930's [4] that H_2 is preferentially evolved on several cathode materials over HD or D_2 during electrolysis of water:

$$H_2O \rightarrow H_2 + 1/2 O_2 \tag{4}$$

thus causing an enrichment of the deuterium content of the remaining electrolyte. The reason for the selective reduction is not totally clear but appears to be a combination of thermodynamic and kinetic effects. Lewis [5] and others [6] used Pd cathodes to measure the selectivity, S, of H to D in the cathodic discharge gas to that in the catholyte. The values of S range from 4 to 10. Separation due to the equilibrium, well catalyzed by Pd:

$$HD + H_2O \rightleftarrows H_2 + HDO \tag{5}$$

would be no greater than that allowed by the equilibrium constant, or about 3, indicating strong kinetic effect.

To improve the power demand of the electrolysis, it was suggested in the open literature in May, 1963 that the H_2-rich cathode effluent gas be recycled to the anode (Fig. 1) [7]. In fact, a complex array was proposed (Fig. 2) to increase the enrichment, the final product recovery as D_2O. It was later reported, however, that no selectivity is found at Pd *anodes*; that is, no selectivity for H_2/HD oxidation [8]. Because of this, the incentive to recycle the evolved cathode gas was weakened and apparently was not applied.

Electrolysis continued to be used for primary enrichment in countries with abundant electric power, such as Iceland and Norway, where the H_2 is used in ammonia manufacture [9]. Molecular deuterium, D_2, is produced in Norway by the electrolysis of D_2O. For heavy water production, the method has, for the most part, been replaced by steam-H_2S exchange columns for heavy water enrichment:

$$H_2O_{(l)} + HDS_{(g)} \rightarrow HDO_{(l)} + H_2S_{(g)} \tag{6}$$

a rapid reaction which does not require a catalyst [10].

Tritium in water can be concentrated using the same electrokinetic enhancement as with deuterium; with H_2O/HTO, the enhancement in cathodic reduction is about 75 [11].

The more standard electrochemical separation of H_2 from gas mixtures was first described in October, 1963, by Langer [12]. Using a simple cell composed of five

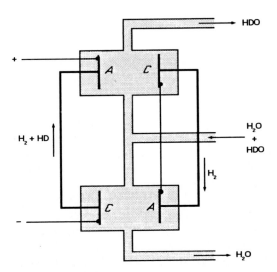

Fig. 1. Recycle scheme for enriching HDO.

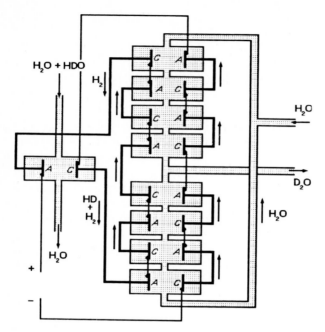

Fig. 2. Multi-stage recycle scheme for D_2O production.

pieces of filter paper saturated with $6 N H_2SO_4$ with Pt gas diffusion electrodes on either side, he was able to obtain nearly pure H_2 from a H_2/N_2 mixture at current densities over $200 mA/cm^2$ with a total voltage of less than $300 mV$. The process is merely the oxidation of H_2 at the positive electrode (anode):

$$H_2 \rightarrow 2H^+ + 2e \tag{7}$$

and, in acid solution, reduction of hydronium at the cathode:

$$2H_3O^+ + 2e \rightarrow H_2 + 2H_2O \tag{8}$$

A similar scheme is operative in alkaline electrolyte:

$$H_2 + 2OH^- \rightarrow 2H_2O + 2e \tag{9}$$

The voltage demand was the sum of the various overpotentials and ohmic loss, since there was little concentration difference in H_2 across the membrane in these experiments. These polarizations were not individually identified.

Significant advances have been made in this decade in electrochemical H_2 separation, mostly through the use of solid polymer electrolytes. Since the overpotentials for H_2 reduction and oxidation are extremely low at properly constructed gas diffusion electrodes, very high current densities are achievable at low total polarization. Sedlak [13] plated thin layer of Pt directly on Nafion® proton conductors 0.1–$0.2 cm$ in thickness, and obtained nearly $1200 mA/cm^2$ at less than $0.3 V$. The current efficiency reaches maximum at relatively low current densities, dropping off as a result of H_2 back-diffusion (Fig. 3), although the H_2 product purity continues to improve with increased H_2 flux (Fig. 4). In addition to producing a nearly pure H_2 stream from a contaminated one, the cathodic product can be maintained at a pressure up to 2,000 kPa above the anode feed, effectively compressing the product as well. Other solid, proton-conducting polymer membranes have been tested [e.g. 14] but their performance is not comparable with that of Nafion.

Table 1. Feed gases for H_2 production.

| H_2 | Volume % (dry basis) | | | |
	CO_2	CO	N_2	H_2S (ppm)
49	11	16	24	800
73	21.4	5.7	—	180

In applications where Nafion is not suitable, at temperatures above 200 °C with feed gas heavily contaminated with CO and sulfur species, a phosphoric acid fuel cell (PAFC)-based concentrator has been effective [15]. Treating the gas shown in Table 1, a H_2 product containing 0.2% CO, 0.5% CO_2 and only 6 ppm H_2S was produced. The anode electrode was formed from a catalyst consisting basically of Pt-alloy mixed with 50% PTFE on a support of Vulcan XC-72 carbon. The cathode was

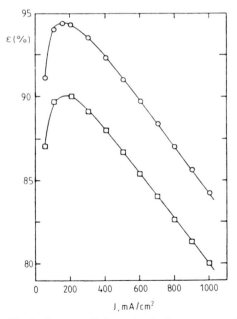

Fig. 3. Current efficiency for hydrogen separation. Calculated overall energy efficiency vs. current density of hydrogen purification for conditions of Table 1: □ including reversible work; ○ excluding reversible work.

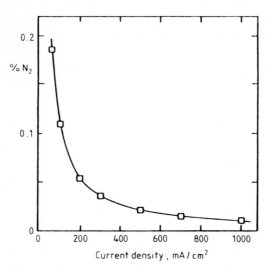

Fig. 4. Nitrogen content of purified hydrogen vs. current density.

10% Pt on Vulcan XC-72 with 50% PTFE; the electrolyte 100% H_3PO_4 immobilized in a SiC matrix.

Despite the technical success of electrochemical H_2 purification, it has not found commercial use. The reason is economic: at a total voltage of 200 mV and an electricity cost of $0.05/kWh, the electric cost alone amounts to $2.36 per million BTU, an unacceptable price for commercial-grade H_2. The real value in this process is the high purity of the product, and if a need for this purity arises the process will be attractive. At present, electrolytically-produced by-product H_2 is often discarded.

3 Oxygen

It is perhaps obvious that oxygen would be the second gas to be purified electrochemically. Langer noted in 1964 that the same sort of apparatus used for H_2 could be used for O_2 [16], with the impure feed admitted to the cathode chamber of the same type of membrane cell, and the pure product obtained at the anode. In alkaline electrolyte,

$$O_2 + 2H_2O + 2e \rightarrow 4OH^- \tag{10}$$

nickel electrodes can be used; in acid electrolyte,

$$O_2 + 4H^+ + 4e \rightarrow 2H_2O \tag{11}$$

noble materials are required.

An improvement in this process was shown by Fujita [17] among others [e.g. 18] who used an ion-exchange membrane such as Nafion as the electrolyte. The cathode was made of graphite powder loaded with 10% Pt and 60% PTFE, which was hot-pressed onto the Nafion. Dissolved Pt was plated on the other side to form the anode (Fig. 5). The cell performance with air as the cathode feed as shown in Fig. 6. When run at low air flow rates, the device can effectively scrub the air of oxygen (Fig. 7) producing a stream with less than 0.02% O_2.

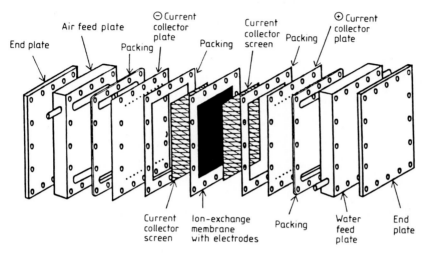

Fig. 5. Exploded view of an ion-exchange membrane electrochemical oxygen separator. Oxygen removal characteristics of the flow-through type oxygen removal system are shown. Air cathode area = 100 cm^2, water temperature = 40 °C.

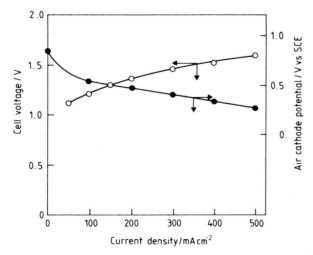

Fig. 6. Polarization of the oxygen separator. Air cathode area = 10 cm^2; water temperature = 40 °C; air feed = 4 dm^3/min.

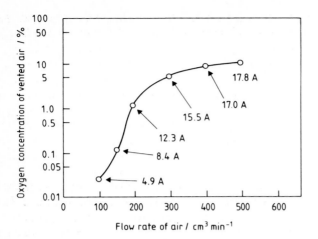

Fig. 7. Oxygen removal characteristics of the flow-through type oxygen removal system. Air cathode area = 100 ²; water temperature = 40 °C.

General Electric built an oxygen supply system for high-altitude aircraft using a similar device [18]. Above 30,000 feet the inhalation rates of pilots, on a mass-basis, are six times what they are at sea level, even though the metabolic use rates are about the same. Thus, a rebreather was designed into a system with an ion-exchange membrane concentrator. (A CO_2 scrubber was also needed in the loop.) Since the oxygen delivery pressure at the anode is limited only by the strength of the membrane, which can be constructed with mechanical support; and a voltage penalty of only:

$$\Delta E = \frac{RT}{4F} \ln \left(\frac{P_{O_2'}}{P_{O_2}} \right) \tag{12}$$

or 45 mV for a partial pressure ratio of 1,000, this system was built to deliver the oxygen at 2.38 MPa (350 psi). It operated at voltages below one volt overall with current densities about 125 mA/cm².

Solid-oxide electrolytes are natural choices for oxygen transport since they transfer oxide ions directly:

$$O_2 + 4e \rightarrow 2O^{2-} \tag{13}$$

Lawless [19] has patented a multi-cell device with various solid-oxide ion-conductors as electrolyte. Some are described to be viable at temperatures as low as 400 °C, but the conductivity is so low ($\sigma T = 0.9$ K/Ω cm) as to make the voltage demand unrealistically high at economic current densities (e.g.: $i > 100$ mA/cm²).

With these solid-oxide electrolytes, designed to operate in relatively O_2-rich feed (e.g. air), gas-diffusion electrodes with their enhanced contact area, are not necessary, and electrode materials can be applied directly onto the electrolyte surfaces in thin films.

Efforts to decrease the power demand of the oxygen transfer process focus, of course, on minimizing the voltage demand by decreasing the electrolyte thickness and improving the electrocatalysis at the electrodes. The four-electron transport mechanism is not immutable, however. On graphite cathodes, in basic solution, oxygen is reduced by the two-electron path:

$$O_2 + H_2O + 2e \rightarrow HO_2^- + OH^- \tag{14}$$

followed by the catalytic decomposition of the peroxide:

$$HO_2^- \rightarrow 1/2\ O_2 + OH^- \tag{15}$$

and then, at the anode,

$$2OH^- \rightarrow 1/2 O_2 + H_2O + 2e \tag{16}$$

completing the process. Tseung [20] has designed an apparatus (Fig. 8) such that the decomposition Eq. (15), occurs in the anode chamber. This is accomplished by using only graphite in the cathode chamber, a poor catalyst for Eq. (15). In the anode chamber, a separate catalytic surface, No. 15 in Fig. 8, is provided for the decomposition. The anode is made of $NiCo_2O_4$, as is the decomposition catalyst. With this arrangement, Tseung was able to purify 1,000 L of O_2 from air at 1 V with the expenditure of 2.7–3 kWh as compared with 4.4 kWh with the 4e path.

Although the two-electron peroxide oxygen transport mechanism appears viable and of some benefit over the four-electron path, the use of superoxide, O_2^-, to provide a one-electron path seems unattained. Gagné [21] has patented such a process, although apparently without actually testing it. It will no doubt be difficult to keep the superoxide from disproportionating:

$$2O_2^- + H_2O \rightarrow O_2 + HO_2^- + OH^- \tag{17}$$

even in basic solution.

A one-electron path, however, has been successfully used in conjunction with organometallic complexes, as used in facilitated transport (non-electrochemical

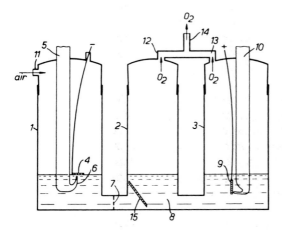

Fig. 8. Three-chamber peroxide-based oxygen separator.

membranes). While the equilibrium across a non-electrochemical membrane cannot be altered from that defined by Eq. (1), the kinetics of transport can be affected significantly by the addition of species-selective carriers. These "facilitated-transport" membranes have been studied since 1960; details of the theory and practice are widely available [e.g. 22]. Oxygen carriers have received the most attention; their function is to react quickly and reversibly with oxygen on the high-pressure side, and to discharge in the same manner on the low-pressure side, as shown in Fig. 9. Most carriers have been developed in an effort to mimic hemoglobin. Thus, iron and cobalt chelated with one or more organic ligands form the bulk of the successful complexes.

The metal ion, e.g. Fe or Co, when in its lower oxidation state can share electron charge with the oxygen molecule adduct. Several iron and cobalt prophyrin derivatives and cobalt-Schiff bases show the necessary reversibility and rates for successful application [e.g. 23]. $\alpha, \alpha', \alpha'', \alpha'''$-*meso*-tetrakis[(*o*-piralamidophenyl)-porphinato] Co(II) has been complexed with 1-methylimidazole to make a complex (CoPIm) which, when mixed with polybutyl methacrylate gave oxygen permeabilities on the order of 10^{-9} sec-cm/(cm^2-s-cmHg) with a selectivity of about 5 over nitrogen [23].

Even better results were obtained by Johnson et al. [24] with the Co compounds shown as Fig. 10. These were dissolved in appropriate solvents and immobilized into standard microporous membranes composed of nylon or PTFE. Oxygen permeabilities were as high as 3×10^{-8} sec-cm/(cm^2-s-cmHg) with selectivity over nitrogen up to 20, leading to product purity up to 85%.

The mechanism of facilitated transport involves using the metal ion only in its reduced state; in the oxidized state the oxygen-carrying capacity is virtually nil. It is thus natural that electrochemical processes should be attempted to improve both the flux and selectivity obtained with the membranes described above by exploiting this O_2 capacity difference. For example, the best of the ultra-thin membranes developed by Johnson et al. [24] delivered oxygen at a rate equivalent to a current density of only 3 mA/cm^2, at least an order lower than that achievable electrochemically. Further, the purity was but 85% and the lifetime of the carrier less than a year.

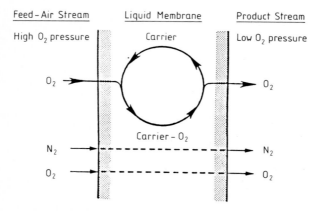

Fig. 9. Facilitated transport of oxygen by a carrier in a liquid membrane.

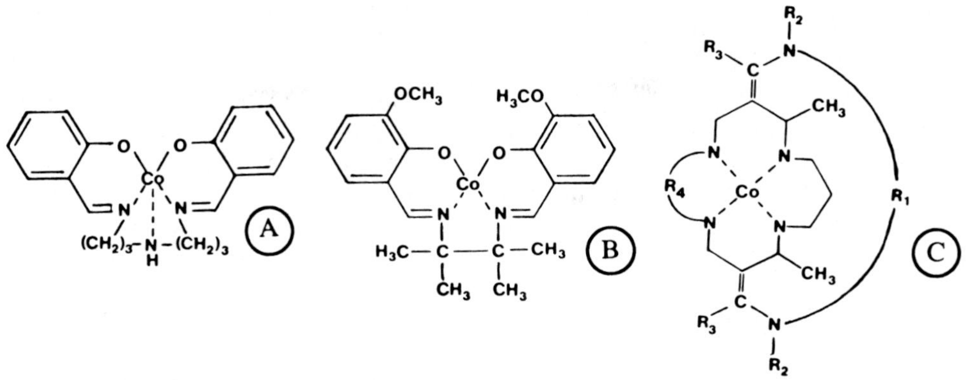

Fig. 10. Chemical structures of some oxygen carriers: (A) five-coordinate cobalt Schiff-base: Co(salPr), (B) four-coordinate cobalt Schiff-base: Co(3-MeOsaltmen), (C) cobalt dry-cave.

Roberts and Laine [25] describe a process for electrochemically transferring oxygen with the scheme shown as Fig. 11. Air or other oxygen-containing gas is drawn into the left side of the vessel. The solution contains an organic solvent such N-methyl pyrrolidine with a supporting electrolyte as tetrabutyl ammonium tetrafluroborate. The carrier, one of the two Schiff bases shown as Fig. 12, is dissolved to about 3 volume percent.

The vessel is divided into two compartments of approximately equal volume by a central divider; the lower portion of the divider is a permeable membrane which

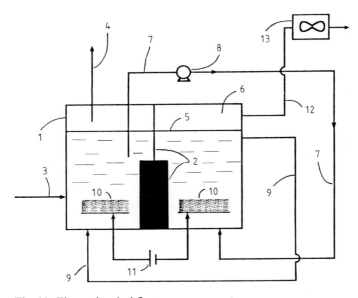

Fig. 11. Electrochemical flow oxygen separator.

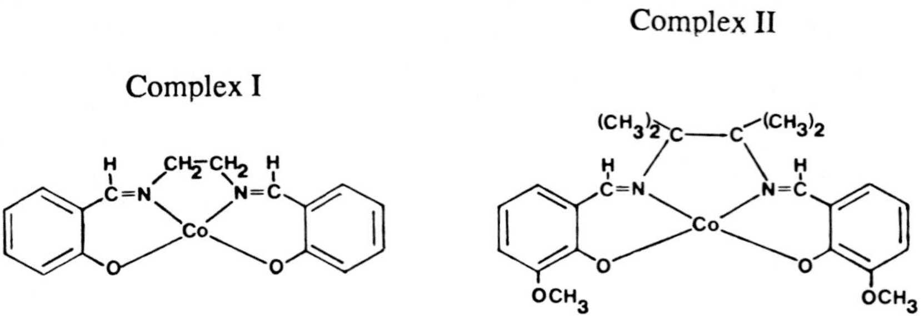

Complex II

Complex I

Fig. 12. Carriers for electrochemical oxygen transport.

prevents intermixing of the liquids in the left-hand and right-hand compartments of the container but provides liquid electrolytic communication between the two compartments. The upper portion of the divider is a metal sheet. Feed gas is introduced into the left-hand compartment of vessel 1. Line 4 is an exhaust line through which the feed gas depleted in oxygen content is removed from the compartment. The upper surface of the solution lies at a level below the top of container 1 and provides a gas space between the upper level of the liquid and the top. Solution is withdrawn from the upper part of the liquid body in the left-hand compartment and is passed into the bottom portion of the right-hand compartment of vessel 1. A pump controls the rate of circulation of the liquid material. Liquid is withdrawn from the upper part of the right-hand compartment and is passed into the bottom part of the left-hand compartment. Gas enriched in oxygen is pulled in by a fan. Metal mesh electrodes are placed in the lower portions of the left-hand and right-hand compartments of the vessel. Cell 11 is connected to the two metal mesh electrodes, the left-hand electrode being the cathode and the right-hand electrode being the anode in the system. Oxygen is taken up by the solution in the left-hand compartment of the vessel and air depleted in oxygen is withdrawn through line 4. The solution-containing carrier bound oxygen is introduced into the lower part of the right-hand compartment where it comes into contact with the anode. Through contact with the anode the *metal* component of the oxygen carrier is *oxidized* to a higher valence, and the oxygen which is bound to the carrier is concurrently released. The released oxygen is withdrawn through line 12. Liquid is withdrawn from the upper portion of the right-hand compartment where the liquid contains the oxygen carrier metal at a higher valence and is passed into the lower part of the left-ahnd compartment where it comes into contact with the cathode. At the cathode the *metal* component of the carrier is *reduced* to a *lower* valence, its capacity to bind oxygen is restored, and further oxygen is picked up from the air introduced through line 3. Operation is continuous. Air is continuously introduced into the left-hand compartment of the vessel. Air depleted in oxygen is continuously withdrawn. Solution-containing oxygen bound to the carrier is continuously passed from the left-hand compartment to the lower part of the right-hand compartment. The oxygen carrier containing bound oxygen is continuously oxidized by contact with the anode, and oxygen is continuously withdrawn as product. Solution containing the metal carrier with its metal at a higher valence level is continuously passed into the lower part of the left-hand compartment, where it is contacted with the cathode and reduced to the lower valence level at which its capacity to bind oxygen is restored.

The reversibility of the carrier was tested by cyclic voltammetry. The scan of the solvent and supporting electrolyte is shown in Fig. 13, with and without dissolved oxygen. The oxygen reduction occurs at about -0.43 V. (vs. SCE). The scan with the complex added, but the solution free of dissolved oxygen is shown as Fig. 14. The carrier is seen to be reduced at about 0.04 V, well within the window of the solvent and electrolyte, and well before reduction of molecular oxygen.

The behavior of the system with the carrier with bound oxygen is shown in Fig. 15. Until the scans are run to potentials positive enough to oxidize the carrier to the non-oxygen-binding state, no oxygen reduction wave is seen. Thus, without electro-chemical oxidation of the carrier, no dissolved molecular oxygen is detectable, but with electrochemical oxidation, bound oxygen is released to the solution. The reduced carrier is free to pick up molecular oxygen once again.

Referring back to Fig. 11, since the solution undergoes no temperature change, or pressure lowering, as in standard facilitated membrane transport, the other dissolved gases (e.g. N_2) are not released into the gaseous product in the anode (right-side) chamber. Product purity is thus inherently much higher in the electrochemical process.

In an attempt to scavenge pure oxygen directly from seawater, Aquanautics Corp. [26] has developed a process similar to that of Roberts. In the Aquanautics scheme, the solution containing the carrier (in this case, a linear pentadentate polyalkylamine chelate of Fe or Co) flows through both anode and cathode of the electrochemical cell.

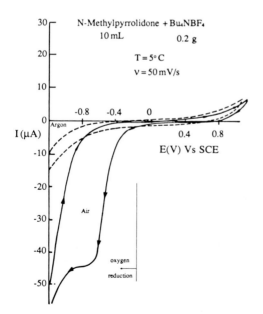

Fig. 13. CV scan of solvent and supporting electrolyte, with and without dissolved oxygen.

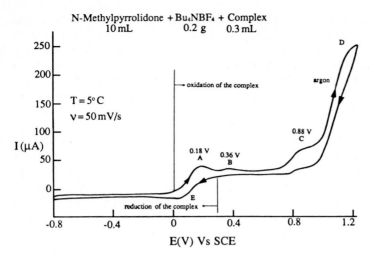

Fig. 14. CV scan with solvent, supporting electrolyte and carrier (oxygen-free).

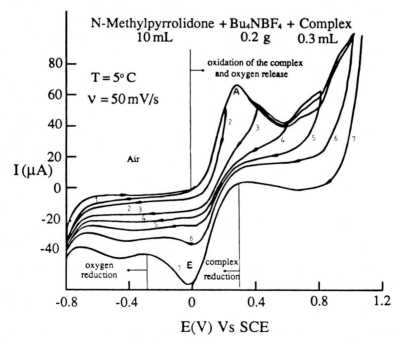

Fig. 15. CV scan with solvent, supporting electrolyte, and carrier with bound oxygen.

The solution carries the metal ion in the oxidized state from the anode with its free, dissolved oxygen to the release chamber, then back to the cathode for reduction, through an exchanger for oxygen pick-up from seawater and back to the anode. The exchanger is a tube and shell device consisting of hundreds of hollow fibers through which the seawater flows. The reduced-carrier solution flows on the shell side, its effective oxygen partial pressure near zero, and oxygen is exchanged from the seawater through the fibers into the carrier.

The main problem to be overcome with these organic-ligand oxygen-transfer processes is the lifetime of the carrier, which is measured in weeks and not months or years, although improvements are continuing.

There seems to be an opportunity to extend the electrochemical process to direct membrane transport; that is, with electrodes plated on either side of a facilitated-transport membrane similar to that of Johnson [24]. The shuttling action of the carrier (Fig. 9) could then be brought about by electrochemical reduction and oxidation instead of pressure difference.

4 Carbon Dioxide

Aqueous, alkaline fuel cells, as used by NASA for supplemental power in spacecraft, are intolerant to CO_2 in the oxidant. The strongly alkaline electrolyte acts as an efficient scrubber for any CO_2, even down to the ppm level, but the resultant carbonate alters the performance unacceptably. This behavior was recognized as early as the mid 1960's as a way to control space cabin CO_2 levels and recover and recycle the chemically bound oxygen. While these devices had been built and operated at bench scale before 1970, the first comprehensive analysis of their electrochemistry was put forth in a series of papers in 1974 [27]. The system comprises a bipolar array of fuel cells through whose cathode chamber CO_2-containing air is passed. The electrolyte, aqueous Cs_2CO_3, is immobilized in a thin (0.25–0.75 mm) membrane. The electrodes are nickel-based fuel cell electrodes, designed to be hydrophobic with PTFE.

At the cathode, oxygen is reduced:

$$O_2 + 2H_2O + 4e \rightarrow 4OH^- \tag{18}$$

followed by the neutralization by CO_2 diffused from the oxidant gas stream:

$$CO_2 + OH^- \rightarrow HCO_3^- \tag{19}$$

At the low CO_2 pressures common to manned spacecraft (0.4 k Pa), the catholyte will be at high pH levels (> 12) so that bicarbonate quickly reacts:

$$HCO_3^- + OH^- \rightarrow CO_3^{2-} + H_2O \tag{20}$$

Moist H_2 is passed over the anode; it reacts with the anloyte, lowering the pH:

$$2H_2 + 4OH^- \rightarrow 4H_2O + 4e \tag{21}$$

This shifts Eq. (20) to the left, producing bicarbonate, which then dissociates:

$$HCO_3^- \rightarrow CO_2 + OH^- \tag{22}$$

and also:

$$HCO_3^- + H_2O - OH^- \rightarrow H_2CO_3 \rightarrow CO_2 + H_2O \tag{23}$$

The net reaction is the transfer of CO_2 at a rate close to 1 mole per 2 Faradays, and the production of water. The process is quite complex; the detailed analysis showed that cathode-side, gas-phase mass transfer of CO_2 was totally controlling only at the lowest CO_2 levels and high current densities. At other conditions chemical reaction rates and transport through the membrane became important; representative results

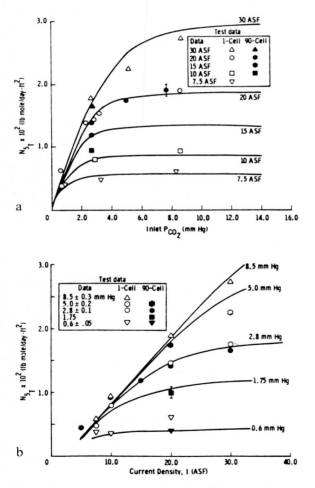

Fig. 16. (a) CO_2 removal as a function of inlet CO_2 pressure at various applied current densities. (b) CO_2 removal as a function of current density at various CO_2 pressures. Curves are calculated [27b].

are shown as Fig. 16. Hydrogen flow-rate did not affect the performance, either in laboratory tests or in the simulations. The simulations did predict an unexpected result: the CO_2 transport decreases as the electrolyte matrix decreases in thickness beyond a certain point (Fig. 17). The reason is the transport of bicarbonate toward the cathode, as shown in the computed concentration gradients, e.g. Fig. 18.

The current efficiency (Fig. 19) (100% = 1 mol/2 F) also shows a maximum with current density. At high currents, insufficient CO_2 is available to neutralize the hydroxide; at low currents, the catholyte pH drops, lowering the rate of CO_2 absorption. This maximum is seen to shift with CO_2 pressure, as expected [28]; the behavior is strikingly similar to that seen in H_2 separation (Fig. 3).

This process can be modified slightly to function as a true concentration cell. If electrical power is applied instead of supplying H_2 to the anode, the carbonate and bicarbonate will be directly oxidized, as shown in Fig. 20 [29]. The advantage to this mode of operation is that the mixture of CO_2 and O_2 from the anode can be delivered to the cathode of an *acid*-electrolyte cell (Fig. 21) which will act as an oxygen concentrator, rejecting the CO_2. The purified oxygen is returned to the cabin; the CO_2, containing only 2 or 3% O_2, is dumped overboard.

Just as the aqueous, alkaline fuel cell can be adopted to CO_2 separation and concentration, the molten carbonate fuel cell (MCFC) can function in this application as well. Recall that the MCFC cathode operates with the *net* reaction

$$CO_2 + 1/2 O_2 + 2e \rightarrow CO_3^{2-} \qquad (24)$$

The mechanism is quite complex. In free electrolyte the reaction order in CO_2 is actually negative [30]; the order in a functioning fuel cell, with gas-diffusion electrodes, rises to near zero [31]. This low order in CO_2 is essential in the efficient operation at very low CO_2 pressures as would be encountered in life-support. The MCFC has been

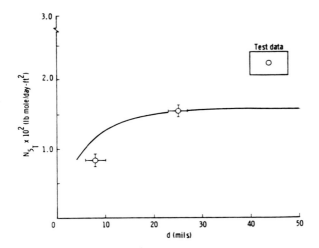

Fig. 17. CO_2 removal as a function of membrand thickness. Current density = 20 A/sq. ft. CO_2 pressure = $4 \cdot 10^2$ pa Curve is calculated [27b].

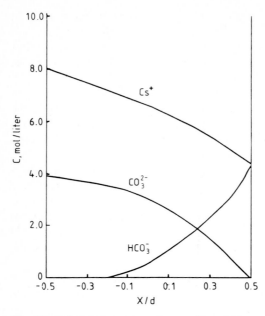

Fig. 18. Calculated concentration profiles within membrane. $x/d = -0.5$ at cathode interface; $+0.5$ at anode interface.

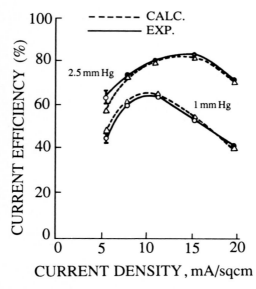

Fig. 19. Current efficiency for CO_2 removal as a function of current density. Curves are calculated [28].

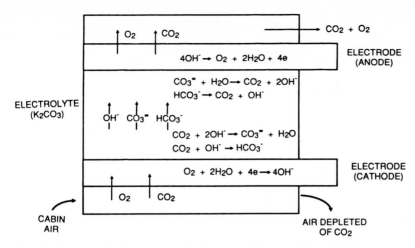

Fig. 20. Stage I cell schematic for simultaneous CO_2 scrubbing and O_2 recovery.

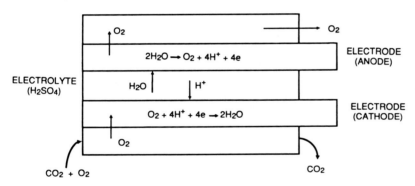

Fig. 21. Stage II cell schematic.

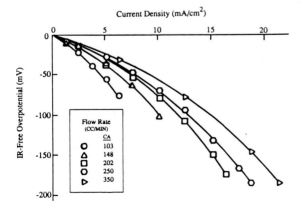

Fig. 22. Cathode overpotential for MCFC-based CO_2 separator. Inlet $CO_2 = 0.50\%$, temperature $= 915$ K.

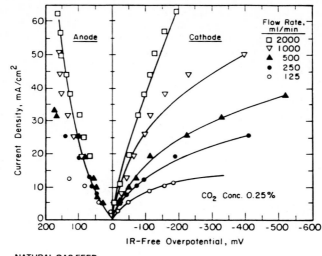

a

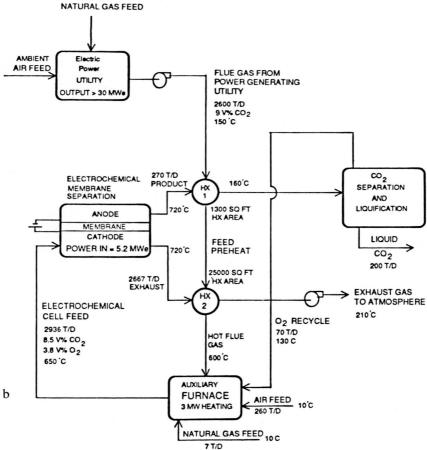

b

operated with CO_2 inlet levels as low as 0.3% in air; the cathode outlet could be reduced below 300 ppm with reasonable polarization, Fig. 22 [32a]. The current densities shown are similar to those in the aqueous device [27]; however, the polarization at both cathode and anode is far lower.

The MCFC has also been tested as a true concentrator, with electric power supplied instead of H_2 [32]. With inlet CO_2 at 0.25% (in air), the outlet could be brought as low as 75 ppm, albeit with the rather low current efficiency of 40% (based on 1 mol $CO_2/2$ F). The parasitic current is due to transport of oxyanions (O_2^{2-} or O_2^-) which are discharged as molecular oxygen at the anode. For lower CO_2 utilization, say below 80%, the polarization is quite acceptable, as seen in Fig. 23. Application to manned spacecraft, however, is handicapped by the high temperatures needed ($> 500\,°C$).

An interesting concept combining H_2 and CO_2 transfer was proposed some time ago [33]. In it a MCFC-based electrochemical cell is fed streams rich in H_2 and CO_2 (CO and H_2O will also be present) to both anode and cathode. Electric power is supplied so that at the cathode:

$$2e + CO_2 + H_2O \rightarrow CO_3^{2-} + H_2 \tag{25}$$

while at the anode:

$$H_2 + CO_3^{2-} \rightarrow H_2O + CO_2 + 2e \tag{26}$$

resulting in no net reaction, hence $E°$ is zero; but there is transfer of H_2 from anode to cathode and CO_2 from cathode to anode. This may have value for a situation where both H_2-rich and CO_2-rich streams are needed.

To date, electrochemical CO_2 transport has not been applied to situations other than manned spacecraft life support. It appears possible that commercial MCFC's can provide an economical means for providing food-grade CO_2 [34]. In this scheme (Fig. 23b) the MCFC acts as a concentrator for CO_2 (and some O_2) from the flue gas emitted from a power plant burning relatively clean fuel (e.g. natural gas). Electric power input is required, as is considerable regenerative heat exchange. Despite these charges, the process would be economical if the MCFC stacks become commercially available at projected prices [34].

5   Chlorine

Chlorine, when contaminated with CO_2 or other relatively inert gases, can, in principle, be purified by electrochemical concentration. At a graphite cathode, Cl_2 is reduced to chloride ions:

$$Cl_2 + 2e \rightarrow 2Cl^- \tag{27}$$

◀ **Fig. 23.** (a) Experimental IR-free overpotentials in MCFC-based separator. Cell performance: 0.25% CO_2 Feed. All curves calculated [32]; (b) CO_2 production scheme using molten carbonate fuel cell stack.

In the conception of Bjorkman [35], an extended area graphite cathode is comprised of a packed, flow-through bed. The electrolyte, in this case aqueous HCl, was immobilized in an asbestos matrix. Chloride ions are transported to the graphite anode where pure Cl_2 is liberated.

This process was developed to transfer essentially all the chlorine from the impure stream. Since this will require, in general, orders of magnitude lowering of the Cl_2 concentration, a series of cells will be required to operate at current densities commensurate with the gas phase transport limitation. If this limiting current density is exceeded, H_2 will be liberated as the electrolyte is reduced.

This device has not reached commercialization, no doubt in part because bulk electrochemical transport of major gaseous components will rarely be economical compared with more standard separation processes. It is in the transport of minority species from low partial pressure to high (e.g. O_2 from seawater, CO_2 from air) where the benefits of the electrochemical driving force, as detailed at the outset of this chapter can best be exploited. Two final examples of contaminant control of great commercial interest demonstrate this principle.

6   Hydrogen Sulfide

Sulfur, found to varying degrees in all fossil fuels, is always present as H_2S after gasification. It is a common contaminant in natural gas, to as high a concentration as 80%. Its presence, even at ppm levels is hazardous to plant and animal life as well as to materials of construction. The standard methods for removing it during fuel processing are complex and expensive; with gas or high-sulfur coal, up to 30% of the cost of the fuel [36].

Standard processing combines sorption in alkaline solution, regeneration of the solvent with concurrent release of high-concentration H_2S, and finally conversion of the H_2S to elemental sulfur through the Claus reaction scheme:

$$H_2S + 3/2\ O_2 \rightarrow SO_2 + H_2O \tag{28}$$

$$2H_2S + SO_2 \rightarrow 3S + 2H_2O \tag{29}$$

Pressure-driven membrane processes to replace the absorption-sequence are under development, but the separated H_2S (and other components which co-diffuse) will have to be treated with Claus or similar technology.

An electrochemical membrane process can, in principle, perform the entire sequence in a single step while enriching the process gas slightly with H_2. If the H_2S could be electronated at a suitable cathode:

$$H_2S + 2e \rightarrow H_2 + S^{2-} \tag{30}$$

and the sulfide ions transported across a suitable membrane to an anode where they would be preferentially oxidized:

$$S^{2-} \rightarrow 1/2\, S_2 + 2e \tag{31}$$

the separation would be achieved. To remove the sulfur from the anode efficiently, it would be necessary to operate above the boiling point of sulfur at the operating pressure (which would be essentially constant across the membrane). A lower operating temperature limit of nearly 500 °C would be set, even at ambient pressure.

Early tests [37] utilized a cell design similar to that of early MCFC experiments. The assembled cell, machined from graphite blocks, is shown as Fig. 24. The electrodes and current collectors were machined from graphite and dense carbon, respectively. The electrolyte was a mixture of 63% Na_2S, 37% Li_2S, believed to melt near 850 °C; the melting point after several days of operation was below 700 °C, probably because of polysulfide formation. The electrolyte was immobilized in a matrix of MgO, the whole formed by hot-pressing a mixture of electrolyte and ceramic powders.

While some tests with this set-up were run with N_2 as the anode sweep gas, elemental sulfur was found to occlude the exit tube. For this reason most tests were run with H_2 at the anode, the product recovered as H_2S:

$$S^{2-} + H_2 \rightarrow H_2S + 2e \tag{32}$$

The removal efficiency was quite good. The first gases tested had about 1% H_2S in a H_2/N_2 carrier; over 98% was removed as shown in Fig. 25. The dashed lines in this figure are calculated, incorporating the effect of back-leakage through cracks in these first crude membranes.

The cell voltage was acceptable: at the higher temperature of 820 °C about 600 mV at 30 mA/cm^2 (Fig. 26). At the lower temperature, 710 °C, the higher overpotential probably indicates partial freezing of the electrolyte.

In a study of the fundamentals of the process, White [38] used free electrolyte with dense graphite electrodes (Fig. 27). He analyzed the electrokinetics at a number of

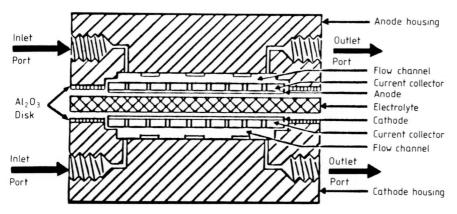

Fig. 24. Laboratory cell for H_2S removal study. Schematic diagram for assembled cell with Al_2O_3 ring.

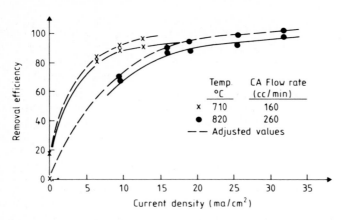

Fig. 25. H$_2$S removal efficiency from inert gas as a function of current density. Dashed lines are adjusted for back-leakage of H$_2$S from anode to cathode.

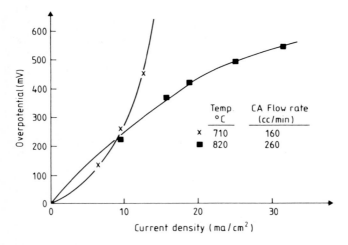

Fig. 26. Overpotential in laboratory cell.

cathode gases (Table 2). Analysis of chronoamperometric data (Fig. 28) showed quite high exchange currents, averaging near 40 mA/cm^2. These studies also indicated the electroactive species in solution to be S$_2^{2-}$ and that the reaction order in H$_2$S was quite low, probably 0.5.

Banks [39] later verified S$_2^{2-}$ as electroactive with quantitative analysis of cyclic voltammograms (Fig. 29a,b) in free electrolyte with varying gas composition equilibrated with the melt. From Banks' analysis of the distinctive current function dependence on scan rate (Fig. 29c) it was concluded that the cathodic reaction was "catalytic" in nature:

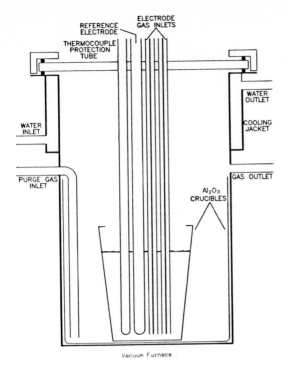

Fig. 27. Free-electrolyte experimental apparatus.

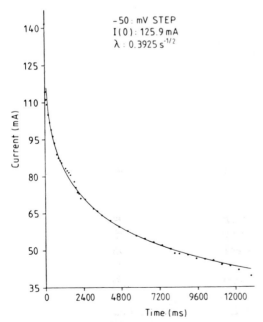

Fig. 28. Chronoamperometry in free-electrolyte (sulfide). Comparison of experimental data with quantitative theory [38].

Table 2. Test parameters for experimental runs.

Data series	Working electrode gas composition	Counter electrode gas
1	6500 ppm H_2S, 50% H_2	H_2
2	6500 ppm H_2S, 50%H_2	He
3	6500 ppm H_2S, 8%CO_2, 20%H_2, 25%CO	He
4	3000 ppm H_2S, 50%H_2	H_2
5	1062 ppm H_2, 82%H_2	H_2
6	336 ppm H_2S, 94%H_2	H_2
7	1.5% H_2S, 50% H_2	H_2
8	6500 ppm H_2S, 8% CO_2, 20% H_2, 25%CO	H_2

(a)

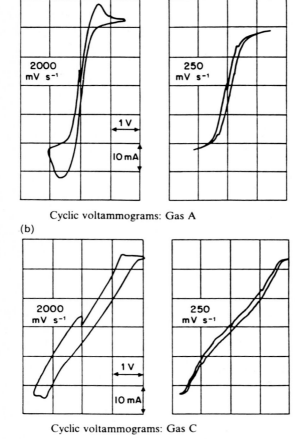

Cyclic voltammograms: Gas A

(b)

Cyclic voltammograms: Gas C

Fig. 29. Cyclic voltammetry in free-electrolyte (sulfide). (a) Gas: 1.5% H_2S, 50% H_2, bal. N_2. (b) Gas: 0.65% H_2S, 20% H_2, 5.8%CO_2, 25% CO, bal. N_2. (c) Current function dependence on scan rate, showing typical ECE behavior.

$$(H_2S) \text{ dissolved} + S^{2-} \rightarrow H_2 + S_2^{2-} \tag{33}$$

$$S_2^{2-} + 2e \rightarrow 2S^{2-} \tag{34}$$

Since application to commercial separations will frequently require treatment of gases containing CO, CO_2 and H_2O, a comprehensive electroanalytical study was undertaken with gases of varying carbon content. Analysis of the cyclic voltammograms showed sharply higher chemical reaction rates with the carbonated gas. This gas also supported higher currents and gave better H_2S removal efficiencies, probably because of an additional reaction pathway of sulfide:

$$CO_2 + H_2O + 2e \rightarrow CO_3^{2-} + H_2 \tag{35}$$

followed by:

$$CO_3^{2-} + H_2S \rightarrow CO_2 + H_2O + S^{2-} \tag{36}$$

These preliminary tests showed the process to be basically sound but left questions concerning, especially, electrode materials. For use in a carbonated gas environment, carbon elements would be unsuitable. At the cathode, carbon would be removed by:

$$C + CO_2 \rightarrow 2CO \tag{37}$$

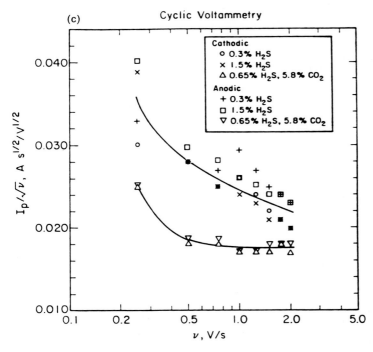

Fig. 29. Continued.

or

$$C + H_2O \rightarrow CO + H_2 \tag{38}$$

Weaver [40] studied alternate cathode materials at 650 °C, finding several that performed well. Steady-state polarization on Ni, Co and Fe porous electrodes operating as cathodes in a MCFC, with a standard $(Li/K)_2 CO_3$ "tile" is shown in Figs. 30–32. Note that the oxidant gas fed to these cathodes is, in normal MCFC operation, the fuel, composed of 32.5% H_2, 17.5% CO_2, 17.5% H_2O, the balance N_2. Polarizations were first taken with this "clean" gas where the only reaction can be Eq. (35). After steady-state was attained, 0.65% H_2S was added and sufficient time allowed for the electrode to convert to the sulfides. After 24 hours, the outlet H_2S reached the inlet level and polarizations were measured. Note in Figs. 30–32, that the performance with H_2S is significantly improved over the clean gas. (The Ni sample was a commercial (Gould) MCFC electrode; the Co and Fe were pressed from powders. Each gas was 8 sq cm in superficial area). The improvement is probably due to a catalytic mechanism involving sulfur interactions with the electrode, as, for Co:

$$Co_9S_8 + 16e \rightarrow 9Co + 8S^{2-} \tag{39}$$

$$9Co + 8H_2S \rightarrow Co_9S_8 + 16H_2 \tag{40}$$

These reactions were seen during cyclic voltammetry on all three cathodes, (Fig. 33). Note in Fig. 33a the relatively small currents representing the reduction and oxidation of Eq. (35). But in Fig. 33b the much larger currents are due to the electrode sulfidation and reduction, Eq. (39). All sulfided electrodes were analyzed by X-ray

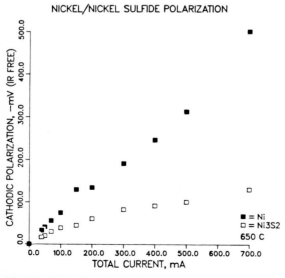

Fig. 30. Cathodic polarization of a nickel gas-diffusion electrode before and after sulfidation with H_2S.

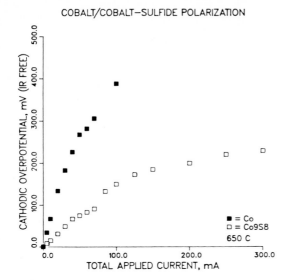

Fig. 31. Cathodic polarization of a cobalt gas-diffusion electrode before and after sulfidation with H$_2$S.

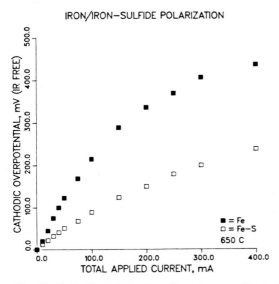

Fig. 32. Cathodic polarization of an iron gas-difusion electrode before and after sulfidation with H$_2$S.

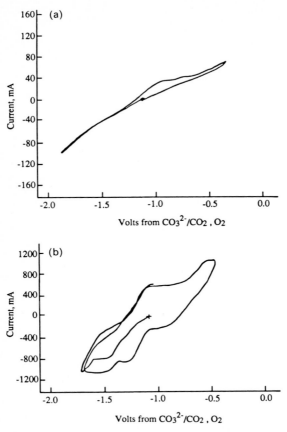

Fig. 33. (a) CV scan with carbonate membrane. Cobalt electrode. Gas: 36% H_2, 5.5% CO_2, 5% H_2O, 9.2% CO. Scan rate = 10 mV/s. (b) CV scan with sulfided membrane. Cobalt electrode. Gas: 36% H_2, 5.5% CO_2, 5% H_2O, 9.2% CO, 2000 ppm H_2S. scan rate = 2.5 mV/s. (c) Cyclic voltammetry of cobalt in 36.0% H_2, 5.5% Co_2, 5.0% H_2O, 9.2% CO, bal. N_2 at 650 °C at (a) 10 mV/s and no H_2S; (b) 2.5 mV/s with 2.000 ppm H_2.

diffraction to identify the solid phase. Cyclic voltammetry was also done with carbon electrodes to verify the mechanism of the reaction in the immobilized electrolyte, as determined by Banks [39] in free electrolyte. Fig. 34a shows the broad window of the clean electrolyte tile in N_2 gas; after partial sulfidation of the tile in Fig. 34b. The reduction and oxidation wave in Fig. 34b must be due to Eq. (34). When H_2S contaminated reducing gas is supplied, Fig. 34c results. Note the large increase in current due to the H_2S involvement.

The H_2S removal was also verified by effluent gas analysis as a function of applied current. Fig. 35 shows representative H_2S outlet concentration as a function of current density. It is seen that the cathode outlet can be brought to near zero H_2S content with sufficient applied current. Measurements at the anode outlet showed very little

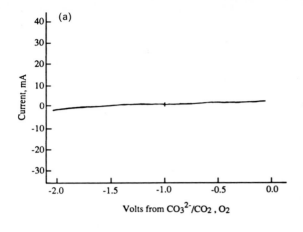

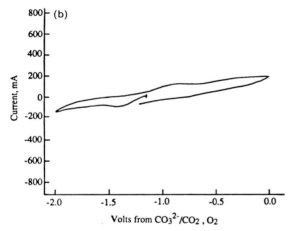

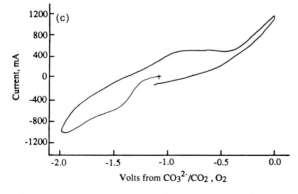

Fig. 34. CV scans on carbon electrode, 650 °C, 500 mV/s. (a) Nitrogen, fresh MCFC membrane. (b) Nitrogen, sulfided MCFC membrane. (c) Coal gas with 1% H_2S.

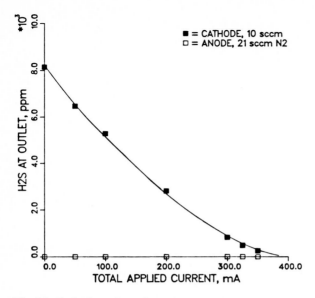

Fig. 35. Cathode and anode outlet H_2S concentration at $700\,°C$. Gases: 0.8% H_2S in coal gas at cathode. Nitrogen at anode.

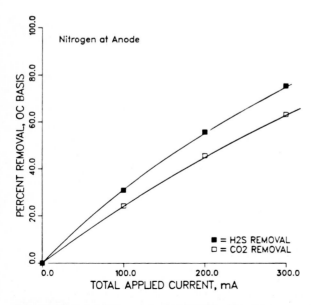

Fig. 36. H_2S and CO_2 removal at $700\,°C$ with nitrogen anode sweep gas.

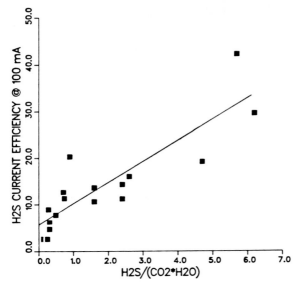

Fig. 37. H_2S removal current efficiency at 100 mA current as a function of reactant-gas ratio. 650 °C.

H_2S in the sweep gas, as shown by the open symbols. However, this fuel gas, containing 17.5% CO_2 and 17.5% H_2O, offers a strongly competitive reaction to H_2S reduction, namely Eq. (35). Since the $E°$ for this reaction is nearly equal to that for Eq. (30), it would be expected to occur to a much greater degree than (30). However, the oxidation reaction at the anode,

$$CO_3^{2-} \rightarrow CO_2 + 1/2 O_2 + 2e \tag{41}$$

occurs some 800 mV positive of sulfide oxidation, Eq. (31). If no fuel, such as H_2, is available at the anode to aid carbonate transfer, it will merely act as supporting electrolyte for the sulfide.

Fig. 36 shows that CO_2 is, in fact, removed along with the H_2S; however, in this and similar runs H_2 leaked through the membrane as indicated by H_2S at the anode exhaust. Despite this difficulty, at lower CO_2 and water levels, a greater fraction of the current was utilized in H_2S removal (Fig. 37). The main obstacle to a truly selective high-temperature H_2S electrochemical process appears to be development of a gas-tight membrane.

7 Sulfur Dioxide

The manner in which CO_2 was scrubbed by carbonate electrolyte in the cells described in Section 4 led Townley to attempt the same scheme for SO_2 in flue gas

[41]. That is, in contact with a sulfate-based electrolyte, at the cathode:

$$SO_2 + O_2 + 2e \rightarrow SO_4^{2-} \tag{42}$$

and at the anode:

$$SO_4^{2-} \rightarrow SO_3 + 1/2 O_2 + 2e \tag{43}$$

In principle, this process should be feasible. SO_2 is present in flue gas to about the same level as the CO_2 in manned spacecraft atmospheres, and that of H_2S in coal gas (300–5000 ppm). SO_2 is much more acidic than CO_2, the equilibrium:

$$CO_2 + SO_4^{2-} \rightarrow CO_3^{2-} + SO_3 \tag{44}$$

lying far to the left, with an equilibrium constant of about 10^{-17} at 500 °C [42]. Further, the electrochemistry had been previously explored. Salzano and Newman [43] had seen the reduction,

$$SO_3 + 1/2 O_2 + 2e \rightarrow SO_4^{2-} \tag{45}$$

in molten sulfate. Flood and Boye [44] had noted the reaction, in melts with small amounts of oxide:

$$SO_2 + 1/2 O_2 + O^{2-} \rightarrow SO_4^{2-} \tag{46}$$

Burrows and Hill [45] had identified the oxidation, Eq. (43), also in molten sulfate. It remained to be seen if the entire process, including the oxidation of sulfur to the +6 state in the flue gas, could be conducted through an immobilized-electrolyte membrane.

Stable, conductive electrodes would also be a problem. Preliminary experiments, were carried out in a cell, using simulated flue gas nearly identical to that shown in Fig. 24. In these tests, the membranes were hot-pressed from mixed powders of electrolyte (ternary eutectic of [Na, Li, K]$_2$ SO$_4$) with LiAlO$_2$ as matrix. The electrodes were constructed of cold-pressed Li$_2$O$\cdot$9Cr$_2$O$_3$, partially sintered to give a highly-porous gas-diffusion structure. The tests were encouraging; up to 50% of the SO_2 was removed from the simulated flue gas with the application of current. Simultaneously, a stream of concentrated SO_3 and O_2 was evolved at the anode.

Projected economics were also highly promising [41]; capital and operating costs would be a fraction of those required by standard methods, e.g. scrubbing. Furthermore, no chemical reagents would be required and no waste stream produced. However, the high melting points of the alkali-metal sulfates (T > 512 °C) offered severe limitations to application, especially for use in power plants, where the flue gas typically is unavailable for treatment at temperatures below 400 °C.

For this reason a lower melting electrolyte was sought, one which would melt below 350 °C. Potassium pyrosulfate, $K_2S_2O_7$, was a natural choice; though thermally unstable above 300 °C in the absence of SO_3,

$$K_2S_2O_7 \rightarrow K_2SO_4 + SO_3 \tag{47}$$

it forms stable solutions, with small amounts of K_2SO_4, under low SO_3 pressures, as would be encountered in flue gas treatment.

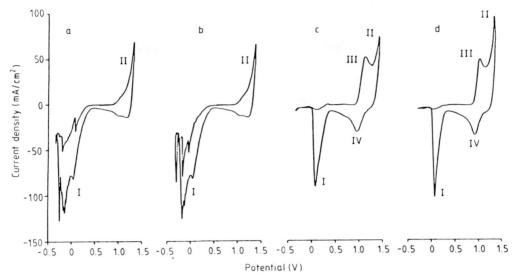

Fig. 38. Series of stabilized cyclic voltammograms of $K_2S_2O_7$ under various gas environments, 400 °C, 100 mV/s. (a) Nitrogen, (b) air, (c) 1.0% SO_2 in N_2, (d) 1.0% SO_2, 10.0% O_2, bal. N_2.

The electrochemistry of the pyrosulfate was clearly more complex than that of sulfate. For example, at a negative electrode Durand [46] had shown it is reduced, actually evolving SO_2:

$$2S_2O_7^{2-} + 2e \rightarrow SO_2 + 3SO_4^{2-} \tag{48}$$

However, with O_2 available and with an approximate oxidation catalyst, the SO_2 should be oxidized to SO_3, the equilibrium:

$$SO_2 + 1/2O_2 \rightleftarrows SO_3 \tag{49}$$

lying well to the right at these temperatures.

Franke [47] undertook a comprehensive electroanalytical study of $K_2S_2O_7$ mixtures with K_2SO_4, which is formed by Eqs. (47) and (48); and V_2O_5, a widely-used oxidation catalyst for SO_2. Pure pyrosulfate under N_2 or air (Fig. 38a,b) shows only the reduction to SO_2 and sulfate, Eq. (48) (all potentials are vs. Ag/Ag$^+$). When SO_2 is added, a new reduction and oxidation peak appear (Fig. 38c,d). When the electrolyte was pre-saturated with K_2SO_4 (ca. 4 wt.%) (Fig. 39) the gas composition had no direct effect on the voltammetry. Although the equilibrium for Eq. (49) lies well to the right at this temperature, 400 °C, the kinetics are quite slow in the absence of a catalyst. The equilibrium between pyrosulfate and sulfate, Eq. (47), lies well to the left ($K = 2 \times 10^{-6}$), but will proceed to the right in the absence of SO_3. Thus, the new peaks are sulfate oxidation, Eq. (43), and SO_3 reduction to sulfite:

$$SO_3 + 2e \rightarrow SO_3^{2-} \tag{50}$$

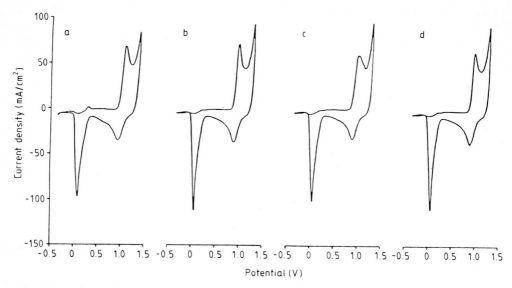

Fig. 39. Series of stabilized cyclic voltammograms with 95% $K_2S_2O_7$, 5% K_2SO_4, 400 °C, 100 mV/s. (a) Nitrogen, (b) air, (c) 1.0% SO_2 in N_2 (d) 1.0% SO_2, 10.0% O_2, bal. N_2.

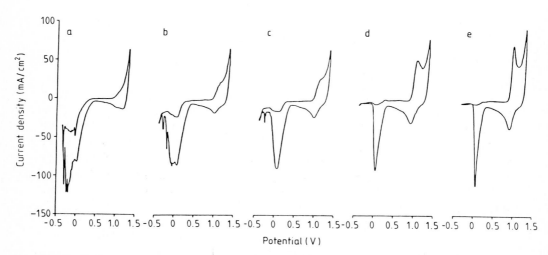

Fig. 40. Series of stabilized cyclic voltammograms of $K_2S_2O_7$ with various sulfate levels, in air, 400 °C, 100 mV/s. (a) 1% $K_2S_2O_4$, (b) 2%, (c) 3%, (d) 5%.

which then reacts with the dissolved O_2 and pyrosulfate:

$$SO_3^{2-} + 1/2 O_2 \rightarrow SO_4^{2-} \tag{51a}$$

$$SO_3^{2-} + S_2O_7^{2-} \rightarrow 2SO_4^{2-} + SO_2 \tag{51b}$$

Voltammetry with varying initial sulfate content, Fig. 40, is in keeping with this scheme; increasing sulfate content lowers the pyrosulfate reduction peak while increasing the peaks identified with sulfate. The oxidation catalyst V_2O_5 was added to the melt (1 wt.%) for the next series. (The mechanism of SO_2 oxidation in the presence of $K_2S_2O_7$ is not totally clear [48, 49], but it does involve the shuttling of vanadium between $+4$ and $+5$ states.) Two more peaks, a reversible couple, V and VI on Figs. 41 and 42 are now present, representing the vanadium-complex redox system:

$$[K_2O - V_2O_5 - 4SO_3] + 2e \rightarrow [K_2O - V_2O_4 - 3SO_4] + SO_4^{2-} \tag{52}$$

the compounds in Eq. (52) having been identified by power X-ray diffraction analysis of the melts [47].

While the peaks associated with sulfate and pyrosulfate are essentially the same with and without vanadia, the equilibrium potentials in the melts are not. In the vanadia-free melts (Table 3) the equilibrium potentials are well-correlated with Eq. (53) [50]:

$$S_2O_7^{2-} + 2e \rightleftarrows SO_4^{2-} + SO_3^{2-} \tag{53}$$

But in the melts with 1% vanadia (Table 4) the equilibrium is more determined by Eq. (52); in an SO_2-rich gas environment, the V_2O_4 complex dominates. Addition of sulfate increases the trend to more negative potentials. With O_2 present, the equilibria;

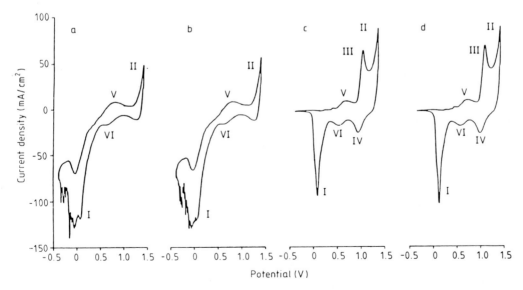

Fig. 41. Series of stabilized cyclic voltammograms with 99% $K_2S_2O_7$, 1% V_2O_5, 400 °C, 100 mV/s. (a) Nitrogen, (b) air, (c) 1.0% SO_2 in N_2, (d) 1.0% SO_2, 10.0% O_2, bal. N_2.

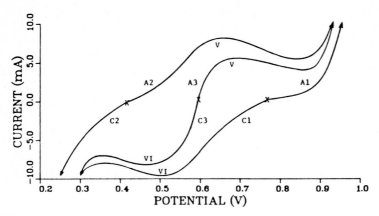

Fig. 42. Single positively and negatively going scans through the vanadia redox region. The electrolyte is $K_2S_2O_7$ with 4.0% K_2SO_4, 1.0% V_2O_5, 370 °C, 100 mV/s. The open circuit potentials are denoted by X, symbols are as follows (A) positive, (C) negaitve, (1) air environment, (2) 1.0% SO_2 in N_2, (3) 1.0% SO_2, 10.0% O_2 in N_2, and IV and V represent the vanadia complex oxidation and reduction as before.

Table 3. Comparison of experimental and calculated equilibrium potentials for various gas environments and electrolyte compositions, 400 °C, relative to Ag/Ag^+ reference. Results are reported on the basis of changes in the equilibrium potential relative to a base case of pure potassium pyrosulfate under an air environment.

Electrolyte Composition	Gas Environment	E_c Calculated (V)	E_e Experimental (V)
$K_2S_2O_7$	air	+0.940	+0.990
		Change in the Equilibrium Potential	
		Calculated	Experimental
$K_2S_2O_7$	air	0.000	0.000
$K_2S_2O_7$ with 0.5% K_2SO_4	air	−0.060	−0.060
$K_2S_2O_7$ with 1.0% K_2SO_4	air	−0.100	−0.090
$K_2S_2O_7$ with 2.0% K_2SO_4	air	−0.140	−0.140
$K_2S_2O_7$ with 4.0% K_2SO_4	air	−0.180	−0.170
$K_2S_2O_7$	N_2	−0.110	−0.120
$K_2S_2O_7$	1.0% SO_2 in N_2	−0.480	−0.470
$K_2S_2O_7$	1.0% SO_2, 10.0% O_2 in N_2	−0.120	−0.130
$K_2S_2O_7$ with 4.0% K_2SO_4	N_2	−0.320	−0.300
S_2O_2 with 4.0% K_2SO_4	1.0% SO_2 in N_2	−0.520	−0.570
$K_2S_2O_7$ with 4.0% K_2SO_4	1.0% SO_2, 10.0% O_2 in N_2	−0.180	−0.170

$$[K_2O - V_2O_5 - 2SO_3] + SO_2 \rightleftarrows [K_2O - V_2O_4 - 3SO_3] \tag{54a}$$

$$[K_2O - V_2O_5 - 2SO_3] + 1/2 O_2 \rightleftarrows [K_2O - V_2O_5 - 2SO_3] + SO_3 \tag{54b}$$

becoming controlling, with SO_2 or O_2 able to shift the potential negative or positive

Table 4. Comparison of the equilibrium potential of non-vanadia and vanadia containing electrolytes (versus Ag/Ag^+) under different gas environments and sulfate compositions, 370 °C.

Base Electrolyte Composition	Gas Environment	Experimental E_e(V)	
		0.0% V_2O_5	1.0% V_2O_5
$K_2S_2O_7$	air	+0.990	+0.870
$K_2S_2O_7$	N_2	+0.870	+0.840
$K_2S_2O_7$	1.0% SO_2 in N_2	+0.520	+0.560
$K_2S_2O_7$	1.0% SO_2, 10.0% O_2 in N_2	+0.860	+0.700
$K_2S_2O_7$ with 4.0% K_2SO_4	air	+0.820	+0.760
$K_2S_2O_7$ with 4.0% K_2SO_4	N_2	+0.690	+0.660
$K_2S_2O_7$ with 4.0% K_2SO_4	1.0% SO_2 in N_2	+0.420	+0.380
$K_2S_2O_7$ with 4.0% K_2SO_4	1.0% SO_2, 10.0% O_2 in N_2	+0.820	+0.600

respectively. The vanadia has no effect on the pyrosulfate or sulfate electrochemistry; its role is to air-oxidize the SO_2, that in the inlet gas as well as that obtained in reduction of pyrosulfate, Eq. (48), to SO_3. These SO_3 molecules are neutralized by the sulfate produced, also in Eq. (48):

$$2SO_3 + 2SO_4^{2-} = 2S_2O_7^{2-} \tag{55}$$

The net cathodic reaction in the presence of vanadia is then Eq. (42); the anodic reaction, oxidation of sulfate, Eq. (43), was already substantiated. The basis was thus established for a membrane process:

$$(SO_2 + O_2) \text{ cathode} \rightarrow (SO_3 + 1/2O_2) \text{ anode} \tag{56}$$

Tests of a pyrosulfate membrane were recently conducted [51]. The membrane was fabricated from MgO and $K_2S_2O_7$ and V_2O_5 in a multi-step procedure. The electrodes were constructed from partially-sintered perovskite ($La_{0.8}Sr_{0.2}CoO_3$) powder [52]. Using the standard cell design (Fig. 24) it was found that, at 400 °C or below, insufficient SO_2 in the inlet gas (3,000 ppm SO_2, 3% O_2) was oxidized by the vanadia in the membrane. For this reason a pre-reactor was added; a packed bed of catalyst giving effectively complete conversion to SO_3. With this arrangement, exact correspondence with theory was obtained (Fig. 43). Removal of over 98% of the inlet SO_x was seen, at 100% current efficiency (1 mol/2F); simultaneously, SO_3 and oxygen were emitted at the anode at the same rate.

One problem remaining with this concentration cell, as it is now configured;

$$(SO_3 + 1/2O_2) \text{ cathode} \rightarrow (SO_3 + 1/2 O_2) \text{ anode} \tag{57}$$

is the high polarization at low current density (Fig. 44). The geometric electrode area is 20 cm^2. It is known from tests in free electrolyte [49] that the true exchange current densities are around 30 mA/cm^2, obtained by potential-step experiments; the apparent value obtained from the slope of the data in Fig. 44:

$$i = \frac{di}{d\eta} \frac{RT}{nF} \tag{58}$$

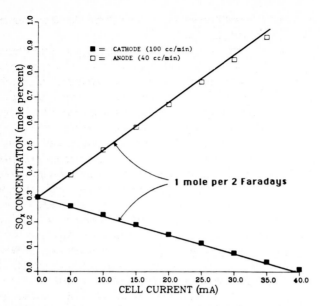

Fig. 43. Full-cell performance with hot-pressed membrane, perovskite electrodes. Cathode removal and anode generation as a function of applied current. Lines calculated from stoichiometry, 1 mol/2 F.

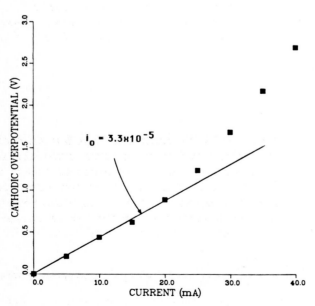

Fig. 44. Cathode polarization. Apparent exchange current density of 3.3×10^{-5} A/cm^2.

is three orders lower. Clearly, with the immobilized membrane there is severe concentration polarization.

The reason appears to be electrolyte flooding from the membrane into the electrodes. The electrochemical reduction of pyrosulfate occurs at the electrolyte-cathode interface, producing sulfate and SO_2. However, with a flooded cathode, there is little dissolved SO_3 available to react. The sulfate then precipitates near the electrode surfaces, raising the overpotential. As expected with this electrode blocking, the open-circuit potential remains highly negative for several hours as the sulfate slowly dissolves. Alternatively, oxidation current is seen to reduce the residual overpotential more quickly. *Post-mortem* analysis indeed gave visual evidence of severe electrode flooding. Improved membrane construction, with higher electrolyte retention is expected to alleviate these problems. A preliminary design of a commercial installation has been made [53]. Economic feasibility is contingent upon achieving current densities similar to those in the MCFC, when used to treat low-reactant-concentration CO_2, as described in Section 4. At 300 ppm SO_2, the typical outlet concentration from a 90%-removal flue gas treatment installation, the gas-phase mass-transfer will be nearly controlling at about 50 mA/cm² [41]; the other resistances, those due to electrode kinetics and solution concentration polarization should be minimal [49]. It is expected that 500 mV total cell polarization is a conservative estimate. At 65% current efficiency, also a conservative estimate, an applied current of 16 MM A is needed for a 500 MW coal-burning plant. This puts the power demand at 8 MW, or less than 2% of the plant output, quite reasonable for flue gas treatment.

The projected size of the installation is also encouraging: Based upon cost estimates for MCFC facilities, including ancillary equipment, this installation is estimated at $5.1 MM, or about $10/kW (installed). This also compares favorably with present treatment techniques, which range from $100 to $250/kW.

8   Conclusions

We have seen several examples of a technique for separation of gas mixtures which, in contrast with most commercial processes, requires no physical transfer of solvent, handling of solids, or cycling of temperature or pressure. The energy requirements can also be far lower: The thermodynamic minimum work of separation is, under isothermal conditions, the free energy difference between the process stream and byproduct, or permeate, stream. When this difference is due only to the partial pressure difference of component 1, it becomes:

$$\Delta G = -RT \ln (P_1/P_1')$$

The theoretical minimum work for electrochemical membrane separation is precisely same,

$$W_{min} = nFEe = RT \ln(P_1/P_1')$$

as seen, for example, in Eq. (12). Of course, the theoretical minimum cannot be achieved because of overvoltages present with current flow; i.e., during any real separation. Yet the actual energy requirements in cases where the transferred component is present in the mixture to 1% or less, are orders lower than those of transfer or acid-base neutralization. These restrictions are seen to be similar to those for electrowinning of metals.

While only a few applications have been tested, there is no fundamental reason to suggest there may not be several others. The separation of HF from process air in phosphoric acid manufacture and HCl from incineration flue gas are but two of these. Although the method will never be as broadly applied as say, distillation or absorption, its simplicity, selectivity and attractive economics make it likely to be employed in several specific situations.

9   References

1. S.A. Newman, "Acid and Sour Gas Treating Processes", Gulf, Houston, 1985, Ch. 20.
2. A.L. Kohl and F.C. Riesenfeld, "Gas Purification", Gulf, Huston, 1985, p. 872.
3. R.W. Rousseau, "Handbook of Separation Process Technology", Wiley, New York, 1987, Ch. 20.
4. M. Benedict, T.H. Pigford, and H. W. Levi, "Nuclear Chemical Engineering", 2nd ed., McGraw-Hill, New York, 1981, p. 751.
5. F.A. Lewis, W.F.N. Leitch, and A. Murray, Surface Technology 7, 385 (1978).
6. J. Brun, W. Gundersen, and T. Varberg, Kgl. Norske Videns. Selsk. Forh. 30, 30 (1957).
7. A. Winsel, Chem. Ing. Tech. 35, 379 (1963).
8. F.A. Lewis, J.H. Kirkpatrick, W.F.N. Leitch, J. Magennis, and A. Obermann, Surface Technology 13, 101 (1981).
9. L, Krogstad, Norsk Hydro, private communication, 1989.
10. M.Benedict, T.H. Pigford, and H.W. Levi, "Nuclear Chemical Engineering", 2nd ed., McGraw-Hill, New York, 1981, p. 765.
11. C.B. Taylor, IAEA – TECDOC – 246, p. 139 (May, 1981).
12. S.H. Langer and R.G. Haldeman, Science 142, 225 (1963).
13. J.M. Sedlak, J.F. Austin, and A.B. La Conti, Int. J. Hydrogen Energy 6, 45 (1981).
14. J.J. Zupancic, U.S. Patent No. 4,664,761, May 12, 1987.
15. J. Giner and N.D. Kackley, U.S. Army Contract DAAK70-86-C-0114, Final Report, November, 1987.
16. S.H. Langer and R.G. Haldeman, J. Phys. Chem. 68, 962 (1964).
17. Y. Fujita, H. Nakamura, and T. Muto, J. Appl. Electrochem. 16, 935 (1986).
18. J.W. Harrison, ASME 75-ENAs-51 (1975).
19. W.N. Lawless, U.S. Patent No. 4,462,891 (July 31, 1984); 4,547,277 (Oct. 15,1985).
20. A.C.C. Tseung and S.M. Jasem, U.S. Patent No. 4,300,987 (Nov.17, 1981).
21. R.R. Gagné and M.L. Marrocco, U.S. Patent No. 4,475,994 (Oct. 9, 1984).
22. D.R. Smith, R.J. Lander, and J.A. Quinn, "Recent Dev'ts in Sep'n. Science," Vol. III B, CRC, Boca Raton, 1977. pp. 225-241; J. S. Schultz, J. D. Goddard and S. R. Suchdeo, AICHEJ 20, 417 (1974).
23. H. Nishide, M. Ohyanagi, O. Okada, and E. Tsuchida, Macromolec. 20, 417 (1987).
24. B.M. Johnson, R.W. Baker, S.L. Matson, K.L. Smith, I.C. Ronan, M.E. Tuttle, and H.K. Lonsdale, J. Membrane Sci. 31, 31 (1987).
25. D.L. Roberts and R.M. Laine, U.S. Patent No. 4,605,475 (Aug. 12, 1986).
26. J. Ciccone, L.Deardurff, E. de Castro, J. Kerr, and B. Zenner, Electrochemistry Soc. Extended Abst. No. 897, Fall, 1987.

27. (a) J. Winnick, R.D. Marshall and F.H. Schubert, I. and E.C. Proc. Des. Dev't. 13, 59 (1974); (b) C.H. Lin and J. Winnick, ibid, p. 63; (c) C.H. Lin, M.L. Heinemann and R.M. Angus, ibid 13, 261 (1974).
28. O.E. Abdel-Salam and J. Winnick, AICHEJ 22, 1042 (1976).
29. R.G. Huebscher and A.D. Babinsky, S.A.E. Trans. 78, No. 690640, 1969.
30. A.J. Appleby and S. Nicholson, J. Electroanal. Chem. 53, 105 (1977); ibid 83, 309 (1977).
31. J. Winnick and P.N. Ross. J. Electrochem. Soc. 128, 5 (1981).
32. (a) J. Weaver and J. Winnick, J. Electrochem. Soc. 130, 20 (1983).
32. (b) M.P. Kang and J. Winnick, J. Appl. Electrochem. 15, 431 (1985).
33. J. Giner and L. Swette, EPRI 391, Nov. 1985.
34. L. Walke, K. Atkinson, D. Clark, D. Scardeville, and J. Winnick, Gas Sep'n. Purif. 2, 72 (1988).
35. H.K. Bjorkman, U.S. Patent No. 4,256,544, (Mar. 17, 1981).
36. R.F. Probstein and R.E. Hicks, "Synthetic Fuels", McGraw-Hill, New York, 1982.
37. H.S. Lim and J. Winnick, J. Electrochem. Soc. 131, 562 (1984).
38. K.A. White and J. Winnick, Electrochim, Acta 30, 511 (1985).
39. E.K. Banks and J. Winnick, J. Appl. Electrochem. 16, 583 (1986).
40. D. Weaver and J. Winnick, J. Electrochem. Soc. 134, 2451 (1987).
41. D. Townley and J. Winnick, I and E.C. Proc. Des. Dev't. 20, 435 (1981); ibid, Electrochim. Acta 28, 389 (1983).
42. I. Barin and O. Knacke, "Thermochemical Prop. of Inorganic Substances", Springer-Verlag, Berlin, 1973.
43. F.J. Salzano and L. Newman, J. Electrochem. Soc. 119, 1273 (1972).
44. H. Flood and N.C. Boye, Z. Electrochem. 66, 184 (1962).
45. B.W. Burrows and G. Hills, Electrochim. Acta 15, 445 (1970).
46. A. Durand, G. Picard, and J. Vedel, J. Electroanal. Chem. 70, 55 (1976).
47. M. Franke and J. Winnick, J. Electroanal. Chem. 238, 163 (1987).
48. P. Mars and J.G.H. Maessen, J. Catal. 10, 1 (1986).
49. A.R. Glueck and C.N. Kenney, Ch. Eng. Sci. 23, 1257 (1968).
50. K. Scott, T. Fannon, and J. Winnick, J. Electrochem. Soc. 135, 573 (1988).
51. M. Franke and J. Winnick, I. and E. C. Research, 28, 1352 (1989).
52. M. Franke and J. Winnick, J. Electrochem. Soc. 135, 1595, (1988); J. Appl. Electrochem. 19, 10, (1989).
53. K. Scott and J. Winnick, Gas Sep'n. Purif. 2, 23 (1988).

Figure Credits

The following figures have been reproduced with kind permission of the publishers:

Fig. 3: Pergamon Press, Int. J. of Hydrogen Energy 6, 45 (1981).
Fig. 4: Pergamon Press, Int. J. of Hydrogen Energy 6, 45 (1981).
Fig. 5: Chapman and Hall, J. Appl. Electrochem. 16, 935 (1986).
Fig. 6: Chapman and Hall, J. Appl. Electrochem. 16, 935 (1986).
Fig. 7: Chapman and Hall, J. Appl. Electrochem. 16, 935 (1986).
Fig. 9: Elsevier, J. Membrane Sci. 31, 31 (1987).
Fig. 10: Elsevier, J. Membrane Sci. 31, 31 (1987).
Fig. 16: American Chemical Society, Ind. and Eng. Chem. Proc. Des. and Dev't. 13, 63 (1974).
Fig. 17: American Chemical Society, Ind. and Eng. Chem. Proc. Des. and Dev't. 13, 63 (1974).
Fig. 18: American Chemical Society, Ind. and Eng. Chem. Proc. Des. and Dev't. 13, 261 (1974).
Fig. 19: American Inst. of Chem. Engrs. AIChE Jour. 22, 1042 (1976).
Fig. 20: Society of Automotive Engrs., Transactions of the S.A.E. 78, No. 690640 (1969).
Fig. 21: Society of Automotive Engrs., Transactions of the S.A.E. 78, No. 690640 (1969).
Fig. 22: The Electrochemical Society, J. of the Electrochemical Soc. 130, 20 (1983).
Fig. 23: Chapman and Hall, J. Appl. Electrochem. 15, 431 (1985).
Fig. 23a: Butterworth and Co., Ltd., Gas Sep. and Purif. 2, 72 (1988).

Fig. 24: The Electrochemical Society, J. of the Electrochem. Soc. 131, 562 (1984).
Fig. 25: The Electrochemical Society, J. of the Electrochem. Soc. 131, 562 (1984).
Fig. 26: The Electrochemical Society, J. of the Electrochem. Soc. 131, 562 (1984).
Fig. 27: Pergamon Press, Electrochemica Acta. 30, 511 (1985).
Fig. 28: Pergamon Press, Electrochemica Acta. 30, 511 (1985).
Fig. 29: Chapman and Hall, J. Appl. Electrochem. 16, 583 (1986).
Fig. 38: Elsevier, J. Electroanalytical Chem. 238, 163 (1987).
Fig. 39: Elsevier, J. Electroanalytical Chem. 238, 163 (1987).
Fig. 40: Elsevier, J. Electroanalytical Chem. 238, 163 (1987).
Fig. 41: Elsevier, J. Electroanalytical Chem. 238, 163 (1987).
Fig. 42: Elsevier, J. Electroanalytical Chem. 238, 163 (1987).
Fig. 43: Am. Chem. Soc., Ind. Eng. Chem. Res., 28, 1352 (1989).
Fig. 44: Am. Chem. Soc., Ind. Eng. Chem. Res., 28, 1352 (1989).

Electrochemical Aspects of Thin-Film Storage Media

Vlasta Brusic,[1] Jean Horkans,[1] and Donald J. Barclay[2]

[1] IBM, Thomas J. Watson Research Center, P.O. Box 218
Yorktown Heights, New York 10598, USA
[2] IBM Laboratories Ltd., Hursley House, Hursley Park,
Winchester, Hampshire S021 2JN, United Kingdom

Contents

1 Introduction

Since its inception in the 1950s [1], the technology used to store information as magnetized domains in a ferromagnetic material has been in a state of constant vigorous change. There is an ever-continuing need to increase the storage density. Areal densities of 0.002 megabits per square inch (Mb/in^2) were the state of the art in 1957 [2], whereas densities of from 27 to 160 Mb/in^2 are predicted for 1990 [3]. The driving force of increased density will cause the work-horse of the present generation of computers – the particulate disk – to be superseded by new, higher density thin-film storage media. This review will describe the role that electrochemistry has played both in the fabrication of thin-film storage media and in the measurement of their stability.

The structures of three kinds of magnetic media are described in Fig. 1. The particulate disk, represented at the top of the figure, consists of particles of a magnetic material suspended in a binder which is applied to an aluminum alloy substrate. The particles are usually γ-Fe_2O_3, Fe_3O_4, a Co-modified iron oxide, CrO_2, or a ferromagnetic metal. The intrinsic characteristics of the particles in the binder affect the maximum storage density. Since the storage density is limited by the width of the magnetic transitions, which is inversely proportional to the thickness of the medium, high-density storage requires a very thin medium: the main drawback of particulate disks is the practical lower limit of $\sim 0.5~\mu m$ on the thickness of the binder layer. The technology of particulate disks is discussed by Bate [4] and by Koster and Arnoldussen [5].

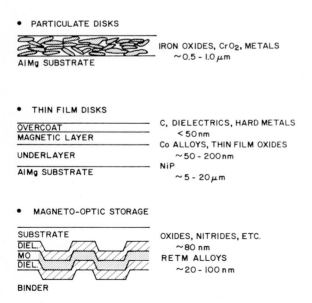

Fig. 1. Typical structures of storage media. Top: a particulate disk. Center: a thin-film disk. Bottom: a magneto-optic disk.

Thin-film disks, which are now appearing on the market, offer higher storage densities than particulate disks. The medium is a thin, continuous layer, which is usually a metal but may alternatively be an oxide, such as sputtered γ-Fe_2O_3. Thin-film disks can be fabricated either by electrochemical deposition or by vacuum deposition. Good reviews of thin-film media exist [4–9], but the technology is changing so rapidly that the preferred compositions of the medium layers and their optimum thicknesses and magnetic parameters continue to evolve even as reviews are being published.

A typical thin-film disk structure is also shown in Fig. 1. There is an underlayer, usually electroless Ni, between the magnetic thin film and the A1 alloy substrate. The electroless Ni underlayer, denoted here as NiP, is deposited from a solution with a hypophosphite reducing agent under conditions resulting in the incorporation of significant amounts of P. It is amorphous to X-rays and is not magnetic. It can be polished to provide a flat surface, thus allowing the head to fly over the disk at the very low flying heights (head-disk separations) required for high density storage. When the magnetic layer is an electroless Co alloy, the NiP also serves as a catalytic surface for the initiation of the deposition reaction. An overlayer, whose main purpose is to serve as a hard, mechanically protecting layer, is deposited on top of the magnetic film. The overlayer must be thin, because high density recording requires a small distance between the head and the medium. The most common overlayer is a sputtered carbon film, but metallic and dielectric overlayers have also been used.

The third structure of Fig. 1 represents the media used for magneto-optical recording, which can afford bit densities in excess of 1 Gbit/in^2. In this structure, the recording medium is a rare earth-transition metal alloy RE-TM (e.g. FeCoTb), usually prepared by sputter deposition. The medium is sandwiched between dielectric layers which can serve to protect the RE-TM from environmental oxidation and which can also enhance the recording performance if they have the appropriate dielectric constants. The grooves on the plastic or glass substrate permit tracking and focusing of the read-write laser beam.

The first requirement of any such disk structure is that it has the necessary magnetic properties. The pertinent magnetic parameters are defined in terms of a hysteresis loop, an example of which is shown in Fig. 2. The induced magnetization M is measured as a function of the applied magnetic field H. The important quantities for media are the coercivity H_c, which is the applied field at which the induced magnetization is zero; the saturation magnetization M_s, which is the maximum magnetization that can be induced in the material; the remanence M_r, which is the magnetization at zero applied field; and the squareness, which can be defined as either a remanence squareness $S = M_r/M_s$ or as a coercivity squareness $S^* = A/H_c$ (where the definition of A is given on the figure). The M_s is an intrinsic property of the material; the other magnetic parameters are dependent on the microstructure.

Information is stored in a magnetic medium as transitions in the direction of the magnetization vector. In conventional magnetic recording, one or more miniature heads are flown a very small distance above the disk and can either write a magnetic flux transition in the medium or can read a signal when passing over a magnetic transition in the medium. The magnitude of the read-back signal depends on the head

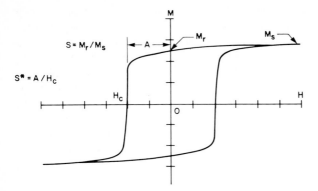

Fig. 2. A representative M-H loop, defining the parameters important for magnetic recording.

flying height and also on M_r. Thus, the medium should have a large M_s (the intrinsic property) and a large S. The coercivity is a measure of the energy required to reverse the direction of magnetization. The H_c should thus be large enough to give permanent storage of data but not so large that the head is unable to write in the medium. The maximum bit density attainable in a material depends on the coercivity.

Recording technologies can be based on media that have the magnetization vector in the plane of the thin film (longitudinal media) or on media having the magnetization vector normal to the film surface (vertical media). Present disk files employ the longitudinal mode, but some researchers believe that vertical recording may potentially offer benefits in future products. In longitudinal media, the bit density is limited by the width of the transition between domains of opposite magnetization; the minimum transition width is $a = M_r\delta/H_c$, where δ is the thickness of the medium [10]. Thus, it is important to have a very thin medium of high coercivity. An upper limit on the coercivity is imposed by the ability of the head to write a bit in the medium. The M_r cannot be reduced to increase storage density, because a high M_r is required for a sufficiently large read-back signal. In the vertical magnetization mode, the recording density is not limited by the transition width between regions of opposite magnetization, which is very small. Geometric factors make it difficult to magnetize thin films normal to the film plane: the self-demagnetizing field is large because of the high energy of the surface magnetic poles. In order to have a vertical medium, the perpendicular anisotropic energy $K_\perp$ must exceed the shape anisotropy $2\pi M_s^2$. This condition can be met by lowering M_s, but since lowering M_s will also reduce M_r, this expedient will have the adverse effect of reducing the read-back signal. Vertical media generally attain the large $K_\perp$ by having a large crystal anisotropy or a large shape anisotropy, or both.

A comparison of horizontal and vertical media is given by Mallinson [11]. He shows that most of the signal from either type of medium comes from the region nearest the surface and that the behaviors of horizontal and vertical media approach each other as the bit frequency becomes high. Although the media themselves may not differ fundamentally in the maximum attainable storage density, there may be differences between the total head-disk assemblies when the head designs for the two storage modes are optimized. The relative merits of horizontal and vertical storage in thin-film disk files are still being debated in some quarters.

Magneto-optic recording uses thermo-magnetic effects to accomplish the writing and erasure of data. The media are generally amorphous RE-TM alloys with the preferred axis of magnetization perpendicular to the plane of the disk. These materials are ferrimagnetic with high coercivities at ambient temperatures and low coercivities at elevated temperatures. During the writing process, the material is locally heated with a laser to a temperature at which its coercivity is sufficiently low to allow the magnetization of the heated spot to align with an applied magnetic field. The written data are read optically, at lower laser power, by sensing the rotation of the plane of polarization that occurs when polarized light is reflected from or transmitted through a magnetic material (i.e. by using the Faraday or polar-Kerr effects). The composition of the alloy is tailored to give relatively high Kerr rotation, which results in high read-back signals. Since the signal is sensed optically, magneto-optic recording has the advantage over conventional magnetic recording that high bit densities do not require a very small flying height between the sensor (the head) and the surface of the medium.

The mechanisms governing the coercivity of a magnetic material are discussed by Livingston [12]. The H_c is determined by the easiest of the two possible mechanisms of reversing the direction of magnetization: either through coherent or incoherent continuous rotation processes or through discontinuous domain processes. Both mechanisms must be impeded in a high-H_c material. Magnetic rotation is impeded through anisotropy, usually crystal or shape anisotropy. Nucleation and growth of domains must also be impeded. Nucleation is limited both by limiting the number of nucleating defects and by limiting the particle to a size in which a single domain is energetically more favorable than two domains. Growth of domains can be prevented by defects, such as second-phase precipitates, that pin domain walls.

Electrochemical deposition techniques are versatile for the fabrication of hard magnetic materials because the incorporation of the precipitated non-magnetic phases giving good magnetic properties is easily achieved. We shall give an overview of electrochemical processes that have been used in the deposition of thin-film magnetic media and shall illustrate the interdependence of structure and magnetic properties. Electroless processes and the structure of electroless deposits will be discussed in the first section of this review, then the deposition and structure of electroplated materials will be reviewed. We shall then turn to the second important electrochemical aspect of thin-film media: their stability. The investment in the file must be protected by assuring that the medium has an adequate lifetime in the environment in which it will be used. We shall review the corrosion of thin-film media. Corrosion mechanisms, means of protection against corrosion, and projected lifetimes will be discussed.

2   Electroless Deposition Processes for Thin-Film Media

Metallic thin-film media almost always are composed of Co alloys. Cobalt is chosen because of its high moment, but pure Co (either plated or vacuum-deposited) has a relatively low coercivity. It is thus necessary to manipulate the film composition or deposition parameters to obtain a microstructure with suitably high coercivity. In vacuum-deposited films, this structure is generally achieved by alloying with metals having a low solubility in Co; the second metal will segregate at grain boundaries and provide the second-phase precipitate that pins domain wall motion. Electrochemical deposition processes give considerable flexibility for the fabrication of materials with second-phase precipitates, as will be described in the following sections. Other film properties must also be controlled; these include grain size and shape, alloy composition. stress, etc.

2.1   Media for Longitudinal Recording

Electroless, or chemical, plating is widely employed in the manufacture of thin-film disks for longitudinal recording; it is commonly used to deposit the NiP underlayer and the CoP or other recording layer [13]. Although functionally equivalent alloys can be prepared by electrodeposition, electroless processes are favored by engineering considerations. For example, no external power supply is required, there is no dependency of deposit thickness on cell geometry or anode-cathode spacing, and a given volume of electrolyte can accommodate more disks than it can for electrodeposition.

Most published work has focused on the deposition of Ni, Co, and NiCo alloys from hypophosphite electrolytes [14], and this part of the review will deal primarily with these alloys. Other Co alloys studied include CoZnP [15, 16], the recording characteristics of which were described by Soraya [17]; CoSnP [18], which is reported to have enhanced corrosion resistance; and the rhenium and manganese alloys used for vertical recording, discussed below. Other reductants, such as hydrazine [19], dimethylamine borane [20–22], pyridine borane [23], and borohydride [24, 25], can be used for the chemical deposition of nickel and cobalt, but to date there has been no significant application of these to the technology of magnetic media.

In 1946, Brenner and Riddell [14] demonstrated that Ni^{2+} cations could be reduced by $H_2PO_2^-$ anions to form sound Ni deposits containing phosphorus. Since then, a great deal of activity has focused on improving and understanding the process, both for nickel deposition and for cobalt deposition, which is similar. Typical electrolytes contain a salt of the metal ion to be plated, hypophosphite in the form of the sodium salt, and a complexant for the metal ion (often a carboxylic acid). Practical plating solutions may also include surface-active chemicals, which induce facile dispersion of hydrogen from the electrode surface, and mediator compounds, which accelerate or inhibit deposition. Nickel electrolytes function over a range of acid and alkaline pHs, the pH of many commercial operations being maintained in the region of 4–6. Cobalt solutions operate in the alkaline region.

The fundamental mechanistic details of the electroless process are not well understood and are a matter of debate in the literature [26]. The overall process can be depicted as

$$Co(II) + P(I) + H(I) \rightarrow Co(0) + P(0) + H(0) + P(III)$$

where Co(0) and Co(II) are cobalt in the oxidation states 0 in the deposit and $+2$ in the solution complex; P(0), P(I), and P(III) are phosphorus in oxidations states 0 in the alloy, $+1$ in $H_2PO_2^-$, and $+3$ in $H_2PO_3^-$; H(0) and H(I) are hydrogen in the oxidation states 0 in H_2 and $+1$ in any of the species H_2O, H^+, $H_2PO_2^-$. Phosphorus is incorporated in the deposit in concentrations determined by the solution composition and by the operating conditions; for magnetic media, there is typically 5–10 a/o P in the CoP.

Proposed intermediates in the above reaction include atomic hydrogen [27, 28], hydride ions [29, 30], metal hydroxides [31], metaphosphites [32, 33], and excitons [34]. In general, the postulated mechanisms are not supported by direct independent evidence for these intermediates. Some authors [35] maintain that the mechanism is entirely electrochemical (i.e. it is controlled by electron transfer across the metal-electrolyte interface), but others [26] advocate a process involving a surface-catalyzed redox reaction without interfacial electron transfer.

An example of the electrochemical mechanism is the following [36, 37]:

$$H_2PO_2^- + H_2O \rightarrow H_2PO_3^- + 2H^+ + 2e^-$$

$$Ni^{2+} + 2e^- \rightarrow Ni$$

$$2H^+ + 2e^- \rightarrow H_2$$

$$H_2PO_2^- + e^- \rightarrow P + 2OH^-$$

The principal evidence in favor of the electrochemical mechanism is the observation that metal deposition can occur, albeit at a much reduced rate, in a two-compartment cell. When one of the compartments contains only the metal ions and the other only the sodium hypophosphite, metal deposition occurs on the electrode in the first compartment, being driven by hypophosphite oxidation at the electrode in the other half-cell [36, 37].

The electrochemical mechanism was rejected by Salvago and Cavallotti [26] on the basis that it does not explain several features of electroless deposition of ferrous metals: it does not account for the isotopic composition of the H_2 gas evolved; it does not explain the effect of the various solution components on reaction rate; and it does not account for the homogeneous decomposition of very active solutions or the fact that they can give deposition on insulating surfaces. These authors put forward a chemical mechanism, involving various "hydrolyzed" nickel species, which they claim explains the observed behavior of the system:

$$Ni(OH)_2 + H_2PO_2^- \rightarrow NiOH + H_2PO_3^- + H$$

$$NiOH + H_2PO_2^- \rightarrow Ni + H_2PO_3^- + H$$

$$NiOH + H_2O \rightarrow Ni(OH)_2 + H$$

$$Ni + H_2PO_2^- \rightarrow P + NiOH + OH^-$$

This mechanism is based on the known importance of hydroxides in other deposition reactions, such as the anomalous codeposition of ferrous metal alloys [38–39]. Salvago and Cavallotti claim an analogy with the mechanism of Ni^{2+} reduction from colloids in support of their proposed mechanism. There is no direct evidence for the hydrolyzed species, however. Furthermore, the mechanism does not explain two experimentally observed facts: Ni deposition will proceed if the Ni^{2+} and the reducing agent are in separate compartments of a cell [36, 37]; and P is not deposited in the absence of Ni^{2+}. The chemical mechanism does not take adequate account of the role of the surface state in catalysis of the reaction. It has no doubt been the extreme oversimplification, by some, of the electrochemical mechanism that has led other investigators to reject it.

Mital et al. [40] studied the electroless deposition of Ni from DMAB and hypophosphite electrolytes, employing a variety of electrochemical techniques. They concluded that an electrochemical mechanism predominated in the case of the DMAB reductant, whereas reduction by hypophosphite was chemically controlled. The conclusion was based on mixed-potential theory: the electrochemical oxidation rate of hypophosphite was found, in the absence of Ni^{2+} ions, to be significantly less than its oxidation rate at an equivalent potential during the electroless process. These authors do not take into account the possible implication of Ni^{2+} (or Co^{2+}) ions to the mechanism of electrochemical reactions of hypophosphite.

Feldstein and Lancsek [30] measured plating rate, potential, and hydrogen evolution rate during the reduction of Ni^{2+} and Co^{2+} with $H_2PO_2^-$ in the presence of various additives. They concluded that the deposition process could be described by a modified hydride mechanism. The basic steps of the process were identified as follows:

$$H_2PO_2^- + OH^- + X \rightarrow [X-H]^- + PO_2^- + H_2O$$

$$PO_2^- + OH^- \rightarrow HPO_3^{2-}$$

$$2[X-H]^- + M^{2+} \rightarrow M + 2[X-H]$$

$$2[X-H] \rightarrow H_2 + 2X$$

$$[X-H]^- + H_2O \rightarrow H_2 + OH^- + X$$

$$3[X-H] + 2H_2O + PO_2^- \rightarrow P + 3[X-H] + 4OH^-$$

$$H_2PO_2^- + OH^- \rightarrow HPO_3^{2-} + H_2$$

This mixed chemical/electrochemical mechanism has been critized by Randin [41], who points out that the main characteristics of the process are accommodated by a number of mechanisms, including the one described.

A new, exciton-like mechanism has been proposed by Demidenko et al. [34] to describe the formation of amorphous NiP films. According to these authors, the electroless process is initiated by the reduction of Ni^{2+} by adsorbed hydrogen to produce clusters of pure Ni. Vacancy sites on these clusters capture a proton (presumably from the $H_2PO_2^-$ anion) to form an "exciton-like" state, which subsequently captures more Ni atoms, and so on. Phosphorus can be captured and fixed at a vacancy site.

Van der Meerakker [28] gives a general mechanism for electroless processes involving various reductants. The mechanism assumes the formation of adsorbed hydrogen atoms, which can then be oxidized or desorbed as H_2 gas. In the case of Co(II) reduction by hypophosphite, the relevant mechanism is as follows:

$$H_2PO_2^- \rightarrow HPO_2^- + H$$

$$HPO_2^- + OH^- \rightarrow H_2PO_3^- + e^-$$

$$H + OH^- \rightarrow H_2O + e^-$$

$$Co(II) + 2e^- \rightarrow Co$$

$$2H_2O + e^- \rightarrow H_2 + 2OH^-$$

$$H_2PO_2^- + e^- \rightarrow P + 2OH^-$$

The most likely approximation of the true mechanism seems to be the Van der Meerakker proposal, which involves both chemical and electrochemical contributions and does not invoke unlikely intermediates. The exciton hypothesis may add a physical description to this chemical picture. The most important need in establishing the mechanism is better experimental verification. The disagreement concerning the mechanism, however, has not kept significant technological progress from being made.

2.2   Media for Vertical Recording

The CoP and CoNiP alloys discussed above have in-plane anisotropy and are employed in the horizontal recording mode. Electroless techniques have also been used to deposit thin films with perpendicular anisotropy for vertical recording media. This application has been pursued primarily by groups in Japan [42–60]. Vertical magnetization can be achieved by modification of a CoNiP alloy through the inclusion of other metals, typically Mn and Re. The mechanism of incorporation of the Mn and Re in electroless CoNiP has not been discussed in the literature. It is known, however, that the complexant plays an important role in determining the deposit properties. Various complexants have been investigated [43, 44, 46, 47, 60]. The optimal systems of complexants were found to be ammoniacal malonate-malate for CoNiMnP, ammoniacal malonate-tartrate-tartronic acid-gluconate for CoNiReP, and ammoniacal malonate-tartrate-succinate for CoNiReMnP. Osaka et al. have shown how the microstructure and the magnetic properties depend on the chemistry of the electroless plating solution, specifically on the complexant used [43]. This example is only one illustration of the strong interdependences between the electrochemistry and the microstructure and properties of the deposit, a subject that will be discussed in more detail below.

3   Structure-Magnetics Interrelationships for Electrolessly Deposited Media

The manner in which a material's structure determines its magnetic properties is of over-riding interest in understanding recording media. A general discussion of this topic is beyond the scope of this review, however. Romankiw and Thompson [61] in 1975 reviewed the magnetic properties of plated magnetically hard and soft films. Here we shall consider only hard films suitable as media, reviewing the role of electrochemistry in determining a deposit's structure and the way in which the electrochemical parameters can be manipulated to give the desired magnetic properties.

3.1   Media for Longitudinal Recording

As discussed above, the two archetypical electrolessly plated alloys of relevance to longitudinal magnetic recording are NiP and CoP. The metallurgical requirements of these alloys are quite different. In the case of NiP, the underlayer on which the recording alloys is deposited, the typical thickness is in the range of 10–25 μm, and the material must not be magnetic. The recording alloy CoP is generaly 0.05–0.10 μm thick; among the magnetic requirements for the CoP are an H_c of $\sim$ 500–1000 Oe and a high squareness of the M-H loop. There is a voluminous literature on the relationships among structure, deposition conditions, and post-treatments of these alloys, (especially of NiP [62, 63], because of the wide range of technological applications of that material). This section of the review will focus on the relationship between structure and processing as it relates to magnetic recording.

The requirement of a non-magnetic underlayer imposes constraints on the composition and structure of the NiP, since unalloyed Ni is a ferromagnetic metal and since NiP can, at elevated temperatures, recrystallize to become magnetic. Speliotis and Fernelius [64] have discussed the conditions under which NiP can be prepared with intermixed ferromagnetic and paramagnetic regions; they have also discussed the consequences to magnetic recording. Depending upon the deposition conditions and post-deposition treatment, NiP and CoP may be amorphous or crystalline. Clements and Cantor [65] studied the recrystallization behavior of Ni, as well as NiCo and NiFeCo, containing 5–20 a/o P. Except for the lowest P concentrations, the alloys were amorphous.

Amorphous NiP alloys with $> \sim$ 10% P (generally obtained by deposition from acidic electrolytes) are non-magnetic (see [66] and references therein), as required of the underlayer for thin-film media. Although the structure of these alloys is generally assumed to be a solid solution of P in Ni, a recent report [67] has suggested that NiP with 7.4–10% P deposited from acid sulfate electrolytes is better represented by a microcrystalline structure composed of 4–5 nm fcc NiP solid-solution grains.

Since the as-deposited NiP is metastable, it should undergo spontaneous crystallization, even at room temperature. Recrystallization is a highly activated process, however, with an activation energy in excess of 1.5 eV [65]. Thus, crystallization under ambient conditions is not of practical consequence. These glassy metastable alloys have been shown to have a layered structure with super-imposed lamellae [68]. They undergo complex crystallization processes upon annealing above 200 °C [69], ultimately leading to a phase separation of fcc Ni and nickel phosphides [65, 66]. The annealed alloys are ferromagnetic, and therefore magnetic disks incorporating NiP must not experience temperatures much above 200 °C, either during processing [62] or in use.

Simpson and Brambley [70] investigated CoP alloys with 5–9 a/o P. The alloys were amorphous if the P content exceeded 8.8%. The CoP for applications in magnetic media is deposited in the crystalline state. It is generally assumed to be in the hcp modification, although Chow et al. [71] concluded that the structure could be hcp or fcc depending on the chemistry of the metallizing solution. Cavallotti and co-workers ([26] and references therein) have studied the preferred orientation PO as a function of electrolyte composition and have shown that the $\{11.0\}$, $\{00.1\}$, and $\{10.0\}$ preferred orientations can be obtained, depending on the acidity and $H_2PO_2^-$ concentration of the solution and on complexing agent used. Similarly, Cortijo and Schlesinger [72] have found that the solution chemistry determines whether films will grow with the c axis perpendicular to the surface plane. These authors found no evidence of an fcc phase.

Of particular interest with respect to magnetic recording is the mechanism whereby high H_c and S are achieved in the CoP alloys. These magnetic parameters have been shown to correlate with the size and nature of the microcrystallites within the film [73]. Electrolytes that yield an average grain size close to the critical size for single-domain particles have been found to result in high-H_c deposits. Additionally, it has been postulated that high S correlates with a small separation between the grains.

Mirzamaani et al. [74, 75] point out that the earlier studies of the interrelationships between structure and magnetics have examined films substantially thicker than those now being used in thin-film disks. These authors have examined very thin CoP films and have studied the relative roles of shape anisotropy, stress anisotropy, and crystal anisotropy in determining the magnetic properties. For their CoP-deposition system, shape anisotropy dominated the other factors in determining the film magnetic properties. The shape anisotropy of a particular deposit was determined by the surface condition of the substrate on which the CoP was deposited.

Nicholson and Khan [76] and Khan and Lee [77] have studied the deposition of CoP and NiCoP on NiP in the pH range 7.8–9.3. Coercivity is markedly dependent on pH for both alloys, as shown in Fig. 3, with uniformly higher H_c being achieved with NiCoP. The authors postulate that Ni initiates deposition in the case of the ternary alloy, resulting in a different nucleation and growth process from CoP. The rapid decrease in the H_c of NiCoP above pH 8.8 occurs as the grain size decreases and approaches the size having superparamagnetic behavior.

Other authors [78, 79], however, report a different pH dependence of the H_c of CoP and CoNiP. Judge et al. [78] deposited CoP from two solutions identical except

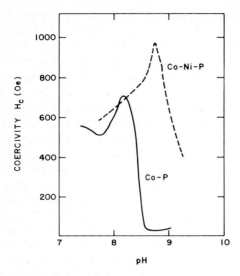

Fig. 3. The coercivity of electroless CoP and CoNiP deposits as a function of the pH of the solution from which they.were deposited [76]. (Reprinted by permission of The Electrochemical Society).

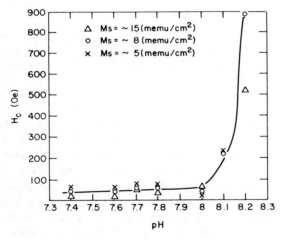

Fig. 4. The H_c of electroless CoP as a function of solution pH. Solution A of Ref. [78]. (Reprinted by permission of the Americal Electroplaters Society).

that the pH of one (solution A) was adjusted with NH_3 solution and that of the other (solution B) with NaOH. The pH dependence of H_c in their solution A is shown in Fig. 4. The H_c is low (< 100 Oe) at the low end of the pH range and increases rapidly above pH ~ 8.8. The coercivity increase corresponds to an increase of crystallinity of the deposit. If instead the pH adjustment is done with NaOH (solution B), H_c is low and nearly constant from pH 8.2 to 8.8. Lawless and Fisher [79] examined the pH dependence of the H_c of electroless CoNiP. The coercivity of this material was low (< 10 Oe), lower than the H_c reported by Judge et al. [78] for CoP, and was essentially independent of pH in the range 7.7 to 8.6.

Nicholson and Khan [76] do not report the composition of their electroless CoP solution, other than describing it as a modified Brenner formulation. Judge [78] and Lawless and Fisher [79] have used similar, citrate-based solutions. The pH ranges examined in the three papers are similar but not identical, and it is conceivable that some trends were missed by examining too narrow a pH range. The discrepancies from author to author not only in the absolute values of the magnetic properties, but also in their trends, pointedly illustrate the difficulties in understanding the relationships between the deposition parameters and the structure and magnetic properties.

An unfortunate factor complicating the understanding of the H_c-pH relations is that H_c is generally reported at constant deposition time. Since deposition rate is also a function of pH, however, the H_c values are being compared for films of varying thickness, even though H_c is also a function of thickness.

Judge et al. [78] examined other magnetic properties of CoP. The saturation moments σ_s were 100 and 116 emu/g for deposits from solutions A and B, respectively. Although the P content of the deposits varied with pH, the σ_s values were essentially independent of pH. The squareness ratio of the M-H loop showed no significant dependence on any quantity other than film thickness; it ranged from ~ 0.5 for thin films to ~ 0.8 for thick deposits.

Other alloys have been examined less extensively than NiP, CoP, and CoNiP. Doss and Phipps [80] discussed the use of NiPB alloys for the non-magnetic underlayer of magnetic disks. The boron greatly enhances the thermal stability of the material, but increasing B content results in decreased hardness and corrosion resistance. During annealing, the precipitation of Ni_3P and Ni_3B proceeds in parallel with the crystallization of Ni.

Films of CoB have been prepared by electroless deposition. Chang et al. [25] deposited magnetically soft amorphous films, which could be annealed to give materials with an H_c of 250 Oe. Depending on the annealing temperature, the films crystallized as the hcp or fcc modifications of Co. Matsui and co-workers [22] obtained crystalline materials in the as-deposited state, the crystalline characteristics being determined by processing conditions. A maximum H_c of 300 Oe was observed for films with $\{10.0\}$ preferred orientation.

Alloys of CoSnP [17] and CoZnP [16] have been prepared by electroless deposition. Both Zn and Sn increase the H_c, and Sn also improves the corrosion resistance and the appearance. For Sn alloys, X-ray diffraction measurements suggest that the Sn addition orients the c axis parallel to the plane of the film and decreases the grain size.

3.2 Media for Vertical Recording

Electrolessly deposited vertical media are generally CoNiP alloys containing Mn or Re or both. The Ni is added to lower the M_s and hence to lower the demagnetizing field $4\pi M_s$. Up to 30% Ni can be added while maintaining the hcp phase of Co. The Re also serves to lower M_s. The earliest descriptions of these media claimed that the addition of a small amount of Mn caused the magnetization to be normal to the surface [43–45]. The c axis of the hcp Co changed from in-plane to perpendicular with increasing addition of $MnSO_4$ to the plating solution [43]. Manganese also increased the H_c of the deposit. Subsequently to these early publications, however, it was discovered that perpendicular orientation could be achieved in the absence of Mn by optimizing the conditions of pH and complexation [48]. The dependence of the magnetic properties of these alloys on deposition conditions has been studied [51–69], and their read/write characteristics as media have been determined [53, 55, 57, 59, 63, 69].

The structures of deposits with the compositions $Co_{40.1}Ni_{44.2}Re_{6.0}P_{9.7}$ and $Co_{23.6}Ni_{58.5}Re_{5.3}Mn_{0.2}P_{12.4}$ have been determined and correlated with the magnetic properties [51]. These alloys have columnar structures, but the columnar nature is less marked than for CoNiMnP. A dependence of magnetic properties on film thickness is caused by structural changes at the interface rather than by compositional variations with film thickness. The CoNiReP has the greater degree of hcp Co c-axis orientation normal to the surface for very thin films.

The origin of the thickness dependence of magnetic properties is discussed by Koiwa et al. [56]. An ammoniacal malonate-tartrate-tartronic acid-gluconate solution was used, with the malonate concentration being varied. For 500 nm films, the c axis of hcp Co is strongly oriented normal to the film; the degree of orientation is independent of malonate concentration. For 100 nm films, however, the c-axis orientation and the crystalline state strongly depend on malonate concentration. Thus, the malonate is important mainly in determining the structure of the interfacial region betwen the magnetic film and the substrate. These authors propose a model of the CoNiReP structure as follows. The structure is columnar with a highly crystalline center surrounded by less crystalline material. The highly crystalline portion is composed mainly of Co and Re, the Co having the hcp structure. The less crystalline surrounding material is primarily Ni and P, the P being insoluble in the Co lattice. The incorporation of Re in the alloy enhances the formation of the less crystalline portion.

Recent studies of CoNiReP media on flexible disks for vertical recording [57] have examined an alternative underlayer, an "almost non-magnetic" NiMoP alloy [58]. The electroless CoNiReP behaves like a so-called double layer film, magnetically soft near the interface but magnetically hard in the bulk. Ouchi and Iwasaki [59] have proposed such double-layer structures for perpendicular recording, but they are usually prepared by depositing two separate hard and soft magnetic materials. The "quasi-soft" initial layer is ~ 150 nm thick in CoNiReP films on NiMoP and has an in-plane anisotropy. Typical $H_c(\parallel)$ values for this layer are 50–85 Oe. As the magnetic film becomes thicker, it attains a high coercivity with a perpendicular anisotropy. The properties of the initial layer are determined by the NiMoP underlayer. The CoNiReP grows epitaxially as the fcc phase on the fcc NiMoP underlayer. The shape anisotropy of this thin film causes it to be oriented in-plane. Away from the influence of the underlayer, however, the fcc phase is replaced by the hcp phase with vertical anisotropy. There is no detectable difference in the crystallinity of the two regions.

3.3 The Effect of Underlayer on the Properties of the Magnetic Layer

The strong dependence of the magnetic properties on the microstructure of the magnetic alloy implies that the substrate will also influence the magnetics, since the substrate will affect the nucleation and growth, the structure, and the PO of the material deposited on it. For electrochemically deposited films, most studies have examined underlayer effects as secondary to other aspects of magnetic thin films. A few relevant papers dealing primarily with the effects of underlayers can nevertheless be mentioned here.

The substrate can have a significant influence on the H_c and S of the M-H loop [74, 75, 81]. Osaka and Nagasaka [81] have examined the substrate material (Cu, Au, NiP, NiCuP, and acetylcellulose), the degree of activation of an insulating substrate by $PdCl_2$, and the surface roughness. The authors concluded that the magnetic properties were controlled by the substrate grain size and the degree of surface activation. (Surface activation by $PdCl_2$ is not used in the fabrication of hard disks, but it is used to initiate the deposition of the magnetic layer of "floppy" disks, which use flexible polymer substrates.) Macroscopic roughness did not have a significant effect.

The influence of the underlayer was demonstrated by Mirzamaani et al. [74] and by DiMilia et al. [75]. Electroless CoP was deposited from a solution yielding an intrinsically isotropic structure onto NiP that had undergone various pretreatments. When the NiP was exposed to a NaOH treatment before CoP deposition, good in-plane magnetic properties resulted, but deposition on a chemically or electrochemically pre-reduced NiP surface gave unacceptable magnetics. These authors found

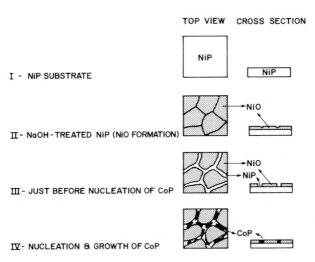

Fig. 5. Model of the origin of shape anisotropy in electroless CoP thin-film media [74]. (Reprinted by permission of The Electrochemical Society).

shape anisotropy to be the biggest factor in determining the film magnetic properties; the results can be explained by the model of Fig. 5. Nucleation is facile on the reduced NiP surface, yielding small isotropic crystallites; there is no source of induced anisotropy. Surfaces pretreated with NaOH, on the other hand, are covered with an oxide film of non-uniform thickness (which has been observed by electron microscopy). Oxide-free surface sites are required for the CoP reaction, but these are limited on the NaOH-pretreated surface. The CoP begins to grow in channels where the oxide was thinnest, and the shape anisotropy of the growing crystallites leads to the desired magnetic properties.

The substrate was also found to influence the properties of the electrolessly deposited vertical media CoNiMnP, CoNiReMnP, and CoNiReP. The c-axis orientation had a larger degree of perpendicular orientation for films deposited on electroless NiP than for those deposited on Cu foil, presumably because of the smaller roughness of the former substrate [43]. The double-layer (magnetically soft interface, magnetically hard bulk) properties of CoNiReP deposited on a NiMoP underlayer [57] have already been discussed.

Electroless CoNiMnP and CoNiP were examined as underlayers for electroless CoP by Matsubara et al. [82]. The CoNiMnP has a c-axis orientation normal to the film plane, whereas the CoMnP has a low degree of orientation. The CoP was found to deposit with a microstructure resembling that of the underlayer. Very thick deposits ($> 0.5 \mu$m) resume the intrinsic structure of CoP (with a low degree of PO). These composite structures have been tested as vertical recording media.

4 Electrodeposition Processes for Thin-Film Media

Electrodeposition is more flexible than electroless deposition, in that it is not limited by the requirement of having a catalytically active surface. Electrodeposition allows a wider variation in the alloy composition and in the deposit properties than does electroless deposition. This flexibility has not been widely exploited, however, and most of the electrodeposited alloys have had compositions similar to those obtained by electroless deposition (i.e. CoP or CoNiP).

4.1 Media for Longitudinal Recording

Koretzky [83] in 1963 published a review of electrodeposited magnetic films, which considers an earlier generation of storage devices. Some magnetic properties of electroplated Co alloys were tabulated by Safranek [84] and by Morral [85].

The discovery that P could be electrodeposited with Ni and Co was made by Brenner and co-workers [86]. Because of the ability to incorporate P over a wide range of composition, a wide range of magnetic properties is available in electro-deposited CoP and NiP. These materials have thus found a variety of applications. Only the deposition of magnetically hard alloys will be considered here.

Whereas the electroless CoP alloys are deposited from basic solutions, electro-deposition is done from acidic solutions. Either sodium hypophosphite NaH_2PO_2 or phosphorous acid H_3PO_3 can be used as the source of P. Those plating solutions containing H_3PO_3 are generally used to deposit high-P, amorphous CoP alloys. References to H_3PO_3-containing solutions for magnetically hard CoP alloys are rare, e.g. [87]. The reduction of the P(I) (or P(III)) in these oxyanions to P(0) does not occur in the absence of the Ni^{2+} or Co^{2+} ions. A CoP medium typically has a P content of a few weight percent [88]. The mechanism of phosphorus incorporation in electroplated alloys has received little attention.

A much greater range of coercivities is available in electroplated CoNiP than in electroplated CoP [88, 89]. The mechanism by which Ni incorporation raises the H_c has not been much examined, nor have the differences between electroplated and electrolessly deposited CoNiP. Phosphorus-free alloys of CoNi also have been deposited [90] but do not appear to be promising because of low coercivities. Bonn and Wendell [91], in a 1953 patent, were the first to suggest electrodeposited CoNiP as a magnetic medium. The essential components of their plating solution were $CoSO_4$, $NiSO_4$, NaH_2PO_2, and NH_4Cl. Similar solutions are still commonly used for deposition of CoNiP. Bonn and Wendell noted that deposition takes place at a current efficiency greater than 100%, showing that there is a parallel electroless reduction proceeding at the cathode.

A review of plated films in early storage applications was given by Sallo [92]. He describes efforts to develop suitable CoNi electrodeposits and gives examples of marketed products incorporating tapes and drums with electrodeposited CoNiP magnetic media. A more recent use of electrodeposited CoNiP for magnetic storage is the plated thin-film disks described by Tago et al. [93] and Pearce et al. [94].

Other high-coercivity alloys of Co have been studied in less detail than CoP and CoNiP. Luborsky [95] has examined the structure and magnetic properties of Co and CoNi electrodeposits with additions of elements from Group VIB (Cr, Mo, and W) and Group VA (P, As, Sb, and Bi). The mechanism of incorporation of the alloying element was not investigated, however. The effect of the VA elements on H_c is shown in Fig. 6. Similar magnetic properties can be attained by additions from either group. The quantity of added element necessary to achieve the maximum in H_c goes in the order P < As < Sb < Bi for Group VA and W < Mo < Cr for Group VIB. These relationships are apparently caused by the dependence of grain size on the incorpor-ated element and by the segregation of this alloying element at grain boundaries. Luborsky suggests that CoNiW compares favorably with CoNiP for storage applications, in that it has similar magnetic properties and is more resistant to corrosion.

Electroplated CoPt [96, 97] and CoSnP [98] have also been suggested for thin-film media. Both alloys are reported to have good corrosion resistance. A plated 2.25″ diameter disk uses Zn in the plating solution to control coercivity [24]; details of the process are not given.

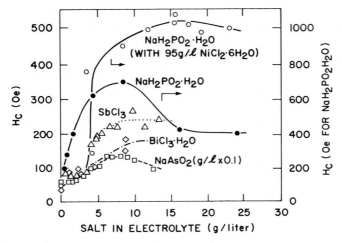

Fig. 6. Effect of additions of Group VA elements on the H_c of Co and CoNi [95]. The thicknesses of the films at the knees of the curves are as follows: CoNiP, 350 nm; CoP, 650 nm; CoAs, 1000 nm; CoSb, 560 nm; CoBi, 220 nm. (Reprinted by permission of The IEEE).

The good magnetic properties of CoMoP electrodeposits have led Shuvalova et al. [99] to study the mechanism of their deposition. These authors conclude that the molybdenum is discharged through the MoO_2^{2+} ion, which may be complexed with ligands like citrate. Increasing the Mo content of the deposit lowers its P content. This alloy also reportedly affords better corrosion resistance than CoP.

Manganese is yet another element that can be deposited to yield hard materials with good magnetic properties. An investigation of CoMnP electrodeposition was described by Bondar' et al. [100], who also reviewed earlier reports of CoMn alloys. The Mn content of these alloys is on the order of 1%; increased concentrations of $MnCl_2$ in the electrolyte lead to a greater incorporation of P of the deposit. The H_c of the CoMnP is higher than that of either CoP or CoMn deposited under the same conditions. Coercivity goes through a maximum as a function of the solution concentration of Mn(II). The magnetic properties are strongly dependent on the solution pH and temperature.

4.2   Media for Vertical Recording

Since vertical recording is not yet being employed in disk files, the electrochemical deposition of perpendicular materials is still at an exploratory stage. Electroplated vertical media are uncommon. In some cases, the vertical media are modifications of materials used as horizontal media. In others, the vertical medium is produced by introducing a strong shape anisotropy.

To produce a continuous thin-film vertical medium, the electroplating process must yield a structure with high coercivity and a vertical anisotropy. One way of achieving this goal is through the electrodeposition of unalloyed Co under the appropriate conditions [101–107]. These conditions result in the incorporation of non-metallic "impurities", which are thought to be precipitated cobalt hydroxides. Chen and Cavallotti [101, 104–107] have plated Co from acidic sulfate and sulfamate solutions of relatively high pH. The Co^{2+} was shown to discharge through a hydrolyzed intermediate [105]. The extent of hydroxide incorporation depends strongly on the current density and on the temperature.

Cobalt-platinum alloys with vertical magnetization have been plated by Baumgaertner et al. [108]. A series of alloys was produced with a wide range of Pt contents (up to ~ 70 a/o). The physical and magnetic properties are strongly dependent on the processing parameters. Deposits are generally highly stressed. The alloy with ~ 30 a/o Pt is considered by the authors to be a potential vertical medium.

Another potential vertical medium is not a continuous thin film, but rather an assembly of metal particles deposited in well defined pores in an alumina film on Al [109–117]. The shape anisotropy of the particles gives the desired vertical anisotropy. In producing such structures, the proper conditions for the anodization of the Al disk and for the subsequent control of pore size are as important as the conditions of metal deposition. The structures of such disks are discussed in detail below.

5 Structure-Magnetics Interrelationships for Electrodeposited Alloys

5.1 Media for Longitudinal Recording

The structures of electroplated hard alloys have been less extensively studied than those of similar electrolessly deposited materials. Sallo and co-workers [118–120] have investigated the relationship between the structure and the magnetic properties of CoP and CoNiP electrodeposits. The structures and domain patterns were different for deposits with different ranges of coercivity. The lower-H_c materials formed lamellar structures with the easy axis of magnetization in the plane of the film. The high-H_c deposits, on the other hand, had a rod-like structure, and shape anisotropy may have contributed to the high coercivity. The platelets and rods are presumed to be isolated by a thin layer of a nonmagnetic material.

Tago et al. [93] have described complete thin-film disks with electrodeposited CoNiP layers. The magnetic properties of the CoNiP vary with its thickness and also depend on the substrate (Au, Cu, or NiP). The grain size of the CoNiP is large when it is plated on Cu foil. On Au, the deposit is finer grained and more uniform in composition. On electroless NiP, the CoNiP is very fine grained. The composition of the magnetic layer is different at the substrate interface and in the bulk of the film, and the compostion distribution depends on the substrate. The deposition of the hard overcoat can alter the magnetic properties through such factors as stresses in the overcoat film.

Another electroplated thin-film disk has been fabricated and tested by Yoshida et al. [121]. This disk is unusual in that the non-magnetic underlayer is electroplated CuSn. A thin NiP layer is interposed between this underlayer and the electroplated CoNiP magnetic layer. The overcoat is a layer of Co_3O_4 formed by selective oxidation of the magnetic layer. The heat treatment necessary for selective oxidation is the process step that obviates the use of a thick NiP underlayer; the NiP will recrystallize, roughen, and become magnetic during the 50 min exposure to air at 300 °C. The overall structure of this electroplated disk is then

$$\text{Al substrate/CuSn 25 } \mu\text{m/NiP} \quad 0\text{–}120 \text{ nm/Co}_{84.2}\text{Ni}_{15}\text{P}_{0.8} \text{ 50 nm/Co}_3\text{O}_4 \text{ 50 nm.}$$

These authors [121] discuss mechanisms of media noise in thin-film disks in relationship to their structures. High intrinsic noise levels in thin-film media make high signal-to-noise ratios difficult to achieve. The noise originates in so-called zig-zag transitions between bits. The bulk magnetic properties and the micromagnetic properties have been determined for disks with varying thicknesses of electroplated and electrolessly deposited NiP interlayers. The bulk magnetic properties of the electroplated CoNiP depends on the thickness of the NiP interlayer. The M_s values of disks also depend on the method of depositing the NiP. The interlayer dependence of the bulk magnetics was postulated to be caused by processes occurring during the heat treatment: the magnetization of the NiP and the diffusion of elements from the CuSn underlayer into the magnetic film. The micromagnetic properties of the structure were also dependent on the interlayer thickness. Media noise increases with increasing thickness of the NiP. There is less noise in disks with electroplated NiP layers.

The structures of CoX and CoNiX, where X is an element from Group VA or Group VIB, were compared by Luborsky [95]. He concluded that the origin of the magnetic properties is similar in the two cases: the deposits act like an array of strongly interacting single-domain crystals. The CoW and CoP deposits were shown to be metastable solid solutions except at the highest concentrations of W and P. The saturation moment σ_s increases with atomic number for the Group VIB elements and decreases for Group VA, a reflection of the trend in solubility down the groups.

5.2   Electroplated Perpendicular Media

Pure Co is not generally suitable for vertical storage, because its H_c is low. It is expected to have its easy axis of magnetization in-plane, because the magneto-crystalline anisotropy $K_\perp$ is less than $2\pi M^2$ [101]. The conditions for electroplating "pure Co" films for perpendicular media [101–107] change the structure of the material so as to raise the H_c and to induce a vertical anisotropy. Croll [102, 103] has examined the dependence of the X-ray diffraction of electroplated Co on the pH of the plating solution. A large reflection intensity from the (00.2) plane of hcp Co indicates

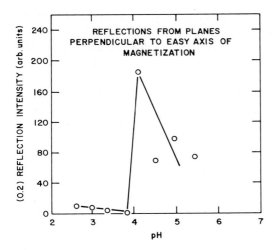

Fig. 7. The intensity of the (00.2) peak for X-ray diffraction measurements of electrodeposited Co as a function of the pH of the plating solution [102]. (Reprinted by permission of The IEEE).

that the c axis, the easy axis, is normal to the surface. The orientation of the Co easy axis depends on the chemistry of the plating solution and on the deposition conditions. Croll's data are shown in Fig. 7. A moderately high pH results in the Co easy axis normal to the film surface. Yet higher pH values result in disruption of the {00.2} PO, presumably through the incorporation of large quantities of precipitated hydroxides.

Chen and Cavallotti [101, 104–107] have proposed the use of such Co with hydroxide inclusions as perpendicular media. Under the appropriate plating conditions, a so-called cellular growth of the Co deposit will occur. The cellular growth is characterized by an hcp Co structure with the basal plane parallel to the surface (i.e. the c axis normal to the surface). The crystallites consist of sharp needles isolated from each other by precipitated basic salts, $Co(OH)_2$. The volume fraction of basic salts can be as high as 22%. The magnetic properties of these cellular Co deposits are the result of a combination of crystalline anisotropy and shape anisotropy. The cellular growth requires a relatively high solution pH. Low current density also favors cellular growth. At a high current density, dendrite formation occurs, and the preferred perpendicular orientation is disrupted. The degree of {00.2} PO also is a strong function of the temperature [101]. Strong {00.2} PO can be obtained over a much wider range of current densities as the temperature of plating is increased [101]. A large degree of PO is found at large deposit thicknesses [105]. At the interface, where the deposit starts to grow, there is less preferred orientation. The PO of very thin films also depends on the substrate, and the effect of substrate roughness is much more important for thin films [107].

The H_c of the cellular Co films depends both on the current density and on the temperature of deposition [101]. The H_c does not correlate directly with the cell dimension, and the maximum H_c does not occur at the maximum in {00.2} PO. A complex interaction among grain size, the PO, and the interparticle separation determines H_c. The magnetization increases with increasing plating temperature because of a decrease in the ratio of the amount of nonmagnetic to magnetic phases with increase in temperature.

Media with a high degree of perpendicular anisotropy can be made by plating a ferromagnetic material into pores formed by anodizing an Al substrate. An early proposal of fabricating media by plating into alumina pores was made by Brownlow [109]. The concept was reduced to practice primarily by groups in Japan [110–116]. The anodization of Al under appropriate conditions will produce a system of alumina cells, each having a central pore, Fig. 8 [114]. Uniformity of the cells requires a very high purity Al substrate [114]. The cell size and pore size are dependent on anodization conditions and can be modified by post-anodization treatments [114–116]. Once the cell structure is formed, a ferromagnetic material can be electrodeposited into the pores [110–116]. Structures have been made with electrodeposited Co, Ni, Fe, CoNi, FeNi, and FeCo. These electrodeposits have a strong shape anisotropy because of the shape of the pores, and thus the structures tend to have a vertical magnetization. The shape anisotropy is not a sufficient condition for obtaining vertical magnetization, however; certain CoNi compositions were shown to grow with the Co hcp c axis parallel to the surface [112] and to have an in-plane magnetic anisotropy despite the shape of the deposits.

Vertical media with very high coercivities can be produced by plating into alumina pores [112]. Some of these media are too hard to be easily written with present heads. Tailoring of the pore size can be used to obtain structures with the desired H_c [115, 116], however. Recording characteristics of disks have been determined [112–114, 116]; such media show excellent promise as vertical recording media. In addition, structures with electrodeposited Fe in the pores were tested in life-tests at elevated temperatures and humidity and in corrosive atmospheres. They were found to perform satisfactorily.

The structures of these disks were examined by Kawai and co-workers [110, 111] and by Tsuya et al. [113–116]. The packing factor, the square of the ratio of the pore

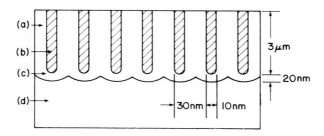

Fig. 8. Schematic representation of a vertical media fabricated by electrodeposition of a ferromagnetic metal into the pores of alumina cells formed by anodization of an Al disk [107]. (Reprinted by permission of The Electrochemical Society).

size to the cell size, is governed by the anodization conditions and the post-anodization pore widening. This factor can be experimentally measured by a re-anodization of the disk. Typically, the cells were 40–140 nm in diameter and the pores 20–120 nm. The Fe electrodeposits into the pores as single crystals with a [110] growth direction. The M_s of the films was directly proportional to the packing factor. An H_c of 1000 Oe was obtained in films with pore diameters > 40 nm; lower pore diameters resulted in high-coercivity disks impossible to write with existing heads.

Further examination of the relationship between structure and properties has been carried out by Arai et al. [117]. Iron was again used as the ferromagnetic material plated in the pores. The length of the Fe particles was varied in two ways: in pores of constant length, the deposition time was varied; and completely filled pores were polished to varying thicknesses. The trends were the same for the two ways of controlling the Fe particle length. For very thin films, $H_c(\parallel) > H_c(\perp)$. The $H_c(\perp)$ increased rapidly with an increase of the ratio of the length to the diameter until a constant value was reached; $H_c(\perp) > H_c(\parallel)$ for thicker films. These results are explained by the geometry of the initial portion of the deposit (i.e. there is no shape anisotropy to induce perpendicular magnetization when the length is not large compared to the diameter) and by the observation that the initial portion of the deposit consists of small crystallites separated by voids. The $H_c(\perp)$ was almost independent of packing density. The M_r and its angular dependence were strongly governed by the packing density.

6   Corrosion of Thin-Film Disks

Corrosion is one of the primary factors determining reliability of thin-film and MO disks. Not only is the reactivity of the magnetic layer higher in thin-film disks than in particulate disks, but also the file environment is less controlled. As pointed out by Koster and Arnoldussen [5], the most significant problems in thin-film disks stem from microscopic, localized, and statistically sparse corrosion sites. Such corrosion can produce magnetic errors, as well as hazardous mechanical defects, and yet be undetectable by conventional macroscopic corrosion measurements [5]. Although corrosion is clearly an electrochemical issue, only a few electrochemical studies have been conducted with magnetic media. The testing is precise and fast and offers mechanistic understanding of the processes involved. In prediction of atmospheric corrosion, however, electrochemical testing will not provide simple acceleration factors. One can, however, attempt to predict the life of the product through a number of accelerated corrosion tests. Most researchers have chosen to expose the media (with or without overcoats) to a temperature/humidity (T/H) test accompanied by surface analyses and functional tests. Disk corrosion will be reviewed with attention given to the corrosion of the magnetic layers, to the possible effects of other layers of the disk structure (notably overcoats) on the corrosion kinetics, and to the projected lifetime.

6.1 CoP and Co Alloys

Judge et al. [122] investigated the corrosion, stress, and magnetic properties of CoP films exposed to elevated temperatures (43.5 to 65 °C) and humidity (80%). The corrosion product was determined by electron diffraction to be Co_3O_4 and was shown by stress tests to have a greater specific volume than the metal. The growth law was parabolic or logarithmic. Although the results indicate some protectiveness of the corrosion product, the remanent moment, which was used to follow the progress of corrosion, decreased about 12% in the 140 hrs of the test. Dubin et al. [123] observed a similar degradation of Co-based films at 85 °C in a humid contaminated atmosphere. In dry contaminated air, no corrosion was observed. The dominance of humidity underlines the electrochemical nature of the corrosion processes.

Helfand et al. [124] observed (by electrochemical techniques) that the corrosion of amorphous CoP_{20} in a dilute solution of NaCl is lower than that of Co alone. This is explained by formation and adsorption of the hypophosphite anion (detected by Auger spectroscopy) resulting from a reaction of P and water. The model of the protective layer is shown in Fig. 9. The formation of such a film could also explain the results of Brusic et al. [125] on magnetic, crystalline CoP films with 8–10% P, given in Fig. 10. A decrease of the average corrosion rate of CoP with time (measured electrochemically in a droplet of DI water) can be due to the time-dependent build-up of a hypophosphite layer. Even in water without any aggressive ions, however, the protectiveness of the P-containing film is marginal. The film allows a significant dissolution of the parent alloy at the open circuit potential: the spontaneous dissolution rate of CoP levels off to a value of 3×10^{-6} A/cm^2, or approximately 0.06 nm/min, only 3X lower than measured on sputtered Co films (Fig. 10). At higher potentials, the film completely loses its protectiveness, as shown by the potentiodynamic polarization curve in Fig. 11. The anodic Tafel slope is very low, approximately 44 mV/decade [125, 126], and it extends throughout several decades. (The reversal in the current-potential curve at about -0.75 V vs. the mercurous sulfate electrode is the result of the complete dissolution of the CoP film.) The behavior of CoP and other

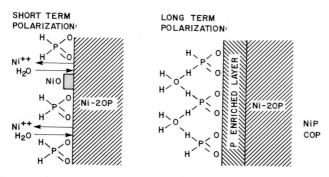

Fig. 9. Model of chemical passivity for CoP_{20} in NaCl [124]. (Reprinted by permission of The Electrochemical Society).

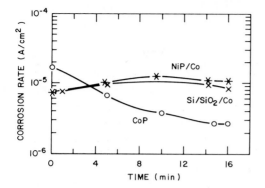

Fig. 10. Time dependence of corrosion rate of Co_8P and Co films, electrolessly plated and sputter-deposited onto NiP substrates, in a droplet of DI water. The results were obtained by a repeated application of the polarization resistance technique with the potential scanned at 1 mV/sec in a potential range 15 mV above and below the corrosion potential [125]. (Reprinted by permission of The Electrochemical Society).

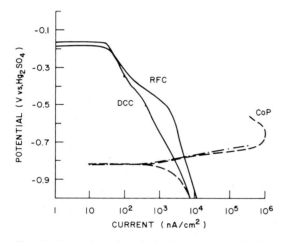

Fig. 11. Potentiostatic polarization curve on Co_3P, measured in a droplet of DI water. The Tafel region is marked [125]. (Reprinted by permission of The Electrochemical Society).

magnetic layers at higher potentials is not only of fundamental interest, it also has a practical value for predicting and assessing the effects of more noble overlayers (such as carbon, rhodium, and others). One can predict that CoP is vulnerable to galvanic attack, since it has a fairly low corrosion potential in DI water (-0.84 V vs. Hg_2SO_4), a low anodic Tafel constant, and no indication of self-passivation [125]. Galvanic attack will be discussed in detail below in the section on overcoats.

A systematic evaluation of the corrosion properties of Co media is not available. From a variety of published works, one can deduce some general trends, as will be discussed below.

6.2 Alloys with Oxide Formers

It is not surprising to find that alloying of Co with the oxide-formers from the transition metal group invariably decreases the corrosion rate [125, 127–134]. The effect is similar to the improvement of the corrosion resistance of steel by addition of Cr. The basic trend, observed by many, is that the corrosion resistance increases from Co $\simeq$ CoP $<$ CoNi $<$ CoCrTa $\simeq$ CoCr. The alloys were exposed to a variety of tests, and the progress of corrosion was measured in a variety of ways. Most often, the films to be tested were exposed to 80–90% relative humidity at elevated temperatures in T/H tests. Sugita et al. [127] exposed e-beam deposited CoCr films with Cr contents $>$ 16% to 90% RH and 60 °C for six months. Optical microscopy detected little change; Auger electron spectroscopy detected a very thin oxidation layer. The authors suggest that the films will have "enough corrosion resistance for the practical use." Yamada et al. [128] compared corrosion resistance of sputtered Co, CoNi, and CoNiCr films on disks subjected to corrosion tests at 90% RH and 60 °C for 26 days. The magnitude of the magnetic changes occurring during the corrosion test demonstrated that the CoNiCr alloy is the most corrosion resistant of the three. Fisher et al. [129] compared the magnetic properties of sputtered CoNi, CoCr, and CoCrTa after a 2-week exposure to 80% RH at 65 °C. The CoNi alloy was deemed inferior to CoCr and CoCrTa, the last two being similar.

Some authors have exposed the films to a humid atmosphere contaminated with ppb amounts of Cl_2, SO_2, NO_2, and H_2S. Examples are given by Phipps et al. [130] for sputtered FeCoCr and Smallen et al. [131] for electroless CoP and sputtered CoNi and CoCr.

Environmental tests have been combined with conventional electrochemical measurements by Smallen et al. [131] and by Novotny and Staud [132]. The first electrochemical tests on CoCr thin-film alloys were published by Wang et al. [133]. Kobayashi et al. [134] reported electrochemical data coupled with surface analysis of anodically oxidized amorphous CoX alloys, with $X =$ Ta, Nb, Ti or Zr. Brusic et al. [125] presented potentiodynamic polarization curves obtained on electroless CoP and sputtered Co, CoNi, CoTi, and CoCr in distilled water. The results indicate that the thin-film alloys behave similarly to the bulk materials [133]. The protective film is less than 5 nm thick [127] and rich in a passivating metal oxide, such as chromium oxide [133, 134]. Such an oxide forms preferentially if the Cr content in the alloy is, depending on the author, above 10% [130], 14% [131], 16% [127], or 17% [133]. It is thought to stabilize the non-passivating cobalt oxides [123]. Once covered by stable oxide, the alloy surface shows much higher corrosion potential and lower corrosion rate than Co, i.e. it shows more noble behavior [125].

Tada et al. [135] reported that additions of small amounts of Pr to evaporated CoNi produced a fine-grained alloy with greatly improved magnetic properties and corrosion resistance. Polarization resistance and T/H tests (50 °C, 90% RH) were used, both showing a significant decrease of the corrosion rate with alloys containing 1 to 4% Pr. XPS and SIMS results suggested that the additions of Pr served to prevent oxidation of metallic Co and Ni through the predominant oxidation of Pr metal.

The corrosion of Re-containing alloys, such as CoNiMnReP, has not been reported, but in disk structures with an SiO_2 overcoat they perform well in corrosion tests [55]. Rhenium itself is very reactive and prone to corrode [136]. Small amounts of Re in magnetic alloys are most likely in an oxidized state and play a role similar to that of Mo in the protection of stainless steel.

6.3 Addition of Noble Metals

According to the thermodynamics of Co corrosion [136], a galvanic contact of Co and Pt or any other noble metal should result in the catastrophic dissolution of Co. Rhodium-overcoated media have been fabricated by exchange, and high-coercivity CoAgP has been fabricated by immersion of CoP films in a solution of Ag^+ ions. The fact that such structures can be fabricated by exchange processes in itself demonstrates the vulnerability of CoP to a galvanic attack [137]. Once cobalt alloys with noble metals are formed, however, they can achieve a high corrosion resistance [96, 97]. Yanagisawa et al. [138] have developed sputtered CoPt alloys as new materials of "high corrosion resistance" for use in high density perpendicular recording. Films of CoPt and CoNiPt were immersed in water for a month. The measured changes of the magnetic moment with time show that both alloys have corrosion resistance superior to Co. Hoshi et al. [139] examined the corrosion of sputtered $CoPt_{11}$, also by immersion in water. They confirmed that this alloy has good corrosion resistance, which was shown to depend on alloy deposition parameters and is the best for films deposited with bias and at low Ar partial pressures. A mechanistic explanation of this enhanced corrosion resistance has not yet been published. Since it has been repeatedly shown that Co oxides do not provide significant corrosion protection, the effect of Pt cannot be explained by "anodic protection". Rather, the original film is apparently de-alloyed, leaving a Pt-rich, corrosion-resistant surface.

6.4 Effects of Overcoats

Overcoats are an integral part of thin-film disk structures. Their primary role is to provide wear protection. The most common overlayers, such as Rh, plasma-polymerized coatings, SiO_2, and carbon, are all chemically stable; if they were fully to cover the disk surface, they would provide good corrosion resistance. The thinness of the overcoats and the roughness of the surface preclude "perfect" coverage and open up the path for localized corrosion at the sites where the magnetic layer is exposed to the environment.

Hattori et al. [140] investigated various overcoats for the thin sputtered γ-Fe_2O_3 and plated CoNiP films. Rhodium overcoats exhibit defects caused by corrosive attack of the plating solution on the magnetic layer (prior to the corrosion test);

plasma polymerized coatings lack adhesion; SiO_2, either sputtered or spin-coated, gives the best results. A galvanic enhancement of localized corrosion by Rh is expected and was observed in the work of Garrison [141], Rossi et al. [142], and Doss et al. [143, 144].

The most common overcoats, however, are sputtered carbons. Their role in disk corrosion has been described in contradicting ways. Whereas Garrison [141] clearly observed that carbon, like Rh, can enhance galvanic corrosion, Smallen et al. [131] believe that carbon decreases corrosion by preventing lateral growth of corrosion products. Results of similar tests are sometimes contradictory: Nagao et al. [145] have shown an improvement of the corrosion resistance of carbon-coated CoCr alloys on T/H test (with either SO_2 gas or NaCl mist), whereas Black [146] finds that pyrolitic carbon over a CoCrMo alloy results in elevated corrosion rates.

Brusic et al. [125] have shown that the galvanic effects of DC- or RF-sputtered carbon overcoats are predictable from electrochemical data. Galvanic attack is the prevailing corrosion mechanism for carbon-overcoated disks in accelerated corrosion tests. Potentiodynamic polarization curves of carbon films on glass show that a typical DC-sputtered C is conductive, fairly noble, and catalytically active for the reduction both of oxygen and of hydrogen ions, Fig. 11. Assuming from this figure that the more noble carbon would be the cathode for the anodically dissolving magnetic layer, carbon in contact with an equal area of CoP would cause the localized corrosion rate of the CoP to be increased by up to 6X. Increases of many orders of magnitude would result from higher C:CoP area ratios, such as would be typical of disk structures.

Furthermore, T/H tests, with or without corrosive gasses, also demonstrate the galvanic effects of C overcoats [125]. An uncoated CoP film exposed to 70% RH and 10 ppb of Cl_2 at room temperature for 24 hrs forms a uniform corrosion product about 2.5 nm thick. Carbon-coated CoP films have very low corrosion rates, on the average, but very high rates on the sites of galvanic contact. Consequently, there is barely any change in the ellipsometric parameters of C-coated CoP (the average change corresponding to a film 0.25 nm thick). The SEM micrographs in Fig. 12, however, show an abundance of corrosion sites with the localized corrosion product reaching a height of several micrometers. In contrast to carbon overcoats, overlayers of dielectric glassy oxides on CoP and Co cause a decrease of the average corrosion rate and show no signs of enhanced localized attack, even if the coverage is incomplete. The galvanic action of sputtered carbon on other Co alloys, such as CoCr, is less pronounced, since they are more noble and thus are electrochemically more compatible with carbon.

Yamashita et al. [147] have reported that a sputtered zirconia offered corrosion protection much superior to that obtained with a hard carbon overcoat. The mechanical performance and corrosion protection of 20–25 nm overcoats on CoNiPt magnetic layers were examined. The overcoats studied were sputtered SiO_2, Al_2O_3, ZrO_2, and DC-sputtered carbon. Corrosion resistance was evaluated by a glide-height test and by defect mapping of disks exposed to elevated temperature and humidity. The superior behavior of the ZrO_2 overcoat, in particular one stabilized with Al_2O_3 and Y_2O_3, was attributed to its near-amorphous structures, providing a

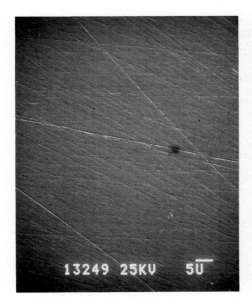

Fig. 12. SEM photographs of CoP and CoP/C surfaces after 24 hrs exposure to Cl$_2$ and 70% RH at RT. Mag 1200X [125]. (Reprinted by permission of The Electrochemical Society).

dense and pore-free overcoat, and to its non-stoichiometry, speculated to have as much as 10% oxygen deficiency. The unassociated Zr was thought to contribute to electrochemical passivation of the medium.

These results with zirconia overcoats would seem to confirm the galvanic-action model of disk corrosion, but the same authors [147] also observed that sputtered SiO$_2$, which is expected to be galvanically inactive, behaves worse than carbon, showing an enhanced corrosion in pinholes. These observations contradict those of Hattori et al. [140], who showed that a sputtered SiO$_2$ overcoat provided good corrosion protection. Yamashita et al. [147] did not elaborate as to why enhanced degradation should occur. One plausible explanation might be the occurrence of crevice attack around the pinholes. Another might be connected with possible non-stoichiometry of the thin, normally insulating layer, which could result in a partially conductive, galvanically active layer. In any case, these results are an example of diverse behavior noted with nominally the same material, showing the influence of the deposition parameters on a material's performance.

In the recent work by Khan et al. [148], the DC-sputtered carbon overcoat on a CoCr disk was evaluated as a function of the processing parameters, such as argon pressure, target power, substrate texture, substrate-to-target distance, and substrate bias. Although the process variations were examined in order to prepare carbon with improved wear characteristics, the changes in the carbon are also interesting from the corrosion point of view: the layer has acquired a more homogeneous grain size and grain distribution, a smooth work function distribution, and a higher percentage of

sp^3-bonded carbon compared with the carbon prior to process tuning [148]. Such a film, more homogeneous and more diamond-like (having sp^3 bonding), could also provide better corrosion protection.

6.5   Enhancement of Corrosion Resistance

If one could choose the disk materials to minimize the risk of corrosion, one might select an alloyed magnetic layer like CoCr and an adherent dielectric overcoat having uniform coverage. The latter could be SiO_2 or some other oxide or carbide that would give protection against wear [149]. Product designers have not relied on such choices for corrosion protection, but instead have built files with control of the environment against dust and moisture. The files take a "conservative approach to corrosion", using absorptive filters to limit the effect of contaminating gasses and to reduce humidity levels [150–152].

6.6   Life Projections

Even though filters do not eliminate corrosion concerns, they narrow the range of field conditions the product will experience, limiting them to a range of humidities but eliminating aggressive gasses. In other words, filters provide assurance that the humidity and temperature are the only variables in the working environment of the disk. Temperature/humidity tests, the most widely used, are thus valid in determining acceleration factors and predicting the disk life. Such tests are simpler, both in theory and in practice, than those involving a multitude of possible reactions with gaseous contaminants.

Sharma et al. [153] have devised a "gentle accelerated corrosion test" using a kinetic rate equation to establish appropriate acceleration factors due to relative humidity and thermal effects. Using an estimate for the thermal activation energy of 0.6 eV and determining the amount of adsorbed water by a BET analysis on Au, Cu and Ni, they obtain an acceleration factor of 154 at 65 °C/80% RH with respect to 25 °C/35–40% RH.

Although such acceleration factors were applied by Fisher et al. [129] in their discussion of the corrosion of Co alloys, there are only a few papers trying to establish the experimental acceleration factors for disk materials. One of these is the work of Novotny et al. [154]. These authors observed that carbon-coated CoCr, CoPt, or CoP disks, when exposed to elevated humidity and temperature, show an accumulation of a corrosion product, similar to those seen on uncoated alloys, on the top surface of the carbon. For both overcoated and uncoated disks, oxides and hydroxides of cobalt were detected on the surface, but there was no detectable Cr, Pt, or P. The amount of Co was on the average less on the C-covered disks. The surface Co resulted from a preferential dissolution from the magnetic layer and diffusion of the dissolved species into a layer of adsorbed water. The Auger signal of Co was used to monitor the progress of corrosion with time, temperature, and humidity.

Examples of their results [154] are shown in the set of curves in Fig. 13. At a given humidity, the Co concentration increases with T; the thermal activation energy is about 0.4 eV. At a given temperature, the corrosion increases with an increase in humidity. As the humidity changes from 30 to 90%, the corrosion rate increases about an order of magnitude. The data allow a calculation of the acceleration factors for a variety of conditions. For example, the acceleration factor for 90 °C/90% RH with respect to 30 °C/40% RH is calculated to be ∼ 150. If the product passes a 2-week exposure to 90 °C/90% RH, the test indicates that it will survive in excess of 6 years at 30 °C/40% RH. The values of the acceleration factors, however, may vary from film to film.

The work of Tagami et al. [155] on similar CoCr/sputtered-C disks shows a dependence of the thermal activation energy on the conditions of CoCr deposition, specifically the Ar partial pressure. The activation energy varied from 0.07 to 0.3 eV. If the thermal acceleration is indeed that low, the predominant factor accelerating the

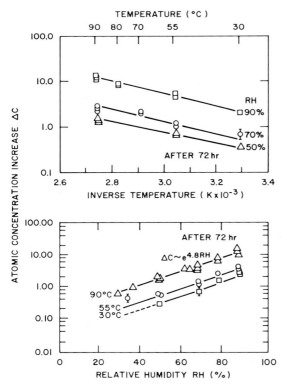

Fig. 13. Increase of Co concentration above the carbon overcoat on CoCr/C disks exposed to temperatures up to 90 °C and RH up to 90% [154]. (Reprinted by permission of The IEEE).

corrosion is humidity. The variation of humidity within reasonable limits, from 30 to 90%, results in acceleration factors that could be as low as 10X. In a view of this, many tests performed at lower temperature and humidity and for a relatively short time (e.g. 2 weeks) do not provide reliable information about the life of the product. There are reports that a $CoNiP/SiO_2$ disk survived a T/H test for 7 months at 40 °C, 80% RH [156] and that Fe plated in alumina pores and overcoated with SiO_2 survived 1000 hrs on test at 40 °C, 90% RH [157]. Despite optimistic interpretations of these results, there is an understandable drive to develop a corrosion-free disk like one incorporating γ-Fe_2O_3 [157].

7   Corrosion of Magneto-optical Disks

7.1   FeTb and Alloys

The thermodynamic instability of the amorphous rare-earth transition metals is reflected in their ease of oxidation and susceptibility to corrosion. The reversible potential for oxidation of Tb is extremely low, below − 2.4 V vs. NHE [136]. Like all other lanthanides, Tb is a powerful reducing agent. It has a great propensity to decompose water (at any pH and concentration) with the evolution of hydrogen. Terbium alloys have similar reactivity. Hatwar et al. [159] performed the first electrochemical measurements of FeTb, using a pH 3.1 Cl^--containing solution. From a typical potentiodynamic polarization curve, Fig. 14, one can extrapolate an average corrosion rate of 7×10^{-5} A/cm^2, which is about 10X higher than that

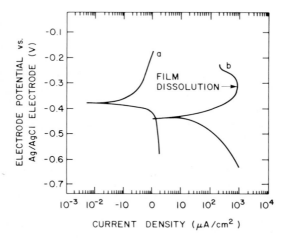

Fig. 14. Potentiodynamic polarization curves obtained on FeTb and FeTb/SiO$_2$ samples in pH 3.1 Cl^--containing solution [159]. (Reprinted by permission of The IEEE).

reported for Co in a similar solution [126]. The film is completely dissolved at anodic potentials, resulting in the decrease of the current and a loss of the reflectance, which is measured simultaneously [160]. When immersed in a NaCl solution, FeTb and similar films tend to pit [161]. Farrow and Marinero [162] observed pitting even in a DI water droplet, but their findings could not rule out the probability of pit initiation by some accidental impurity. They employed AES, XPS, SIMS, and electrochemical techniques to determine the corrosion reactions. The pit-growth rate was found to be controlled by the dissolution of iron and its transport away from the pit area, the pit being filled by an insoluble terbium oxide [162].

Additions of small amounts of Pt to FeTb greatly reduce its pitting tendency in NaCl; additions of Al, Ti, and Co reduce pitting to a lesser extent [161]. Additions of Pt to a ternary alloy, FeCoTb, also suppress pit formation and reduce the average corrosion rate by 2.5X [163]. A similar observation was made in T/H tests, with additions of Al, Ti, and Cr suppressing the selective oxidation of TbFe and TbCo alloys whereas Ni and Co have undetectable effects [164]. The change of the magnetic properties with the course of corrosion of FeTb alloys was reported to be slower if some of the Tb is replaced by Gd or Dy [165]. Good results are reported for FeCoTb alloys with Be [166] and for FeTb with In [167]. Additions of In have caused a significant increase of the thermal activation energy, changing it from 0.3 eV for FeTb to 1.3 eV for the ternary alloy. Thus one can speculate that oxidized In hinders the diffusion of Fe and Tb ions and thereby the growth of surface oxides.

Many researchers have found that differences in film deposition techniques and deposition parameters profoundly influence the oxidation and stability of MO layers. For example, increases of Ar pressure and bias voltage, leading to more porous structures, result in easier oxidation and quicker changes in aging tests [168–172]. Such films are particularly corrosion-prone.

7.2 MO Layers and Overcoats

In practical applications, the magneto-optical alloys require underlayers and over-layers [173]. These have several functions, not the least being an assurance of stability of these highly reactive materials. Luborsky et al. [174] reviewed some of the work relevant to stability of CoGdTb overcoated with SiO_2, starting with the first annealing experiments in 1977. They have concluded that isothermal annealing causes film degradation through surface oxidation and consequently results in a decrease of perpendicular anisotropy. The degradation can be minimized through the use of metallic overcoats, such as Au, Al, and rare earth elements. Hartman et al. [175] tested Al, Cr, CrNi, or Cu as protective films over GdTbFe annealed in air at temperatures of 65–160 °C. Aluminum and Cr gave the best performance. Although the results clearly show some protectiveness of the metallic overcoats during annealing, their application in conditions relevant to corrosion has to be carefully reviewed, since many of them can clearly introduce unwanted galvanic effects at points of incomplete coverage. A number of dielectric oxide and nitride layers has been suggested and tested. Some of these were deposited as single layers such as SiO_2 [174], AlN [159, 160, 176], Al_2O_3 [168], silicon nitride [177], and Al-Si nitride and Al-Si oxynitride [178]. Also evaluated were a two-layer structure of Al_2O_3/SiO_2 [168], mixtures of aluminum oxide and tantalum oxide [179], SiO_2 codeposited with Tb to trap the active oxygen in SiO_2 [180], and layers of Tb/SiO_2 [181]. In general, all but AlN are found to be protective. Grundy [176] has reported that AlN over TbFeCo films allows catastrophic oxidation at higher humidities on T/H tests. In contrast, AlN is reported to be superior to SiO_2 in providing protection to FeTb on electrochemical tests [159, 180]. Hatwar et al. [159] have concluded that AlN has higher chemical stability than SiO_2, but this interpretation has to be taken with caution. The test only shows that this film provides better coverage than SiO_2; surface coverage, and ways to improve it, were not discussed.

It is difficult to provide an objective ranking of the layers, from the point of view of corrosion protection, as long as the deposition processes are not optimized and the films are not evaluated on the same test. From electrochemical principles, films of SiO_2 with Tb [180] and Tb/SiO_2 layers [181] are very interesting, since the preferential oxidation of Tb should slow diffusion of O_2 through SiO_2 and provide galvanic protection for the underlying alloy.

7.3   Life Projection

The thermal stability for erasable MO media has been projected from the degradation of their magnetic properties during annealing in the absence of humidity. Since RE-TM alloys are amorphous materials, they undergo structural relaxation and consequent changes in perpendicular anisotropy, carrier-to-noise ratio, and coercivity. In addition, as predicted from the free energy change, Tb will reactively extract oxygen from all of the oxygen-containing overcoats tested to date. In spite of these atomic movements and reactions within the medium, some authors have projected useful lifetimes in excess of 300 years, or at least more than 10 years [174, 182]. Environmental stability is more difficult to project. Using bit error rate (BER) as a measure of the corrosion progress on various T/H tests and 3X increase of BER as an end-of-life criterion, Freese et al. [182] and Iijima [167] predicted a lifetime for MO disks of about 10 years at 30 °C and 90% RH.

8   Summary

Electrochemical science and technology have played an important role in all aspects of the new technology of the thin-film disk, which is capable of higher storage density than the particulate disk. The first disks, at the beginning of the technology more than

20 years ago, were produced by electroless deposition. With these media in hand, scientists could start to understand how the structure determined the magnetic properties of continuous thin films. Electrolessly deposited alloys and those deposited by other techniques were selected and optimized primarily because of their magnetic properties. Their reactivity is a reminder that Mother Nature has given these devices only a temporary stay in terrestrial environments. Through the work of many, the thermodynamic stability of these magnetic alloys has been increased, and magnetic media have been built in a fashion that drastically reduces the rate of the corrosion reactions. After many years of development in the laboratory, thin-film and magneto-optic disks are being announced [183] or are appearing in commercially available files. It will be interesting to follow the performance of these files in light of the optimistic projections of the accelerated tests. It is expected that, as the technology matures and improves with time, metallic thin-film disks will remain important in the technology of magnetic storage.

9 References

1. L.D. Stevens, IBM J. Res. Develop. 25, 663 (1981).
2. J.M. Harker, D.W. Brede, R.E. Pattison, G.R. Santana, and I.G. Taft, IBM J. Res. Develop. 25, 677 (1981).
3. J.C. Mallinson, IEEE Trans. Magn., MAG-21, 1217 (1985).
4. G. Bate in Ferromagnetic Materials, Vol. 2, E.P. Wohlfarth, ed. (North Holland, New York, 1980) p. 381.
5. E. Koester and T.C. Arnoldussen in Magnetic Recording, Vol. I: Technology, C.D. Mee and E.D. Daniel, eds. (McGraw Hill, NY, 1987) p. 98.
6. A.H. Eltoukhy, J. Vac. Sci. Tech. A 4, 539 (1986).
7. G. Bate, J. Appl. Phys. 37, 1164 (1966).
8. T.C. Arnoldussen and E.-M. Rossi, Ann. Rev. Mat. Sci. 15, 379 (1985).
9. J.S. Judge, Ann. N.Y. Acad. Soc. 189, 117 (1972).
10. J.K. Howard, J. Vac. Sci. Technol. 4, 1 (1986).
11. J.C. Mallinson, IEEE Trans. Magn., MAG-20, 461 (1984).
12. J.D. Livingston, J. Appl. Phys. 52, 2544 (1981).
13. D.W. Baudrand and M. Malik, Met. Finish. 84(3), 15 (Mar., 1987).
14. A. Brenner and G.E. Riddell, J. Res. Natl. Bur. Standards 3, 31 (1946).
15. R.D. Fisher, IEEE Trans. Magn., MAG-2, 681 (1966).
16. H. Matsuda and O. Takano, J. Jpn. Inst. Met. 52, 414 (1988).
17. M. Soraya, Plating 54, 549 (1967).
18. H. Matsuda and O. Takano, Proc. Interfinish 142 (1980).
19. G.S. Alberts, R.H. Wright, and C.C. Parker, J. Electrochem. Soc. 113, 7, 687 (1966).
20. M. Lelental, J. Electrochem. Soc. 120, 1650 (1973).
21. H. Narcus, Plat. 54, 380 (1967).
22. K. Matsui, Y. Suzuki, T. Yamade, S. Maruno, and T. Kawaguchi, Bull. Nagoya Inst. Technol. 36, 301 (1984).
23. M. Matsuoka and T. Hayashi, Plat. 68(7), 66 (Jul., 1981).
24. K.M. Gorbunova, M.V. Ivanov, and P. Moiseev, J. Electrochem. Soc. 120, 613 (1973).
25. Y.H. Chang, C.C. Lin, M.P. Hung, and T.S. Chin, J. Electrochem. Soc. 133, 985 (1986).
26. P. Cavallotti and G. Salvago in "Proc of Symposium on Electrodeposition Technology," L.T. Romankiw and D.R. Turner, eds. (The Electrochemical Society, Pennington, NJ, 1987) p. 327.

27. G.A. Sadakov and K.M. Gorbunova, Sov. Electrochem. 16, 199 (1980).
28. J.E.A.M. Van der Meerakker, J. Appl. Electrochem. 11, 395 (1981).
29. R.M. Lukes, Plat. 51, 969 (1964).
30. N. Feldstein and T.S. Lancsek, J. Electrochem. Soc. 118, 869 (1971).
31. G. Salvago and P. Cavallotti, Plat. Surf. Finish. 59, 665 (1972).
32. G. Gutzeit, Plat. 46, 1158, 1275, 1379 (1959).
33. G. Gutzeit, Plat. 47, 63 (1960).
34. V.S. Demidenko, A. Szasz, and M.A. Aysawi, Phys. Stat. Sol. (B) 140, 121 (1987).
35. F.M. Donahue and C.U. Yu, Electrochim. Acta 15, 237 (1970).
36. B.D. Barker and D. Taberner, Surf. Tech. 12, 103 (1981).
37. C.H. deMinjer, Electrodep. Surf. Treat. 3, 262 (1975).
38. H. Dahms, J. Electroanal. Chem. 8, 5 (1964).
39. H. Dahms and I.M. Croll, J. Electrochem. Soc. 112, 771 (1965).
40. C.K. Mital, P.B. Srivastava, and R.G. Dhaneshwar, Met. Finish. 84(10), 67 (Oct., 1987).
41. J.-P. Randin, J. Electrochem. Soc. 118, 1969 (1971).
42. T. Osaka, F. Goto, N. Kasai, and Y. Suganuma, Denki Kagaku 49, 792 (1981).
43. T. Osaka, N. Kasai, I. Koiwa, F. Goto, and Y. Suganuma, J. Electrochem. Soc. 130, 568 (1983).
44. T. Osaka, N. Kasai, I. Koiwa, and F. Goto, J. Electrochem. Soc. 130, 790 (1983).
45. F. Goto, T. Osaka, I. Koiwa, Y. Okabe, H. Matsubara, A. Wada, and N. Shoita, IEEE Trans. Magn., MAG-20, 803 (1984).
46. T. Osaka, I. Koiwa, Y. Okabe, H. Matsubara, A. Wada, and F. Goto, Denki Kagaku 52, 197 (1984).
47. T. Osaka, I. Koiwa, Y. Okabe, H. Matsubara, A. Wada, F. Goto, and N. Shiota, Bull. Chem. Soc. Jpn. 58, 414 (1985).
48. T. Osaka, I. Koiwa, Y. Okabe, H. Matsubara, F. Goto, and N. Shiota, IEEE Trans. J. Magn. Jpn., TJMJ-1, 27 (1985).
49. T. Osaka, I. Koiwa, M. Toda, T. Sakuma, T. Namikawa, and Y. Yamizaki, IEEE Trans. J. Magn., Jpn., TJMJ-1, 977 (1985).
50. I. Koiwa, M. Toda, and T. Osaka, J. Electrochem. Soc. 133, 597 (1986).
51. I. Koiwa, M. Matsubara, T. Osaka, Y. Yamazaki, and T. Namikawa, J. Electrochem. Soc. 133, 685 (1986).
52. T. Osaka, I. Koiwa, M. Toda, Y. Yamazaki, T. Namikawa, and F. Goto, IEEE Trans. Magn., MAG-22, 1149 (1986).
53. T. Osaka, H. Matsubara, K. Yamanishi, H. Mizutani, and H. Okabe, IEEE Trans. Magn., MAG-23, 1935 (1987).
54. T. Osaka, I. Koiwa, Y. Okabe, and K. Yamanishi, Jap. J. Appl. Phys. 26, 1674 (1987).
55. F. Goto, H. Tanaka, M. Yanagisawa, N. Shiota, S. Ito, M. Kimura, Y. Suganuma, and T. Osaka in "Electrochemical Technology in Electronics," L.T. Romankiw and T. Osaka, eds. (The Electrochemical Society, Pennington, NJ, 1988), p. 377.
56. I. Koiwa, T. Osaka, Y. Yamazaki, and T. Namikawa, IEEE Trans. Magn., MAG-23, 2800 (1987).
57. H. Matsubara, H. Mizutani, S. Mitamura, T. Osaka, and F. Goto, IEEE Trans. Magn., MAG- (in press).
58. I. Koiwa, M. Usuda, K. Yamada, and T. Osaka, J. Electrochem. Soc. 135, 718 (1988).
59. K. Ouchi and S.I. Iwasaki, IEEE Trans. Magn., MAG-20, 99 (1984).
60. A. Takano and H. Matsuda, Met. Finishing 83(1), 63 (Jan., 1985).
61. L.T. Romankiw and D.A. Thompson in "Properties of Electrodeposits," R. Sard, H. Leidheiser, and F. Ogburn, eds. (The Electrochemical Society, Princeton, NJ, 1975), p. 389.
62. M. Pushiavanam and B.A. Shenoi, Finish. Ind. 1, 48 (1977).
63. G.D.R. Jarrett and L.D. Brown, Trans Ins. Met. Finish. 49, 1 (1971).
64. D.E. Speliotis and N.C. Fernelius, J. Appl. Phys. 61, 3831 (1987).
65. W.G. Clements and B. Cantor, Rapidly Quenched Metals, 2nd Int. Conf. Proceedings, 267 (1976).
66. M. Schwartz and G.O. Mallory, J. Electrochem. Soc. 123, 606 (1976).
67. Su Hoon Park and Dong Nyung Lee, J. Mater. Sci. 23, 1643 (1988).
68. H. Wiegand, F.W. Hirth, and H. Speckhardt, J. Less Common Met. 43, 267 (1975).

69. K. Parker, Plat. Surf. Finish. 68(12), 71 (Dec. 1981).
70. A.W. Simpson and D.R. Brambley, Phys. Stat. Sol. (B) 43, 291 (1971).
71. S.L. Chow, N.E. Hedgecock, M. Schlesinger, and J. Rezek, J. Electrochem. Soc. 119, 1614 (1972).
72. R.O. Cortijo and M. Schlesinger, J. Electrochem. Soc. 131, 2800 (1984).
73. Tu Chen, D.A. Rogowski, and R.H. White, J. Appl. Phys. 49, 1816 (1978).
74. M. Mirzamaani, L. Romankiw, C. McGrath, J. Mahlke, and N.C. Anderson, J. Electrochem. Soc. 135, 2813 (1988).
75. D. DiMilia, J. Horkans, C. McGrath, M. Mirzamaani, and G. Scilla, J. Electrochem. Soc. 135, 2817 (1988).
76. E.L. Nicholson and M.R. Khan, J. Electrochem. Soc. 133, 2342 (1986).
77. M.R. Khan and J.I. Lee, J. Appl. Phys. 57, 4028 (1985).
78. J.S. Judge, J.R. Morrison, D.E. Speliotis, and J.R. DePew, Plating 54, 533 (1967).
79. G.W. Lawless and R.D. Fisher, Plating 54, 709 (1967).
80. S.K. Doss and P.B.P. Phipps, Plat. Surf. Finish. 72(4), 64 (Apr., 1985).
81. T. Osaka and H. Nagasaka, J. Electrochem. Soc. 128, 1686 (1981).
82. H. Matsubara, M. Toda, T. Sakuma, T. Honma, Y. Yamazaki, and T. Namikawa, in "Electrochemical Technology in Electronics," L.T. Romankiw and T. Osaka, eds. (The Electrochemical Society, Pennington, NJ, 1988), p. 433.
83. H. Koretzky, Proc. 1st Australian Conference on Electrochemistry (1963).
84. W.H. Safranek, The Properties of Electrodeposited Metals and Alloys (Elsevier, New York, 1974).
85. F.R. Morral, Plating 59, 131 (1972).
86. A. Brenner, D.W. Couch, and E.K. Williams, Plating 37, 36 (1950).
87. V. Zentner, Plating, 52, 868 (1965).
88. J.S. Sallo and J.M. Car, J. Electrochem. Soc. 109, 1040 (1962).
89. G.W. Reimherr, NBS Technical Note 247, (1964).
90. J.H. Kefalas, J. Appl. Phys. 37, 1160 (1966).
91. T.H. Bonn and D.C. Wendell, U.S. Pat. 2, 644, 787 (1953).
92. J.S. Sallo, Plating 54, 257 (1967).
93. A. Tago, T. Masuda, and T. Taketa, Rev. ECL of NTT (Tokyo) 25, 1315 (1977).
94. D. Pearce, D. Rice, and G. Tang, Solid State Tech. 31(11), 113, (Oct., 1988).
95. F.E. Luborsky, IEEE Trans. Magn., MAG-7, 502 (1970).
96. D.J. Barclay, D.S. Mansbridge, W.M. Morgan, and C.T. Prowting, IBM Tech. Disc. Bulletin 04–74, 3769 (1974).
97. D.J. Barclay and W.M. Morgan, IBM Tech. Disc. Bulletin 11–74, 1591 (1974).
98. D.J. Barclay and W.M. Morgan, IBM Tech. Disc. Bulletin 08–76, 1098 (1976).
99. M.A. Shuvalova, B. Ya. Kaznachei, and G.A. Sadakov, Prot. Met. 13, 522 (1977).
100. V.V. Bondar', M.M. Mel'nikova, and Yu. M. Polukarov, Prot. Met. 1, 467 (1965).
101. Tu Chen and P.L. Cavallotti, IEEE Trans. Magn., MAG-18, 1125 (1982).
102. I.M. Croll, IEEE Trans. Magn., MAG-23, 59 (1987).
103. I.M. Croll and B.A. May, "Proc of Symposium on Electrodeposition Technology", L.T. Romankiw and D.R. Turner, eds. (The Electrochemical Society, Pennington, NJ, 1987), p. 295.
104. Tu. Chen and P. Cavallotti, Appl. Phys. Lett. 41, 205 (1982).
105. P.L. Cavallotti, D. Colombo, E. Gabiati, and R. Martinella, Interfinish 84, A. Israeli, ed. (Jerusalem, 1984), p. 234.
106. P. Cavallotti, D. Colombo, U. Ducati, and A. Piotri, "Proc of Symposium on Electrodeposition Technology," L.T. Romankiw and D.R. Turner, eds. (The Electrochemical Society, Pennington, NJ, 1987), p. 429.
107. P.L. Cavallotti, D. Colombo, E. Galbiati, A. Piotti, and F. Kruger, Plating 74(4), 78 (Apr., 1988).
108. M.E. Baumgaertner, Ch. J. Raub, P. Cavallotti, and G. Turrilli, Metalloberfläche 41, 559 (1987).
109. J. Brownlow, IBM Tech. Disc. Bulletin 11, 238 (1968).
110. S. Kawai and R. Ueda, J. Electrochem. Soc. 122, 32 (1975).
111. S. Kawai, J. Electrochem. Soc. 122, 1026 (1975).
112. S. Kawai and I. Ishigura, J. Electrochem. Soc. 123, 1047 (1976).
113. J.S. Judge, J.R. Morrison, D.E. Speliotis, and G. Bate, J. Electrochem. Soc. 112, 681 (1965).

V. Brusic et al.

114. M. Shiraki, Y. Wakui, T. Tokushima, and N. Tsuya, IEEE Trans. Magn., MAG-21, 1465 (1985).
115. N. Tsuya, T. Tokushima, M. Shiraki, N. Wakui, Y. Saito, N. Nakamura, S. Hayano, A. Furugori, and M. Tanaka, IEEE Trans. Magn., MAG-22, 1140 (1986).
116. N. Tsuya, T. Tokushima, M. Shiraki, and N. Wakui, IEEE Trans. Magn., MAG-23, 53 (1987).
117. N. Tsuya, T. Tokushima, M. Shiraki, N. Wakui, and Y. Umehara, IEEE Trans. Magn., MAG-23, 2242 (1987).
118. K.I. Arai, Y. Wakui, K. Ohmori, and T. Tokushima, IEEE Trans. Magn., MAG-23, 2245 (1987).
119. J.S. Sallo and K.H. Olsen, J. Appl. Phys. Suppl. 32, 203S (1961).
120. J.S. Sallo and J.M. Carr, J. Appl. Phys. 33, 1316 (1962).
121. J.S. Sallo and J.M. Carr., J. Appl. Phys. 34, 1309 (1963).
122. K. Yoshida, T. Yamashita, and M. Saito, J. Appl. Phys. 64, 1 (1988).
123. R.R. Dubin, K.D. Winn, L.P. Davis and R.A. Cutler, J. Appl. Phys. 53, 2579 (1982).
124. M.A. Helfand, C.R. Clayton and R.B. Diegle, Abs. 208, Fall Meeting of the Electrochemical Society, Honolulu (October, 1987).
125. V. Brusic, M. Russak, R. Schad. G. Frankel, A. Selius, D. DiMilia, and D. Edmonson, J. Electrochem. Soc. 136, 42 (1989).
126. J.-B. Ju and H. Smyrl, submitted to J. Electrochem. Soc.
127. R. Sugita, T. Kunieda, and F. Kobayashi, IEEE Trans. Magn., MAG-17, 3172 (1981).
128. T. Yamada, N. Tani, M. Ishikawa, Y. Ota, K. Nakamura, and A. Itoh, IEEE Trans. Magn., MAG-21, 1429 (1985).
129. R.D. Fisher, J.C. Allan, and J.L. Pressesky, IEEE Trans. Magn., MAG-22, 352 (1986).
130. P.B. Phipps, S.J. Lewis, and D.W. Rice, J. Appl. Phys. 55, 2257 (1984).
131. M. Smallen, P.B. Mee, A. Ahmad, W. Freitag, and L. Nanis, IEEE Trans. Magn., MAG-21, 1530 (1985).
132. V. Novotny and N. Staud, J. Electrochem. Soc. 135, 2931, (1988).
133. T.G. Wang and G.W. Warren, IEEE Trans. Magn., MAG-22, 340 (1986).
134. H. Kobayashi, T. Yashiro, A. Kawashima, K. Asami, K. Hashimoto, and H. Fujimori, Abs. 218, Fall Meeting of the Electrochemical Society, Honolulu, October 1987.
135. J. Tada, M. Akihiro, K. Takei, T. Satoh, and T. Suzuki, IEEE Trans. Magn., MAG-22, 343 (1986).
136. M. Pourbaix, Atlas of Electrochemical Equilibria in Aqueous Solution (Pergamon Press, New York, 1966).
137. T. Miyabayashi, H. Wakayama, M. Miyata, K. Kobayashi, M. Yasuda, and F. Hine, Abs. 510, Fall Meeting of the Electrochemical Society, Honolulu, October, 1987.
138. M. Yanagisawa, N. Shiota, H. Yamaguchi, and Y. Suganuma, IEEE Trans. Magn., MAG-19, 1638 (1983).
139. Y. Hoshi, M. Matsuoka, and M. Naoe, J. Appl. Phys. 57, 4022 (1985).
140. S. Hattori, A. Tago, and O. Ishii, Rev. Comm. Lab. 30, 24 (1982).
141. M. Garrison, IEEE Trans. Magn., MAG-19, 1683 (1983).
142. E.M. Rossi, G. McDonough, A. Tietze, T. Arnoldussen, A. Brunsch, S. Doss, M. Henneberg, F. Lin, R. Lyn, A. Ting, and G. Trippel, J. Appl. Phys. 55, 2254 (1984).
143. S.K. Doss and G.A. Condas, Met. Trans. 18A, 158 (1987).
144. S.K. Doss and G.A. Condas, Proceedings ISTFA 1986: International Symposium for Testing.
145. M. Nagao, K. Sano, M. Kojima, H. Iwasaki, A. Nahara, and T. Kitamoto, Digest of the Intermag. Conf. Tokyo, BG-14, April, 1987.
146. J. Black, P. Oppenheimer, and D.M. Morris, J. Biomed. Matter Res. 21, 1213 (1987).
147. T. Yamashita, G.L. Chen, J. Shir, and T. Chen, IEEE Trans. Magn., MAG-24, 2629 (1988).
148. M.R. Khan, N. Heiman, R.D. Fisher, S. Smith, M. Smallen, G.F. Hughes, K. Veirs, B. Marchon, D.F. Ogletree, M. Salmeron, and W. Siekhaus, IEEE Trans. Magn., MAG-24, 2647 (1988).
149. S. Saito, M. Futamoto, Y. Honda, T. Nishimura, and K. Yoshida, Digest of the Intermag. Conf. Tokyo, BG-15, April, 1987.
150. H. Kaneko, IEEE Trans. Magn., MAG-18, 1221 (1982).
151. Y. Mitsuya, K. Kogure, and S. Oguchi, Rev. Elec. Comm. Lab. 30, 46 (1982).
152. C.K. Day, C.S. Harkins, S.P. Howe, and P. Poorman, Hewlett Packard J. 36, 25 (1985).
153. S.P. Sharma, J.H. Thomas III, and F.E. Bader, J. Electrochem. Soc. 125, 2002, 2005 (1978).
154. V. Novotny, G. Itnyre, A. Homola, and L. Franco, IEEE Trans. Magn., MAG-23, 3645 (1987).
155. K. Tagami and H. Hayashida, IEEE Trans. Magn., MAG-23, 3648 (1987).

156. Y. Suganuma, M. Tanaka, M. Yanagisawa, F. Goto, and S. Hatano, IEEE Trans. Magn., MAG-18, 1215 (1982).
157. N. Tsuya, T. Tokushima, M. Shiraki, Y. Wakui, Y. Saito, M. Nakamura, S. Hayano, A. Furugori, and M. Tanaka, IEEE Trans. Magn., MAG-22, 1140 (1985).
158. S. Futami, K. Kawata, and Y. Okamura, in "Electrochemical Technology in Electronics," L.T. Romankiw and T. Osaka, eds. (The Electrochemical Society, Pennington, NJ, 1988), p. 425.
159. T.K. Hatwar, S.C. Shin, and D.G. Stinson, IEEE Trans. Magn., MAG-22, 946 (1986).
160. G.L. McIntire and T.K. Hatwar, submitted to Corr. Sci.
161. N. Imamura, S. Tanaka, F. Tanaka, and Y. Nagao, IEEE Trans. Magn., MAG-21, 1607 (1985).
162. M. Farrow and A. Marinero, J. Electrochem. Soc., in press.
163. Y. Nagao, S. Tanaka, F. Tanaka, and N. Imamura, Digest Intermag. Conf. FC-7, March, 1986.
164. M. Kobayashi, M. Asano, K. Kawamura, and S. Ohno, Trans. J. Magn. Jap., TJMJ-1, 331 (1985).
165. H. Arimune, T. Maeda, and T. Yamada, Trans. J. Mag. Jap., TJMJ-1, 337 (1985).
166. T. Fujii, T. Tokushima, and N. Horiai, Digest Intermag., Tokyo CG-08, 1987.
167. T. Iijima, Appl. Phys. Lett. 50, 1835 (1987).
168. F.E. Luborsky, IEEE Trans. Magn., MAG-22, 937 (1986).
169. M. Tokanuga, M. Tobise, N. Meguro, and H. Harada, IEEE Trans. Magn., MAG-22, 904 (1986).
170. M. Harada, T. Tokanuga, M. Ohkoshi, S. Honda, and T. Kusuda, Digest of Intermag., FC-8, March, 1986.
171. H. Heitmann, M. Hartmann, S. Klahn, M. Rosenkranz, H.J. Tolle, and P. Willich, J. Appl. Phys. 61, 3331 (1987).
172. M. Hong, D.D. Bacon, R.B. van Dover, E.M. Gyorgy, J.F. Dillon, Jr., and S. D. Anderson, J. Appl. Phys. 57, 3900 (1985).
173. M.H. Kryder. J. Appl. Phys. 57, 3913 (1985).
174. F.E. Luborsky, J.T. Furey, R.E. Skoda, and B.C. Wagner, IEEE Trans. Magn., MAG-21, 1618 (1985).
175. M. Hartman, K. Witter, J. Reck and H.J. Tolle, IEEE Trans, Magn., MAG-22, 943 (1986).
176. P.J. Grundy, E.T.M. Lacey and C.D. Wright, Digest Intermag., CG-14, April, 1987.
177. S. Yumoto, K. Toki, Y. Tsukamoto, T. Habara, M. Okada, H. Yokota, and H. Gokan, Digest Intermag., CG-03, April, 1987.
178. M. Kobayashi, M. Asano, Y. Maeno, K. Oishi, and K. Kawamura, Appl. Phys. Lett. 50, 1964 (1987).
179. Y. Watanabe, J. Sasaki, Y. Kobayashi, and T. Yoshitomi, Intermag 87.
180. M. Miyazaki, I. Shibata, S. Okada, K. Itoh, and S. Ogawa, J. Appl. Phys. 61, 3226 (1987).
181. S. Okada, M. Miyazaki, I. Ishibata, K. Naito, K. Itoh, and S. Ogawa, Digest Intermag., DB-05., April, 1987.
182. R.P. Freese, R.N. Gardner, T.A. Rinehart, D.W. Siitari, and L.H. Johnson, Optical Mass Data Storage 529, 6 (1985).
183. J. Markoff, "The PC's Broad New Potential," The New York Times, November 30, 1988.

Index